PROGRESS IN CLINICAL AND BIOLOGICAL RESEARCH

RECENT TITLES

Vol 78: **Female Incontinence,** Norman R. Zinner and Arthur M. Sterling, *Editors*

Vol 79: **Proteins in the Nervous System: Structure and Function,** Bernard Haber, Jose Regino Perez-Polo, and Joe Dan Coulter *Editors*

Vol 80: **Mechanism and Control of Ciliary Movement,** Charles J. Brokaw and Pedro Verdugo, *Editors*

Vol 81: **Physiology and Biology of Horseshoe Crabs: Studies on Normal and Environmentally Stressed Animals,** Joseph Bonaventura, Celia Bonaventura, and Shirley Tesh, *Editors*

Vol 82: **Clinical, Structural, and Biochemical Advances in Hereditary Eye Disorders,** Donna L. Daentl, *Editor*

Vol 83: **Issues in Cancer Screening and Communications,** Curtis Mettlin and Gerald P. Murphy, *Editors*

Vol 84: **Progress in Dermatoglyphic Research,** Christos S. Bartsocas, *Editor*

Vol 85: **Embryonic Development,** Max M. Burger and Rudolf Weber, *Editors*. Published in 2 volumes: Part A: **Genetic Aspects** Part B: **Cellular Aspects**

Vol 86: **The Interaction of Acoustical and Electromagnetic Fields With Biological Systems,** Shiro Takashima and Elliot Postow, *Editors*

Vol 87: **Physiopathology of Hypophysial Disturbances and Diseases of Reproduction,** Alejandro De Nicola, Jorge Blaquier, and Roberto J. Soto, *Editors*

Vol 88: **Cytapheresis and Plasma Exchange: Clinical Indications,** W.R. Vogler, *Editor*

Vol 89: **Interaction of Platelets and Tumor Cells,** G.A. Jamieson, *Editor*

Vol 90: **Beta-Carbolines and Tetrahydroisoquinolines,** Floyd Bloom, Jack Barchas, Merton Sandler, and Earl Usdin, *Organizers*

Vol 91: **Membranes in Growth and Development,** Joseph F. Hoffman, Gerhard H. Giebisch, and Liana Bolis, *Editors*

Vol 92: **The Pineal and Its Hormones,** Russel J. Reiter, *Editor*

Vol 93: **Endotoxins and Their Detection With the Limulus Amebocyte Lysate Test,** Stanley W. Watson, Jack Levin, and Thomas J. Novitsky, *Editors*

Vol 94: **Animal Models of Inherited Metabolic Diseases,** Robert J. Desnick, Donald F. Patterson, and Dante G. Scarpelli, *Editors*

Vol 95: **Gaucher Disease: A Century of Delineation and Research,** Robert J. Desnick, Shimon Gatt, and Gregory A. Grabowski, *Editors*

Vol 96: **Mechanisms of Speciation,** Claudio Barigozzi, *Editor*

Vol 97: **Membranes and Genetic Disease,** John R. Sheppard, V. Elving Anderson, and John W. Eaton, *Editors*

Vol 98: **Advances in the Pathophysiology, Diagnosis, and Treatment of Sickle Cell Disease,** Roland B. Scott, *Editor*

Vol 99: **Osteosarcoma: New Trends in Diagnosis and Treatment,** Alexander Katznelson and Jacobo Nerubay, *Editors*

Vol 100: **Renal Tumors: Proceedings of the First International Symposium on Kidney Tumors,** René Küss, Gerald P. Murphy, Saad Khoury, and James P. Karr, *Editors*

Vol 101: **Factors and Mechanisms Influencing Bone Growth,** Andrew D. Dixon and Bernard G. Sarnat, *Editors*

Vol 102: **Cell Function and Differentiation,** G. Akoyunoglou, A.E. Evangelopoulos, J. Georgatsos, G. Palaiologos, A. Trakatellis, C.P. Tsiganos, *Editors*. Published in 3 volumes. Part A: **Erythroid Differentiation, Hormone–Gene Interaction, Glycoconjugates,**

See pages following the index for previous titles in this series.

LIMB DEVELOPMENT AND REGENERATION PART B

ORGANIZING COMMITTEE
Third International Conference on Limb Morphogenesis and Regeneration

Arnold I. Caplan

Professor of Biology and Anatomy, Case Western Reserve University, Cleveland, Ohio

John F. Fallon

Professor of Anatomy, University of Wisconsin School of Medicine, Madison, Wisconsin

Paul F. Goetinck

Professor of Animal Genetics and Genetics and Cell Biology, University of Connecticut, Storrs, Connecticut

Robert O. Kelley

Professor and Chairman of Anatomy, The University of New Mexico School of Medicine, Albuquerque, New Mexico

Jeffrey A. MacCabe

Associate Professor of Zoology, University of Tennessee, Knoxville, Tennessee

LIMB DEVELOPMENT AND REGENERATION

Part B

Proceedings of the Third International Conference on Limb Morphogenesis and Regeneration University of Connecticut, Storrs, June 27–July 2, 1982

Editors

Robert O. Kelley
Professor and Chairman of Anatomy
University of New Mexico School of Medicine
Albuquerque, New Mexico

Paul F. Goetinck
Professor of Animal Genetics and Genetics and Cell Biology
University of Connecticut
Storrs, Connecticut

Jeffrey A. MacCabe
Associate Professor of Zoology
University of Tennessee
Knoxville, Tennessee

Alan R. Liss, Inc., • New York

Address all Inquiries to the Publisher
Alan R. Liss, Inc., 150 Fifth Avenue, New York, NY 10011

Copyright © 1982 Alan R. Liss, Inc.

Printed in the United States of America

Library of Congress Cataloging in Publication Data

International Conference on Limb Morphogenesis and
 Regeneration (3rd : 1982 : University of Connecticut)
 Limb development and regeneration.

 (Progress in clinical and biological research ;
v. 110)
 Vol. 2 edited by Robert O. Kelley, Paul F. Goetinck,
and Jeffrey A. MacCabe.
 Includes bibliographies and indexes.
 1. Extremities (Anatomy) — Regeneration — Congresses.
2. Morphogenesis — Congresses. I. Fallon, John F.
II. Caplan, Arnold I.
 III. Title. IV. Series. [DNLM: 1. Extremities — Growth
and development — Congresses. 2. Extremities — Physiology
 — Congresses. 3. Regeneration — Congresses. W1 PR668E
v.110 / WE 800 I59 1982L]
QL950.7.I57   1982   596'.031   82-20391
ISBN 0-8451-0110-2 (set)
ISBN 0-8451-0170-6 (v. 1)
ISBN 0-8451-0171-4 (v. 2)

to the memory of
Professor Edgar Zwilling
1913–1971

Participants in the Third International Conference on Limb Morphogenesis and Regeneration, held at the University of Connecticut, Storrs, June 27–July 2, 1982.

Contents of Part B

Contents of Part A

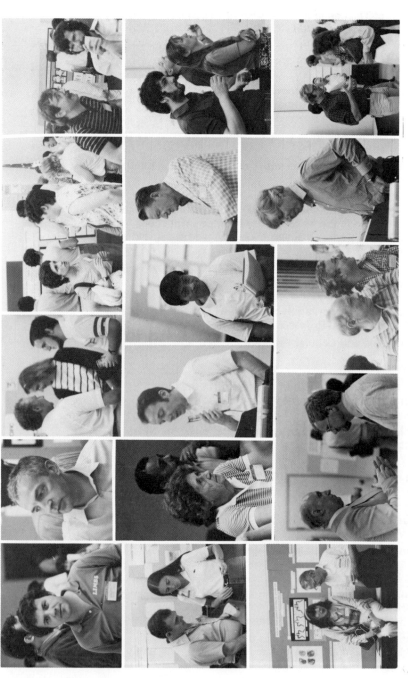

Conference participants. Clockwise from top left: T. Stevens; E. Lheureux; J. Sasse, D. Neubert and Ms. Neubert; B. Vertel; T. Linsenmayer and J. Slack; N. Holder and V. Stirling; R. Kelley and L. Iten; L. Wolpert; L. Rosenberg and D. Fischman; D. Ede and H.J. Jacob; U. Abbott and M. Runner; K. Sparks and P. McKeown-Longo.
Center group, from left: Mrs. E. Zwilling and M. Michael; J. Fallon; M. Yasuda; J.W. Saunders, Jr.

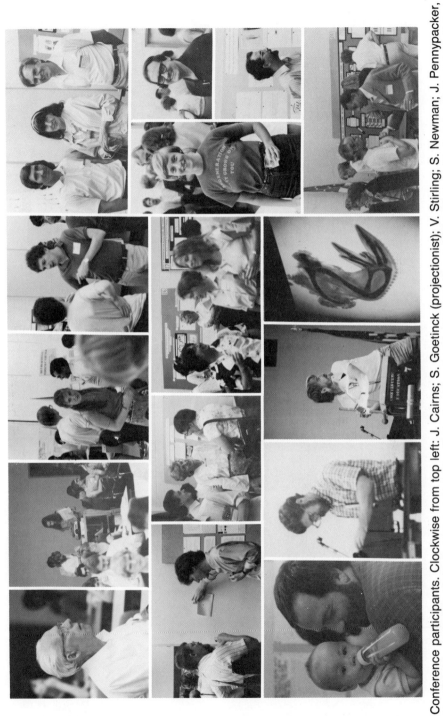

Conference participants. Clockwise from top left: J. Cairns; S. Goetinck (projectionist); V. Stirling; S. Newman; J. Pennypacker; U. Abbott and P. Goetinck; D. Dayton; V. Hascall; A. Caplan and M. Michael; the duplicated limb; D. Summerbell; J. MacCabe; C. Ordahl and offspring; D. Heinegård. Center group, from left: M. Maden and J. McCredie; K. Muneoka, M. Runner, C. Rollman-Dinsmore and N. O'Rourke; C. Olsen.

Participants and Contributors

Ursula K. Abbott [A:13]
University of California, Davis, Davis CA 95616

Thomas R. Adams [B:349]
T.H. Morgan School of Biological Sciences, University of Kentucky,
Lexington, KY 40506

Loulwah Al-Ghaith [A:195]
Department of Biology as Applied to Medicine, Middlesex Hospital Medical
School, London W1P 6DB, United Kingdom

Rodolfo Amprino [A:155]
Institute of Human Anatomy, University of Bari, 70124 Bari, Italy

C.W. Archer [A:267]
Department of Biology as Applied to Medicine, The Middlesex Hospital
Medical School, London W1P 6DB, United Kingdom

John R. Baker [B:17]
University of Alabama in Birmingham, Birmingham, AL 35294

Mark Ballow [B:113]
Department of Pediatrics, University of Connecticut Health Center,
Farmington, CT 06032

J. Biehl [B:271]
Department of Anatomy, Medical School of the University of Pennsylvania,
Philadelphia, PA 19104

David E. Birk [A:245]
Department of Pathology, UMDNJ-Rutgers Medical School, Piscataway, NJ
08854

Philip H. Bonner [B:349]
T.H. Morgan School of Biological Sciences, University of Kentucky,
Lexington, KY 40506

Fred L. Bookstein [A:525]
Center for Human Growth and Development, The University of Michigan, Ann
Arbor, MI 48109

Richard B. Borgens [A:597]
Institute for Medical Research, San Jose, CA 95128

Edward G. Buss [B:85]
Department of Poultry Science, Pennsylvania State University, University
Park, PA 16802

Jo Ann Cameron [A:491]
Department of Anatomical Sciences, University of Illinois, Urbana, IL 61801

The boldface number in brackets following each contributor's name indicates the opening page of that
author's article in either Part A or Part B of these proceedings.

Arnold I. Caplan [A:143, B:229, 379]
Department of Biology, Developmental Biology Center, Case Western Reserve University, Cleveland, OH 44106

Bruce M. Carlson [A:433]
Departments of Anatomy and Biological Sciences, University of Michigan, Ann Arbor, MI 48109

Jill L. Carrington [A:33]
Department of Anatomy, University of Wisconsin, Madison, WI 53706

David A. Carrino [B:379]
Department of Biology, Case Western Reserve University, Cleveland, OH 44106

Bruce Caterson [B:17]
University of Alabama in Birmingham, Birmingham, AL 35294

Matthias Chiquet [B:359]
Department of Embryology, Carnegie Institution of Washington, Baltimore, MD 21210

H. Choi [B:67]
Department of Orthopedic Research, Montefiore Hospital and Medical Center, Bronx, NY 10467

Bodo Christ [B:281,313,333]
Institute of Anatomy, Ruhr-University Bochum, D-4630 Bochum, Federal Republic of Germany

James E. Christner [B:17]
University of Alabama in Birmingham, Birmingham, AL 35294

Sandra B. Conlon [B:417]
Department of Medicine/Oncology, Stanford School of Medicine, Stanford, CA 94305

Thomas G. Connelly [A:525]
Department of Anatomy and Cell Biology, The University of Michigan, Ann Arbor, MI 48109

Thomas Cooper [B:391]
Department of Anatomy, Temple University School of Medicine, Philadelphia, PA 19140

Michael T. Crow [B:417]
Department of Medicine/Oncology, Stanford School of Medicine, Stanford, CA 94305

Ann Dannenberg [B:85]
Department of Animal Genetics, University of Connecticut, Storrs, CT 06268

Charles E. Dinsmore [A:577]
Department of Anatomy, Rush Medical College, Chicago, IL 60612

Albert Dorfman [B:175]
Department of Pediatrics, The University of Chicago, Chicago, IL 60637

D.A. Ede [A:45]
Department of Zoology, University of Glasgow, Glasgow G12 9LU, United Kingdom

William A. Elmer [A:355]
Biology Department, Emory University, Atlanta, GA 30322

John F. Fallon [A:33,119]
Department of Anatomy, University of Wisconsin, Madison, WI 53706

Douglas M. Fambrough [B:359]
Department of Embryology, Carnegie Institution of Washington, Baltimore, MD 21210

Marc Y. Fiszman [B:401]
Department of Molecular Biology, Pasteur Institute, Paris Cedex 15, France

John M. Fitch [B:369]
Department of Medicine, Harvard Medical School of Massachusetts General Hospital, Boston, MA 02114

Jeanne M. Frederick [A:33]
Cullen Eye Institute, Baylor College of Medicine, Houston, TX 77030

Mary T. Gasseling [A:67]
Department of Biological Sciences, State University of New York, Albany, NY 12222

Jacqueline Géraudie [A:289]
Equipe de recherche "Formations squelettiques," Laboratoire d'Anatomie comparée, Université de Paris VII, 75251 Paris Cedex 05, France

Morton Globus [A:513]
Department of Biology, University of Waterloo, Waterloo, Ontario, Canada N2L 3G1

Suresh C. Goel [A:175]
Department of Zoology, University of Poona, Poona 411 007, India

Paul F. Goetinck [B:85,113]
Department of Animal Genetics, University of Connecticut, Storrs, CT 06268

John M. Graham, Jr. [A:413]
Department of Maternal and Child Health, Dartmouth Medical School, Hanover, NH 03755

M. Grim [B:333]
Department of Anatomy, Faculty of Medicine, Charles University Prague, 12800 Praha 2, Czechoslovakia

Brian K. Hall [B:323]
Department of Biology, Dalhousie University, Halifax, Nova Scotia, Canada B3H 4J1

Fiona Harvey [A:109]
National Institute for Medical Research, London NW7 1AA, United Kingdom

Vincent C. Hascall [B:3]
Laboratory of Biochemistry, National Institute of Dental Research, National Institutes of Health, Bethesda, MD 20205

John R. Hassell [B:105]
Laboratory of Developmental Biology and Anomalies, National Institute of Dental Research, National Institutes of Health, Bethesda, MD 20205

Stephen Hauschka [B:303]
Department of Biochemistry, University of Washington, Seattle, WA 98195

L.L. Hearson [A:587]
Department of Biology, Wabash College, Crawfordsville, IN 47933

Dick Heinegård [B:35]
Department of Physiological Chemistry, University of Lund, S-220 07 Lund, Sweden

A. Tyl Hewitt [B:25,105]
Laboratory of Developmental Biology and Anomalies, National Institute of Dental Research, National Institutes of Health, Bethesda, MD 20205

Stuart M. Heywood [B:409]
Biological Sciences Group, Genetics and Cell Biology Section, University of Connecticut, Storrs, CT 06268

J. R. Hinchliffe [A:131]
Department of Zoology, University College of Wales, Aberystwyth, Wales, Dyfed SY23 3DA, United Kingdom

Kenneth S. Hirsch [A:423]
Eli Lilly and Co., Greenfield Laboratories, Greenfield, IN 46140

Nigel Holder [A:477]
Department of Anatomy, King's College University of London, London WC2R 2LS, United Kingdom

Margaret Hollyday [A:183]
Department of Pharmacological and Physiological Sciences, University of Chicago, Chicago, IL 60637

Lewis B. Holmes [A:311,317]
Department of Pediatrics, Embryology-Teratology Unit, Massachusetts General Hospital, Boston, MA 02114

Howard Holtzer [B:159,271]
Department of Anatomy, Medical School, University of Pennsylvania, Philadelphia, PA 19104

S. Holtzer [B:271]
Department of Anatomy, Medical School, University of Pennsylvania, Philadelphia, PA 19104

Lawrence S. Honig [A:57,99]
Laboratory for Developmental Biology, University of Southern California, Los Angeles, CA 90007

Marcia Honig [A:207]
Department of Biology, Yale University, New Haven, CT 06511

David R. Hootnick [A:327]
Department of Orthopedic Surgery, SUNY Upstate Medical Center, Syracuse, NY 13210

Elizabeth A. Horigan [B:105]
Laboratory of Developmental Biology and Anomalies, National Institute of Dental Research, National Institutes of Health, Bethesda, MD 20205

Amata Hornbruch [B:343]
Department of Biology as Applied to Medicine, The Middlesex Hospital Medical School, London W1P 6DB, United Kingdom

Laurie E. Iten [A:77]
Department of Biological Sciences, Purdue University, West Lafayette, IN 47907

Heinz Jürgen Jacob [B:281,313,333]
Institute of Anatomy, Ruhr-University Bochum, D-4630 Bochum, Federal Republic of Germany

Monika Jacob [B:281,313]
Institute of Anatomy, Ruhr-University Bochum, D-4630 Bochum, Federal Republic of Germany

David M. Jargiello [A:143]
Department of Biology, Developmental Biology Center, Case Western Reserve University, Cleveland, OH 44106

Janice A. Jerdan [B:105]
Laboratory of Developmental Biology and Anomalies, National Institute of Dental Research, National Institutes of Health, Bethesda, MD 20205

Thomas Johnson [B:67]
Department of Orthopedic Research, Montefiore Hospital and Medical Center, Bronx, NY 10467

Arnold J. Kahn [B:239]
Department of Biomedical Sciences, Washington University School of Dental Medicine, St. Louis, MO 63110

Gary H. Karpen [A:609]
Department of Genetics, University of Washington, Seattle, WA 98195

Robert O. Kelley [A:119]
Department of Anatomy, University of New Mexico, Albuquerque, NM 87131

Terry Kenny-Mobbs [B:323]
Department of Biology, Dalhousie University, Halifax, Nova Scotia, Canada B3H 4J1

Agnes Kesik [A:513]
Department of Biology, University of Waterloo, Waterloo, Ontario, Canada N2L 3GI

Akbar Khan [A:195]
Department of Anatomy, King's College University of London, London WC2R 2LS, United Kingdom

Madeleine A. Kieny [B:293]
Laboratoire de Zoologie et Biologie animale, Université Scientifique et Médicale, Grenoble 38041, France

D.M. Kochhar [B:203]
Department of Anatomy, Jefferson Medical College, Philadelphia, PA 19107

Robert A. Kosher [A:279]
Department of Anatomy, University of Connecticut Health Center, Farmington, CT 06032

Dean Kravis [B:175]
Department of Pediatrics, The University of Chicago, Chicago, IL 60637

Ralf Krowke [A:387]
Institut für Toxikologie und Embryopharmakologie der Freien Universität Berlin, D-1000 Berlin 33, Federal Republic of Germany

Marilyn Krukowski [B:239]
Department of Biology, Washington University, St. Louis, MO 63130

Gerald Kuncio [B:391]
Department of Anatomy, Temple University School of Medicine, Philadelphia, PA 19140

Thomas E. Kwasigroch [A:335]
Department of Anatomy, Quillen-Dishner College of Medicine, East Tennessee State University, Johnson City, TN 37614

Alan H. Lamb [A:227]
Department of Pathology, University of Western Australia, Nedlands 60009, Western Australia

Karen M. Lamb [B:45]
Department of Anatomy, Cell Biology Division, New Jersey Medical School, Newark, NJ 07103

Lynn T. Landmesser [A:207]
Department of Biology, Yale University, New Haven, CT 06511

Mark E. Lanser [A:33]
Department of Anatomy, University of Wisconsin, Madison, WI 53706

Bernard Lassalle [A:547]
Laboratoire de Morphogenèse animale, Université des Sciences et Techniques de Lille, 59655 Villeneuve d'Ascq Cedex, France

C. W. Leal [A:237]
Department of Zoology, University of Tennessee, Knoxville, TN 37996

K.W. Leal [A:237]
Department of Zoology, University of Tennessee, Knoxville, TN 37996

Claire M. Leonard [A:251]
Department of Anatomy, New York Medical College, Valhalla, NY 10595

E. Mark Levinsohn [A:327]
Department of Radiology, SUNY Upstate Medical Center, Syracuse, NY 13210

Julian Lewis [A:195]
Department of Anatomy, King's College London WC2R 2LS, United Kingdom

Emile Lheureux [A:455]
Laboratoire de Morphogenèse animale, Université des Sciences et Techniques de Lille, 59655 Villeneuve d'Ascq Cedex, France

Thomas F. Linsenmayer [B:369]
Department of Anatomy and Medicine, Harvard Medical School and Massachusetts General Hospital, Boston, MA 02114

Jeffrey A. MacCabe [A:237]
Department of Zoology, University of Tennesse, Knoxville, TN 37996

Thomas McCarthy [B:409]
Biological Sciences Group, Genetics and Cell Biology Section, The University of Connecticut, Storrs, CT 06268

M.E. McGinnis [A:587]
Department of Biological Sciences, Purdue University, West Lafayette, IN 47907

John C. McLachlan [B:343]
Department of Zoology, University of Oxford, Oxford OX1 3PS, United Kingdom

Malcolm Maden [A:445]
Division of Developmental Biology, National Institute for Medical Research, London NW7 1AA, United Kingdom

Robert J. Majeska [B:249]
Department of Oral Biology, University of Connecticut Health Center, School of Dental Medicine, Farmington, CT 06032

J. David Malone [B:239]
Department of Internal Medicine, Jewish Hospital of St. Louis, St. Louis, MO 63110

Klaus von der Mark [B:159]
Max-Planck-Institut für Biochemie, 8033 Martinsried bei München, Federal Republic of Germany

George R. Martin [B:25,105]
Laboratory of Developmental Biology and Anomalies, National Institute of Dental Research, National Institutes of Health, Bethesda, MD 20205

Mary Martini [B:229]
Biology Department, Case Western Reserve University, Cleveland, OH 44106

Pauline Mayne [B:125]
Department of Anatomy, University of Alabama in Birmingham, Birmingham, AL 35294

Richard Mayne [B:125,369]
Department of Anatomy, University of Alabama in Birmingham, Birmingham, AL 35294

Anthony L. Mescher [A:501]
Anatomy Section, Medical Sciences Program, Indiana University School of Medicine, Bloomington, IN 47405

Guy Milton [A:513]
Department of Biology, University of Waterloo, Waterloo, Ontario, Canada N2L 3G1

Jay E. Mittenthal [A:619]
Department of Anatomical Sciences, University of Illinois, Urbana, IL 61801

Didier Montarras [B:401]
Department of Molecular Biology, Pasteur Institute, Paris Cedex 15, France

Barbara Mroczkowski [B:409]
Biological Sciences Group, Genetics and Cell Biology Section, The University of Connecticut, Storrs, CT 06268

Helen Muir [B:55]
Department of Biochemistry, Kennedy Institute of Rheumatology, Hammersmith, London W6 7DW, United Kingdom

Ken Muneoka [A:77]
Developmental Biology Center, University of California, Irvine, Irvine, CA 92717

Douglas J. Murphy [A:77]
Department of Biological Sciences, Purdue University, West Lafayette, IN 47907

Harukazu Nakamura [A:301]
Department of Anatomy, Hiroshima University School of Medicine, Hiroshima 734, Japan

Mark A. Nathanson [B:215]
Department of Anatomy, New Jersey Medical School, Newark, NJ 07103

Diether Neubert [A:387]
Institut für Toxikologie und Embryopharmakologie der Frein Universität Berlin, D-1000 Berlin 33, Federal Republic of Germany

Daniel A. Neufeld [A:407]
Department of Anatomy, University of South Dakota School of Medicine, Vermillion, SD 57069

Stuart A. Newman [A:251]
Department of Anatomy, New York Medical College, Valhalla, NY 10595

Yoshifumi Ninomiya [B:183]
Department of Biochemistry, UMDNJ-Rutgers Medical School, Piscataway, NJ 08854

Michael J. O'Donovan [A:207]
Department of Biology, Yale University, New Haven, CT 06511

Bjorn Reino Olsen [B:183]
Department of Biochemistry, UMDNJ-Rutgers Medical School, Piscataway, NJ 08854

Cherie L. Olsen [A:537]
Department of Zoology, The Ohio State University, Columbus, OH 43210

Pamela S. Olson [B:417]
Department of Medicine/Oncology, Stanford School of Medicine, Stanford, CA 94305

Charles P. Ordahl [B:391]
Department of Anatomy, Temple University School of Medicine, Philadelphia, PA 19140

Philip Osdoby [B:229]
Department of Biomedical Sciences, School of Dentistry, Washington University, St. Louis, MO 63110

Maurizio Pacifici [B:159,271]
Department of Anatomy, Medical School, University of Pennsylvania, Philadelphia, PA 19104

David S. Packard, Jr. [A:327]
Department of Anatomy, SUNY Upstate Medical Center, Syracuse, NY 13210

Subhash Pal [B:67]
Department of Orthopedic Research, Montefiore Hospital and Medical Center, Bronx, NY 10467

Mats Paulsson [B:35]
Department of Physiological Chemistry, University of Lund, S-220 07 Lund, Sweden

R. Payette [B:271]
Department of Anatomy, Medical School, University of Pennsylvania, Philadelphia PA 19104

Isaac Peng [B:391]
Department of Anatomy, Temple University School of Medicine, Philadelphia, PA 19140

John D. Penner [B:203]
Department of Anatomy, Jefferson Medical College, Philadelphia, PA 19107

John P. Pennypacker [B:167]
Department of Zoology, University of Vermont, Burlington, VT 05405

I. Pidoux [B:67]
Joint Diseases Research Laboratories, Shriners Hospital for Crippled Children, Montreal, Quebec, Canada H3G 1A6

A. Robin Poole [B:67]
Joint Diseases Research Laboratories, Shriners Hospital for Crippled Children, Montreal, Quebec, Canada H3G 1A6

A.H. Reddi [B:261]
Bone Cell Biology Section, Laboratory of Biological Structure, National Institute of Dental Research, National Institutes of Health, Bethesda, MD 20205

Charles A. Reese [B:125]
Department of Anatomy, University of Alabama in Birmingham, Birmingham, AL 35294

Agnes Reiner [B:67]
Joint Disease Research Laboratories, Shriners Hospital for Crippled Children, Montreal, Quebec, Canada H3G 1A6

Susan Reynolds [A:477]
Department of Anatomy, King's College University of London, London, WC2R 2LS, United Kingdom

Gideon A. Rodan [B:249]
Department of Oral Biology, University of Connecticut Health Center, School of Dental Medicine, Farmington, CT 06032

P. Rooney [A:267]
Department of Biology as Applied to Medicine, The Middlesex Hospital Medical School, London W1P 6DB, United Kingdom

Marcel André Rooze [A:365]
Laboratory of Human Anatomy and Embryology, Faculty of Medicine, Free University of Brussels, B-1000, Brussels, Belgium

Lawrence C. Rosenberg [B:67]
Department of Orthopedic Surgery, Albert Einstein College of Medicine, Montefiore Hospital and Medical Center, Bronx, NY 10467

Peter Roughley [B:67]
Joint Diseases Research Laboratories, Shriners Hospital for Crippled Children, Montreal, Quebec, Canada H3G 1A6

Meredith N. Runner [A:345]
Department of Molecular, Cellular and Developmental Biology, University of Colorado, Boulder, CO 80309

Richard Rutz [B:303]
Department of Biochemistry, University of Washington, Seattle, WA 98195

Linda Sandell [B:175]
Department of Biochemistry, Rush-Presbyterian-St. Lukes Medical Center, Chicago, IL 60612

J.D. Sandy [B:55]
Department of Biochemistry, Kennedy Institute of Rheumatology, Hammersmith, London W6 7DW, United Kingdom

Joachim Sasse [B:159,271]
Department of Anatomy, Medical School, University of Pennsylvania, Philadelphia, PA 19104

John W. Saunders, Jr. [A:67]
Department of Biological Sciences, State University of New York, Albany, NY 12222

Claire M. Schreiner [A:423]
Division of Pathologic Embryology, Children's Hospital Research Foundation, Cincinnati, OH 45229

Gerold Schubiger [A:609]
Department of Zoology, University of Washington, Seattle, WA 98195

Nancy B. Schwartz [B:97]
Department of Pediatrics and Biochemistry, The University of Chicago, Chicago, IL 60637

William J. Scott [A:423]
Division of Pathologic Embryology, Children's Hospital Research Foundation, Cincinnati, OH 45229

Robert L. Searls [A:165]
Department of Biology, Temple University, Philadelphia, PA 19122

Robert E. Seegmiller [B:193]
Department of Zoology, Division of Genetics and Developmental Biology, Brigham Young University, Provo, UT 84602

Maryam Shamslahidjani [A:45]
Department of Biology, National University of Iran, Teheran, Iran

Val C. Sheffield [B:175]
Department of Pediatrics, The University of Chicago, Chicago, IL 60637

Thomas H. Shepard [A:377]
Central Laboratory for Human Embryology, University of Washington School of Medicine, Seattle, WA 98195

Kohei Shiota [A:377]
Central Laboratory for Human Embryology, University of Washington School of Medicine, Seattle, WA 98195

Allan M. Showalter [B:183]
Department of Biochemistry, UMDNJ-Rutgers Medical School, Piscataway, NJ 08854

Frederick H. Silver [A:245]
Department of Pathology, UMDNJ-Rutgers Medical School, Piscataway, NJ 08854

Michael H. Silver [B:25]
Department of Laboratory Medicine and Pathology, University of Minnesota Medical Center, Minneapolis, MN 55455

B. Kay Simandl [A:33]
Department of Anatomy, University of Wisconsin, Madison, WI 53706

Richard G. Skalko [A:335]
Department of Anatomy, Quillen-Dishner College of Medicine, East
Tennessee State University, Johnson City, TN 37614

Jonathan M.W. Slack [A:557]
Department of Experimental Morphology, Imperial Cancer Research Fund,
London NW7 1AD, United Kingdom

J.C. Smith [A:57]
Imperial Cancer Research Fund, Mill Hill Laboratories, London NW7 1AD,
United Kingdom

Michael Solursh [B:139]
Department of Zoology, University of Iowa, Iowa City, IA 52242

Yngve Sommarin [B:35]
Department of Physiological Chemistry, University of Lund, S-220 07 Lund,
Sweden

Kenneth Sparks [B:113]
Department of Animal Genetics, University of Connecticut, Storrs, CT 06268

Trent D. Stephens [A:3]
Department of Biology, Idaho State University, Pocatello, ID 83209

R. Victoria Stirling [A:217]
Department of Developmental Biology, National Institute for Medical
Research, London NW7 1AA, United Kingdom

Frank E. Stockdale [B:417]
Department of Medicine/Oncology, Stanford School of Medicine, Stanford,
CA 94305

David L. Stocum [A:467]
Department of Genetics and Development, University of Illinois, Urbana IL
61801

Dennis Summerbell [A:109,217]
Division of Developmental Biology, National Institute for Medical Research,
London NW7 1AA, United Kingdom

Gavin Swanson [A:195]
Department of Anatomy, King's College University of London, London WC2R
2LS, United Kingdom

Lih-Heng Tang [B:67]
Department of Orthopedic Research, Montefiore Hospital and Medical
Center, Bronx, NY 10467

Patrick W. Tank [A:565]
Department of Anatomy, University of Arkansas for Medical Sciences, Little
Rock, AR 72205

Roy A. Tassava [A:537]
Department of Zoology, The Ohio State University, Columbus, OH 43210

Steven L. Teitelbaum [B:239]
Department of Pathology, Jewish Hospital of St. Louis, St. Louis, MO 63110

C. Tickle [A:89]
Department of Biology as Applied to Medicine, The Middlesex Hospital
Medical School, London W1P 6DB, United Kingdom

Madeleine Toutant [B:401]
Groupe de Recherches de Biologie et de Pathologie neuromusculaires, CNRS, Paris 75005, France

David P. Treece [A:537]
Department of Zoology, The Ohio State University, Columbus, OH 43210

Robert L. Trelstad [A:245]
Department of Pathology, UMDNJ-Rutgers Medical School, Piscataway, NJ 08854

William B. Upholt [B:175]
Department of Pediatrics, The University of Chicago, Chicago, IL 60637

Joseph W. Vanable, Jr. [A:587]
Department of Biological Sciences, Purdue University, West Lafayette, IN 47907

Hugh H. Varner [B:25,105]
Laboratory of Developmental Biology and Anomalies, National Institute of Dental Research, National Institutes of Health, Bethesda, MD 20205

N.S. Vasan [B:45]
Department of Anatomy, Cell Biology Division, New Jersey Medical School, Newark, NJ 07103

Barbara M. Vertel [B:149]
Department of Biology, Syracuse University, Syracuse, NY 13210

Swani Vethamany-Globus [A:513]
Department of Biology, University of Waterloo, Waterloo, Ontario, Canada N2L 3G1

F. Wachtler [B:281]
Institute of Histology and Embryology, University of Vienna, A-1090 Wien, Austria

Rachel Warga [A:619]
Department of Biology, University of Oregon, Eugene, OR 97403

Cynthia C. Williams [B:125]
Department of Anatomy, University of Alabama in Birmingham, Birmingham, AL 35294

L. Wolpert [A:267]
Department of Biology as Applied to Medicine, The Middlesex Hospital Medical School, London W1P 6DB, United Kingdom

J. Thomas Wright [A:355]
Biology Department, Emory University, Atlanta, GA 30322

B.L. Yallup [A:131]
Department of Zoology, University College of Wales, Aberystwyth, Wales, Dyfed SY23 3DA, United Kingdom

Mineo Yasuda [A:301]
Department of Anatomy, Hiroshima University School of Medicine, Hiroshima 734, Japan

David J. Zaleske [A:317]
Department of Orthopedics, Pediatric Orthopedic Unit, Massachusetts General Hospital, Boston, MA 02114

Preface

The developing vertebrate limb continues to be an excellent model for investigation of morphogenesis, cytodifferentiation, and regeneration. Considerable theoretical and experimental attention has been devoted by investigators around the world to defining the nature of a series of complex morphogenetic interactions that ultimately lead to the development of a definitive structure. These studies have ranged through multiple levels of analysis, from the organismal to the molecular, and attempts are now underway to integrate this vast accumulation of information into meaningful hypotheses that can be tested and that will unify our understanding of limb morphogenesis. In addition, it is hoped that further understanding of basic events in normal limb development will aid in the generation of experimental protocols that will clarify problems of abnormal development and function.

Prior to the meeting on which these volumes are based, investigators interested in the developing limb had participated in two international meetings. The first was held in Grenoble, France, in 1972 and the second in Glasgow, Scotland, in 1976. The major impact of the Grenoble meeting was that it brought together, for the first time, experimentalists in the field with those who were building theoretical models to account for experimental observations. As a result of that meeting, the theoretical constructs generated experimental avenues leading to exploration of fundamental aspects of limb development. The second meeting in Glasgow, Scotland, served two important functions. First, the meeting served as a forum for the consolidation of both theoretical and experimental information that had been generated over the preceding four years. In addition, it drew attention to the importance of directing efforts towards the cellular and molecular aspects of experimentation in this field.

The third international conference was held at the University of Connecticut, Storrs campus, from June 27 through July 2, 1982. The Organizing Committee sought to continue the established emphasis in the field of limb development by bringing together investigators with the latest information in the various experimental aspects of limb development. The meeting provided a unique forum in which focused discussions and interchange of ideas took place between scientists with diverse experimental and theoretical backgrounds, including experimental morphology, cellular biology, and molecular biology.

As an aid to this exchange of ideas and methodological approaches, the Organizing Committee planned two entire sessions in workshop format. In one workshop, the molecular and cellular biologists were requested to emphasize the methodological aspects of their work during a presentation of their recent experimental findings for the explicit benefit of individuals who emphasize the morphological and clinical approaches to the study of the limb. It was anticipated that this type of presentation would demonstrate that molecular techniques of recombinant DNA, monoclonal antibody production, and protein characterization can be successfully applied to the study of the limb at the organized tissue level. The second workshop was designed to analyze and compare current models used to explain limb development and regeneration. Again, the intent was to continue the development of new avenues of communication between theoretical and experimental biologists. This successful combination of workshop format with regular sessions greatly facilitated verbal interchange between audience and speakers, with the result that technological and conceptual strengths of all participants merged and created new experimental approaches.

The Organizing Committee is especially grateful to Mr. John Farling, Assistant Director of Conferences, Institutes and Administrative Services, and Ms. Deborah McSweeney and Ms. Danitza Nall in the School of Extended and Continuing Education at the University of Connecticut for their professional management of the conference. In addition acknowledgment is extended to Ms. Jean O'Neill at the University of New Mexico School of Medicine and to Ms. Sue Leonard and Ms. B. Kay Simandl at the University of Wisconsin School of Medicine for their assistance in the final preparation of these proceedings for publication.

Finally, special acknowledgment is given to the SmithKline Corporation (Smith, Kline and French Laboratories) for financial contributions that defrayed most of the expenses necessitated by a meeting of this size and scope. It is doubtful that the conference would have been possible without this grant assistance. The Organizing Committee also acknowledges support from the National Science Foundation, the International Society for Developmental Biology, the Ortho Pharmaceutical Corporation, and the National Institute of Child Health and Human Development, National Institutes of Health. Collectively, these agencies assured a productive and successful conference.

Considerable effort was expended by members of the Organizing Committee to plan and successfully implement the meeting that formed the basis for these volumes. We share equal responsibility for credit as well as criticism.

The Organizing Committee

SECTION FIVE
EXTRACELLULAR MATRIX

Limb Development and Regeneration
Part B, pages 3–15
Published 1983 by Alan R. Liss, Inc., 150 Fifth Avenue, New York, NY 10011

STRUCTURE AND BIOSYNTHESIS OF PROTEOGLYCANS
WITH KERATAN SULFATE[+]

Vincent C. Hascall

Laboratory of Biochemistry
National Institute of Dental Research
NIH, Bethesda, MD   20205

I. INTRODUCTION[*]

The purpose of this paper is to describe recent work
concerning the structure and biosynthesis of two distinctly
different classes of proteoglycans, (a) the major proteoglycans from cartilaginous tissues and (b) the keratan sulfate-proteoglycans from mature corneal stroma. In both
these tissues, proteoglycans have defined structural and
organizational functions. Further, there are characteristic changes in the structures of these macromolecules
which occur during the sequential developmental stages involved in tissue maturation (Caplan and Hascall, 1980; Hart,
1976). These changes reflect modulations in the complex,
multienzyme biosynthetic pathways involved in the post translation modification steps required to synthesize the final
proteoglycan structures. Hence, a detailed understanding of
structures and of the post translational steps utilized in

+Dr. Karl Meyer has always had a particular fascination and
fondness for keratan sulfate, pioneering many studies on the
chemistry, structure and potential function of this class of
glycosaminoglycan. For this reason, and in admiration of his
many contributions,I dedicate this communication to Karl with
the hope that it provides a few more clues for understanding
"keratan sulfate."

*Because of space limitations, citations are not comprehen-
sive. References cited will, however, contain further references for the various points raised in the discussion.

proteoglycan synthesis is necessary to understand the role
these molecules play in tissue development and for suggesting ways in which cells may regulate biosynthetic pathways
to modulate the final macromolecular structures.

II. CARTILAGE PROTEOGLYCANS

Fig. 1 shows a model for the overall structure of the
major proteoglycan isolated from a transplantable rat chondrosarcoma, which has many structural similarities to those
isolated from chick limb bud chondrocyte cultures. Recent
work (Lohmander et al, 1980; Nilsson et al, 1982) has identified and characterized two classes of oligosaccharides
which are present on the core protein of this proteoglycan.
One class, which contains mannose, is linked to the core via
N-glycoside bonds between N-acetylglucosamine and appropriate asparagine residues in the polypeptide. More than 80%
of the oligosaccharides in this class have three branched
(triantennary) structures of the complex type indicated in
Fig. 1, which are very typical for a variety of glycoproteins
(Kornfeld and Kornfeld, 1980). Most of these N-linked oligosaccharides, about 12 per molecule, are attached to the core
protein in that portion of the polypeptide referred to as
the hyaluronate-binding region which contains the binding
site for hyaluronic acid that is involved in proteoglycan
aggregation (for reviews see Hascall, 1981; Muir, 1981;
Rodén, 1981; Hascall and Hascall, 1981). The second class
of oligosaccharide is linked to the core via O-glycoside
bonds between N-acetylgalactosamine and appropriate serine
and threonine residues in the polypeptide. The structures
of the three major oligosaccharides in this class are shown
in Fig. 1. As indicated below, the largest, oligosaccharide
A, is related to the linkage region for keratan sulfate
chains on proteoglycans isolated from more mature cartilages.
These O-linked oligosaccharides are typical "mucin" type
oligosaccharides, and there are about 150 distributed along
the core protein, approximately 1.5 for each chondroitin
sulfate chain.

Biosynthesis of the proteoglycan by chondrocytes from
the chondrosarcoma involves an orchestrated sequence of steps
from polypeptide synthesis in the rough endoplasmic reticulum
to completion of the complex carbohydrate structures on
the fully processed proteoglycan molecule somewhere in the
Golgi complex. A core protein precursor of about 350,000
molecular weight has been identified in these chondrocytes

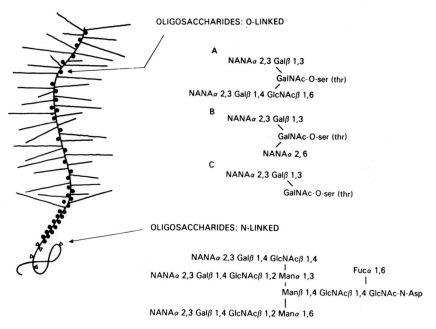

CHONDROSARCOMA PROTEOGLYCAN

OLIGOSACCHARIDES: O-LINKED

A

NANAα 2,3 Galβ 1,3
           \
           GalNAc-O-ser (thr)
           /
NANAα 2,3 Galβ 1,4 GlcNAcβ 1,6

B

NANAα 2,3 Galβ 1,3
           \
           GalNAc-O-ser (thr)
           /
      NANAα 2,6

C

NANAα 2,3 Galβ 1,3
           \
           GalNAc-O-ser (thr)

OLIGOSACCHARIDES: N-LINKED

NANAα 2,3 Galβ 1,4 GlcNAcβ 1,4
                 |
NANAα 2,3 Galβ 1,4 GlcNAcβ 1,2 Manα 1,3   Fucα 1,6
                          |
               Manβ 1,4 GlcNAcβ 1,4 GlcNAc-N-Asp
               |
NANAα 2,3 Galβ 1,4 GlcNAcβ 1,2 Manα 1,6

Fig 1. Schematic model for the structure of the monomer proteoglycan from the Swarm rat chondrosarcoma with the chemical
structures of the O-linked and N-linked oligosaccharides.

(Kimura et al, 1981b). This biosynthetic intermediate has a half life in the cell of about 1 hour before the chondroitin sulfate chains are added, and it has a functional hyaluronate-binding region since it is able to aggregate when reconstituted with carrier proteoglycan aggregate (Kimura et al, 1981a; 1981b). When this core precursor population was isolated from cells labeled with ^{3}H-mannose, it was found to contain ^{3}H-labeled oligo-peptides (Fellini et al, 1982). Therefore, the biosynthesis of the N-linked oligosaccharides begins at the earliest stages, presumably while the polypeptide is still on the ribosome as is the case with the synthesis of these oligosaccharides on normal glycoproteins (Kornfeld and Kornfeld, 1980). Additionally, the synthesis of the N-linked oligosaccharides on the core protein is blocked by tunicamycin (Stevens et al, 1982, Lohmander et al, 1982) which selectively interferes with the synthesis of dolichol-dipospho-N-acetylglucosamine, an obligatory intermediate for this pathway. Therefore, the synthesis of the N-linked oligosaccharides on the proteoglycan almost certainly follows the complex sequence of steps from addition of the high mannose oligosaccharide to an appropriate asparagine residue from a dolichol intermediate, to enzymatic removal of extra mannose and glucose residues, to the eventual addition of the sugars in the complex antennary branches, presumably in the Golgi complex (Kornfeld and Kornfeld, 1980).

The addition of the chondroitin sulfate chains to the core protein precursor occurs in the Golgi complex and is a very rapid step, requiring less than a minute or two (Kimura et al, 1981a). A series of experiments utilizing ^{3}H-glucosamine as a precursor in the chondrosarcoma chondrocyte cultures, has compared the kinetics of synthesis of the O-linked oligosaccharides which contain N-acetylglucosamine and N-acetylgalactosamine, and the chondroitin sulfate chains which contain N-acetylgalactosamine, in purified proteoglycans isolated at labeling times from 30 min to 420 min (Table 1) (Thonar et al, 1982). At all times, the ratio of radioactivity in the O-linked asialo-oligosaccharides to that in the chondroitin sulfate chains was constant within experimental error.* Additionally, the ratio of label in the galactosamine

*The sialic acids were removed since the equilibration of ^{3}H-glucosamine with CMP-sialic acid, the biosynthetic precursor for adding sialate differs from that for the UDP-N-ace-

tylhexosamines, the precursors for all the hexosamines in
the O-linked oligosaccharides and chondroitin sulfate.

Table 1 Biosynthesis of chondroitin sulfate and O-linked
oligosaccharides on proteoglycans from chondrosarcoma chondrocytes*

Labeling time (min)	Chondroitin sulfate (^{3}H-dpm x 10^{-6})	asialo-oligosaccharides (^{3}H-dpm x 10^{-5})	ratio CS/ASO
30	0.140	0.029	48.1
45	0.355	0.076	47.0
60	0.504	0.112	44.9
90	1.111	0.294	37.9
120	1.576	0.374	42.2
180	2.780	0.611	45.5
240	3.747	0.734	51.0
300	4.586	0.951	48.2
420	6.613	1.320	50.1

*Identical cultures of chondrocytes were labeled for the
indicated times with 50 µCi/ml ^{3}H-glucosamine (Lohmander et
al, 1980). The labeled proteoglycans were subsequently
purified and treated with alkaline-borohydride to release
chondroitin sulfate chains and O-linked oligosaccharides.
The O-linked oligosaccharides were purified and treated
with neuraminidase to remove sialic acid (Thonar et al,
1982). The values given are total dpm in purified chondroitin sulfate and O-asialo-oligosaccharides at the indicated
times.

and glucosamine residues in oligosaccharide A, Fig. 1, was
1.0. These results indicate that the entire O-linked oligosaccharide structures are added to appropriate serine and
threonine residues in the core precursor at nearly the same
time that the chondroitin sulfate chains are added, and therefore they are added in the Golgi complex. If the N-acetylgalactosamine residues were added during polypeptide synthesis
in the rough endoplasmic reticulum (Strouss, 1979), there
would be a considerable lag in the entry of the label in this
position into the completed proteoglycan because of the long
half time of the core protein precursor.

As hyaline cartilages mature, the proteoglycan molecules synthesized contain more and more keratan sulfate in addition to chondroitin sulfate (Pal et al, 1981). The most probable structure (see discussion in Nilsson et al, 1982) for the linkage region between the core protein and the keratan sulfate chain is shown in Fig. 2. Its derivation from oligosaccharide A is evident. Thus, the O-linked oligosaccharides are utilized at later stages of cartilage development as primers for synthesis of keratan sulfate chains. The consequence of this would be to introduce more charged glycosaminoglycans within the same domain of the completed proteoglycan molecules (compare Fig. 1 and 2). This might yield macromolecules which provide greater resistance to equivalent compressive load in a normal functional articular cartilage or might influence the relative hydration of the tissue matrix. The regulation of keratan sulfate synthesis, then, may be a critical parameter in tissue maturation. This can operate at the level of synthesis of the O-linked oligosaccharide primer. For example, the tranfer of sialic acid to the 3-position of the galactosyl-β1,4-N-acetylglucosaminyl branch prevents the addition of the next N-acetylglucosamine residue of the growing chain. Regulation could also operate at the level of the N-acetylglucosamine-β-3-galactose transferase required for chain elongation to occur. The comparative structures of the O-linked oligosaccharides and of keratan sulfate suggest that the addition of keratan sulfate to the proteoglycan evolved later, and took advantage of preexisting "mucin" oligosaccharides as chain initiators (Hascall, 1981). Interestingly, a closely related proteoglycan synthesized by smooth muscle cells (see Discussion) contains a large number of O-linked oligosaccharides but no keratan sulfate chains indicating that these cells have not developed this related biosynthetic capacity.

III.   KERATAN SULFATE-PROTEOGLYCAN FROM CORNEA

The keratan sulfate-proteoglycan in the mature corneal stroma is thought to be essential for maintenance of the highly organized structure of the collagen lattice in the extracellular matrix of this tissue, and this organization is probably required for optical transparency. A deficiency in the synthesis of keratan sulfate in the human genetic anomaly, corneal macular dystrophy, eventually requires a corneal transplant to prevent blindness due to corneal clouding (Hassell et al, 1979). Recent work (Nakazawa et al., 1982, Nilsson et al, 1982), has provided many details for the

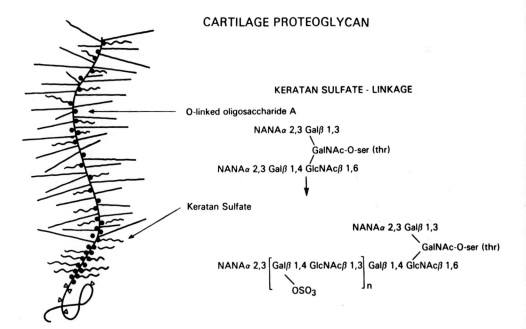

Fig. 2. Schematic model for the structure of a typical monomer
proteoglycan from a mature hyaline cartilage depicting the
relationship between O-linked oligosaccharide A of Fig. 1 and
the most probable structure of the linkage region of the
keratan sulfate chains.

structure of this class of proteoglycans isolated from monkey cornea, Fig. 3. Unlike the cartilage proteoglycans, which have large hydrodynamic sizes and molecular weights, 2-3 million, the corneal keratan sulfate-proteoglycans are small, having molecular weights of about 70,000. The hydrodynamic dimensions of the molecules are such that they can fit within the spaces between the lattice of the collagen fibrils in the stroma without distorting its regularity (Axelsson and Heinegård, 1978).

Oligosaccharide-peptides were separated from keratan sulfate-peptides following papain digestion of the purified proteoglycan isolated from monkey corneas. Only N-linked oligosaccharides were present, and structural analyses using selective chemical degradation procedures and direct gas chromatography-mass fragmentation analyses identified the two high mannose-type structures shown in Fig. 3. They contain the core structure of 3 mannoses and two N-acetylglucosamines with either one or three additional exterior mannoses. These two structures have been identified as intermediates in the processing steps involved in converting the original high mannose-glucose structure to the final complex structures characteristic of this class of oligosaccharides, such as that described above for the cartilage proteoglycan. In this case for the corneal proteoglycan, however, none of the oligosaccharides were processed to the complex form.

The linkage region between the keratan sulfate and the protein core was isolated after removing the backbone glycosaminoglycan chains with endo-β-galactosidase. Its structure, determined by similar procedures used for the oligosaccharides, is clearly related to the two branched (biantennary) form of the N-linked complex oligosaccharides. The endo-β-galactosidase digestion left the first N-acetylglucosamine of the repeating keratan sulfate backbone with the linkage region, and each linkage region originally contained two keratan sulfate chains, Fig. 3. If these two, non-reducing terminal N-acetylglucosamines were replaced by sialic acids, the structure would be a very common one found on many glycoproteins. As with the O-linked oligosaccharide A, Fig. 2, each branch of the linkage structure, Fig. 3, contains a galactosyl-β-1,4-N-acetylglucosaminyl moiety to serve as a primer for keratan sulfate chain elongation when exposed to a proper N-acetylglucosamine transferase. Almost certainly, then, the corneal keratan sulfate-proteoglycan has evolved from a glycoprotein precursor, and the enzyme machinery for

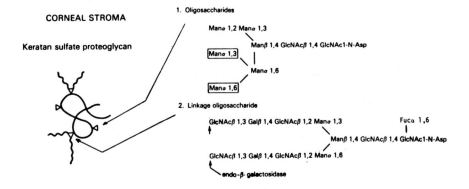

Fig 3. Schematic model for the structure of keratan sulfate-proteoglycan from monkey cornea showing the structures of the oligosaccharides and of the linkage region of keratan sulfate chains. The mannose residues in boxes are absent on about 50% of the oligosaccharides, and the arrows indicate where the endo-β-galactosidase removed the rest of the two keratan sulfate chains originally on the linkage structure.

adding keratan sulfate chains uses an entirely different oligosaccharide structure for initiating synthesis. The linkage region structure is entirely consistent with the observation that keratan sulfate synthesis in corneal tissue is particularly sensitive to tunicamycin treatment (Hart and Lennarz, 1978). As with embryonic cartilage, embryonic corneal tissue contains less and undersulfated keratan sulfate compared with that which appears at later, more mature stages in corneal development (Hart, 1976). It will be interesting to determine how keratan sulfate synthesis is regulated during development in cartilage and cornea and what parts of the pathway are analogous.

IV CONCLUDING REMARKS

Proteoglycans are defined as macromolecules having at least one covalently bound glycosaminoglycan chain. From the discussion above, it seems quite probable that different classes of proteoglycans will contain entirely different core proteins which are not related gene products. However, within a given class there may well be several closely related types. For example, recent work has shown that smooth muscle cells derived from monkey aorta synthesize several classes of proteoglycans, one of which can interact with hyaluronic acid and link protein to form link-stabilized aggregates (Wight and Hascall, 1982; Chang et al, 1982). While this proteoglycan must therefore have a hyaluronic acid-binding region which is similar or identical to that of cartilage proteoglycans, it has distinctly different characteristics in the remainder of the core structure. These proteoglycans have far fewer chondroitin sulfate chains (about 25 per core) and more 0-linked oligosaccharides (about 250 per core) than the cartilage prototype discussed above. Since it is likely that the amino acid sequence specificity around serine residues which accept xylose and initiate chondroitin sulfate synthesis is different from that around serine and threonine residues which accept N-acetylgalactosamine and initiate 0-linked oligosaccharide synthesis, it is very likely that the cartilage proteoglycan differs significantly from the smooth muscle proteoglycan in primary sequence in the region outside the hyaluronic acid-binding region. These core proteins, then, may well be related gene products analogous to the different α-chains of collagen. It is likely that the application of the exciting methodology developed for sequencing DNA to the analysis of proteoglycan core structures as well as conventional peptide mapping will be vital to begin to understand the relationships of different classes of proteoglycan at the molecular and genetic levels. Such information will be needed to answer critical quéstions relating to the regulation of synthesis of proteoglycans during tissue development and maturation.

REFERENCES

Axelsson I, Heinegård D (1978). Characterization of the Keratan Sulphate Proteoglycans from Bovine Corneal Stroma Biochem J 169:517-530.
Caplan AI, Hascall VC (1980). Structure and Developmental Changes in Proteoglycans. In Naftolin F, and Stubblefield PG (eds): "Dilatation of Uterine Cervix," New York: Raven Press, pp 79-98.
Chang Y, Yanagishita M, Hascall VC, Wight TN (1982). Proteoglycans Synthesized by Smooth Muscle Cells Derived from Monkey (Macaca nemestrina) Aorta, submitted.
Fellini SA, Kimura JH, Hascall VC (1982). Localization of Core Protein in Subcellular Fractions of Chondrocytes from the Rat Chondrosarcoma. Submitted.
Hart GW (1976). Biosynthesis of Glycosaminoglycans during Corneal Development. J Biol Chem 251:6513-6521.
Hart, GW, and Lennarz, WJ, (1978). Effects of Tunicamycin on the Biosynthesis of Glycosaminoglycans by Embryonic Chick Cornea. J. Biol Chem. 253:5795-5801.
Hascall VC (1981). Proteoglycans: Struction and Function. In Ginsburg V, Robinson P (eds): "The Biology of Carbohydrates, Vol. I," New York: John Wiley, pp 1-49.
Hascall VC, Hascall GK (1981). Proteoglycans. In Hay ED (ed): "Cell Biology of Extracellular Matrix," New York: Plenum, pp 39-63.
Hassell JR, Newsome DA, Krachmer J, Rodriguez M (1980). Corneal Macular Dystrophy: A Possible Inborn Error in Corneal Proteoglycan Maturation. Proc Natl Acad Sci USA 77:3705-3709.
Kimura JH, Caputo CB, Hascall VC (1981a). Effect of Cycloheximide on Synthesis of Proteoglycans by Cultured Chondrocytes. J Biol Chem 256:4368-4376.
Kimura JH, Thonar EJ, Hascall VC, Reiner A, Poole AR (1981b). Identification of Core Protein, an Intermediate in Proteoglycan Biosynthesis in Cultured Chondrocytes from the Swarm Rat Chondrosarcoma. J Biol Chem 256:7890-7897.

Kornfeld R, Kornfeld S (1980). Structure of Glycoproteins and their Oligosaccharide Units. In Lennarz WA (ed): "Biochemistry of Glycoproteins and Proteoglycans," New York: Plenum, pp 1-34.

Lohmander LS, De Luca S, Nilsson B, Hascall VC, Caputo CB, Kimura JH, Heinegård DK (1980). Oligosaccharides on Proteoglycans from the Swarm Rat Chondrosarcoma. J Biol Chem 225:6084-6091.

Lohmander LS, Fellini SA, Stevens RL, Hascall VC, Kimura JH (1981). Effect of Tunicamycin on the Formation of Proteoglycan Aggregates in Cultures of Chondrocytes from the Swarm Rat Chondrosarcoma. In Yamakawa T, Osawa T, Handa, S (eds): "Glycoconjugates, Proceedings of the Sixth International Symposium on Glycoconjugates," Tokyo: Japan Scientific Societies Press, pp 443-444.

Muir IHM (1980). The Chemistry of the Ground Substance of Joint Cartilage. In Sokolof L (ed): "The Joints and Synovial Fluid II," New York: Academic Press, pp 27-94.

Nakazawa K, Nilsson B, Hassell JR, Hascall VC, Newsome DA (1982). Isolation of Keratan Sulfate Linkage Region Oligosaccharide and Mannose Containing Oligosaccharides from Monkey Corneal Keratan Sulfate Proteoglycan, submitted.

Nilsson B, De Luca S, Lohmander LS, Hascall VC (1982). Structures of N-linked and O-linked Oligosaccharides on Proteoglycan Monomer Isolated from the Swarm Rat Chondrosarcoma. J Biol Chem, In press.

Nilsson B, Nakazawa K, Hassell JR, Newsome DA, Hascall VC (1982). Structure of Oligosaccharides and the Linkage Region Between Keratan Sulfate and the Core Protein on Proteoglycans from Monkey Cornea, submitted.

Pal S, Tang L-H, Choi H, Habermann E, Rosenberg L (1981). Structural Changes During Development in Bovine Fetal Epiphyseal Cartilage. I. Isolation and Characterization of Proteoglycan Monomers and Aggregates. Collagen and Rel Res 1:151-176.

Rodén L (1980). Structure and Metabolism of Connective Tissue Proteoglycan. In Lennarz WA (ed): "Biochemistry of Glycoproteins and Proteoglycans," New York:Plenum, pp 267-371.

Stevens RL, Schwartz LB, Austen KF, Lohmander LS, Kimura JH (1982). Effect of Tunicamycin on Insulin Binding and Proteoglycan Synthesis and Distribution in Swarm Rat Chondrosarcoma Cell Cultures. J Biol Chem, in press.

Strouss GJAM (1979). Initial Glycosylation of Proteins with Acetylgalactosaminylserine Linkages. Proc Natl Acad Sci USA 76:2694-2698.

Thonar EJ, Lohmander LS, Kimura JH, Fellini SA, Hascall VC
(1982). Biosynthesis of O-linked Oligosaccharides on
Proteoglycans by Chondrocytes from the Swarm Rat Chondrosarcoma,
submitted.
Wight TN, Hascall VC (1982). Proteoglycans in Primate Arteries: III Characterization of the Proteoglycans Synthesized by Arterial Smooth Muscle Cells in Culture. J Cell
Biol, in press.

Limb Development and Regeneration
Part B, pages 17–24
© 1983 Alan R. Liss, Inc., 150 Fifth Avenue, New York, NY 10011

THE LINK PROTEINS: SOME ASPECTS OF THEIR STRUCTURE

John R. Baker, Bruce Caterson and
James E. Christner
University of Alabama in Birmingham

Birmingham, Alabama 35294

Link proteins are, together with hyaluronic acid and proteoglycan monomers, components of proteoglycan aggregates, the predominant form in which cartilage proteoglycans occur in vivo. The link proteins are believed to stabilize the interactions between the components of the proteoglycan aggregate. The structure and function of cartilage proteoglycan aggregate components have been reviewed recently (Hascall, 1981).

There are two major link proteins (LP), LP1 and 2, of molecular weights 49K and 45K, respectively. Trypsin and clostripain treatment of the aggregate causes limited cleavage of LP1 and 2 to give a single link protein (LP3) of molecular weight 43K, which remains associated with the proteoglycan aggregate. LP1 and 2 have blocked or buried N-termini whereas LP3 can be sequenced. Therefore, trypsin and clostripain remove one or more peptides from the N-terminal end of LP and may additionally give some cleavage at the C-terminus. To determine what peptide or peptides are indeed removed by clostripain, the strategy outlined in Fig. 1 was employed. LP isolated from bovine nasal cartilage and chondrosarcoma were labeled with ^{125}I using the Bolton Hunter reagent. A sub-equimolar amount of the ^{125}I-LP was mixed with proteoglycan and hyaluronic acid in 4 M guanidine hydrochloride, pH 7, the mixture was brought to associative conditions by dialysis and the reconstituted proteoglycan aggregate purified by chromatography on Sepharose CL 2B (more than 99% of the ^{125}I-LP was incorporated into the proteoglycan aggregate).

Fig. 1

Reconstitution of proteoglycan aggregate (AI) incorporating ^{125}I–link protein, and subsequent digestion by clostripain.

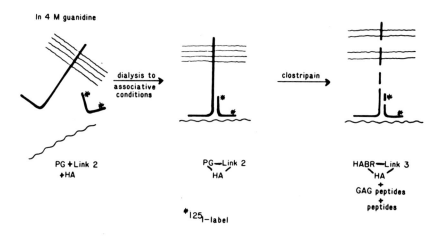

In 4 M guanidine

dialysis to
associative
conditions

clostripain

PG +Link 2
+HA

PG—Link 2
HA

HABR—Link 3
HA
+
GAG peptides
+
peptides

#^{125}I–label

^{125}I–proteoglycan aggregate was digested with clostripain, the digest was fractionated by acrylamide gel electrophoresis in SDS on a 15% gel and ^{125}I products located by autoradiography. Short time exposure revealed clostripain treatment had converted all rat chondrosarcoma LP2 to LP3, (Fig. 2, left side), and longer time exposure showed only one peptide formed in the conversion of LP2 to LP3 (Fig. 2, right side, lane 2). Two small peptides, presumably one from LP1 and the other from LP2, are formed upon cleavage of ^{125}I–proteoglycan aggregate from bovine nasal cartilage with clostripain (Fig. 2, lane 3). Therefore, it may be concluded that in the proteolytic conversion of LP1 or 2 to LP3 only single N-terminal peptides are released.

Some sequence data on the N-terminal of bovine nasal cartilage LP3 has been published by Périn, et al. (1980). We have sequenced through the first twenty residues of rat chondrosarcoma LP3, and compared the results (Fig. 3). One preparation was obtained through the use of trypsin (Périn, et al., 1980) and the other with clostripain, but it is evident that cleavage

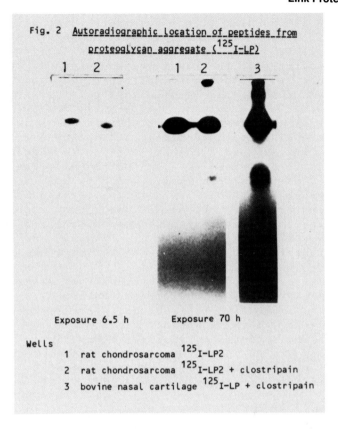

Fig. 2 <u>Autoradiographic location of peptides from proteoglycan aggregate (^{125}I-LP)</u>

Exposure 6.5 h   Exposure 70 h

Wells
1 rat chondrosarcoma ^{125}I-LP2
2 rat chondrosarcoma ^{125}I-LP2 + clostripain
3 bovine nasal cartilage ^{125}I-LP + clostripain

Fig. 3 <u>N Terminal Sequence of Link Protein 3</u>

Bovine Nasal (trypsin)
1   5   10
Leu-Leu-Val-Glu-Ala-Glu-Gln-Ala-Lys-Val-Phe-Ser-Arg-Arg-
15   20
Gly-Gly-X-Val-Thr-Val-

Rat Chondrosarcoma (clostripain)
1   5   10
Lys-Leu-Val-Glu-Ala-Glu-Gln-Ala-Lys-Val-Phe- X - X -Val-
15   20
Gly-Gly-X-Val-Thr-Leu-

occurred at the same site. Considerable homology can be seen. Although the determined sequences represent only approximately 5% of link protein primary structure, one might predict that the remaining sequences may also be homologous, as LP3 possesses the specific hyaluronic acid- and proteoglycan-binding regions of native LP. This prediction is only partially fulfilled by the evidence of peptide mapping on LP3 from both sources. LP3 from bovine nasal cartilage and rat chondrosarcoma (Fig. 4, lanes 1 and 2 respectively) were digested with S. aureus V8 protease (40 µg/ml 20° 16 h). The resulting peptides were fractionated by acrylamide gel electrophoresis in SDS (15% gel) and located by staining with Coomassie blue. Prominant bands from rat chondrosarcoma LP3 are seen at approximately 25K, 20K, 17K, 10K and 7K (Fig. 4, lane 2). The 25K and 10K bands are not shared with bovine nasal cartilage LP3 (lane 1).

It is evident from both biochemical (Baker and Caterson, 1979) and immunological (Baker, Caterson and Christner, 1982) data that LP1 and 2 are structurally similar. To determine whether these two major link proteins are products of a single biosynthetic precursor, and to answer many other important questions concerning link protein biosynthesis, requires study of the translation products of their mRNA. This we have done for the products of rat chondrosarcoma mRNA (Baker, unpublished findings). In this case, a single link protein of molecular weight slightly larger than LP2 was identified by immunoprecipitation with a rabbit anti link protein antiserum (R #13) (Baker, Caterson and Christner, 1982). Unfortunately, the antiserum has low affinity for the translation product and at least five days autoradiography exposure time is required for location. Therefore, we have begun to characterize the determinants recognized by this polyclonal antibody, and by two monoclonal antibodies, termed 6A1 and 8A4, which we might be tempted to employ for recognition of cell-free-synthesized link proteins.

A duplicate V8 protease digest of rat chondrosarcoma LP3 was fractionated by acrylamide gel electrophoresis in SDS (as for Fig. 4) and electroblotted to nitrocellulose (Burnette, 1981). Treatment of this Western blot with polyclonal antibody R #13 followed by ^{125}I-protein A was employed to

Fig. 4 <u>V8 protease digestion of LP3 from bovine nasal
cartilage and rat chondrosarcoma</u>

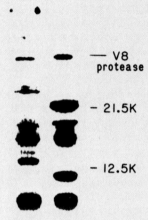

Well
1   bovine nasal cartilage LP3 + V8 protease
2   rat chondrosarcoma LP3 + V8 protease

determine which peptides possessed antigenic
determinants. All peptides reacted although to a
different relative degree than by Coomassie blue
staining. The duplicate Western blot was reacted with
^{125}I-wheat germ agglutinin, whereupon a similar profile
of located peptides was found. Therefore,
oligosaccharide-bearing peptides are antigenic, and
indeed oligosaccharides may form part of the determinant
sites. This conclusion is supported by the finding that
glycopeptides from LP, isolated following pronase

digestion, partially inhibit the binding of this
antiserum (R #13) to link protein.

Both monoclonal antibodies were examined for
reactivity with link proteins from a number of sources:
bovine nasal cartilage, rat chondrosarcoma, human
articular cartilage and chicken sternal cartilage. 8A4
reacted with LP1, 2 and 3 regardless of source and 6A1
reacted only with LP2 from rat chondrosarcoma.
Experiments to determine which peptides, derived from
LP2 of rat chondrosarcoma by V8 protease digestion,
would bind to the two monoclonal antibodies revealed
different patterns of reactivity. The data, leading to
the conclusion that 6A1 recognizes a determinant near
the N-terminal of rat chondrosarcoma LP2, and 8A4 binds
to a determinant on the C-terminal half of the molecule,
have been described by Caterson, et al., 1982.

To test whether oligosaccharides are involved in
determinant regions, samples of rat chondrosarcoma LP2
were digested with neuraminidase, β-galactosidase,
β-N-acetylglucosamindase and endoglycosidase D or with
endoglycosidase H (Muramatsu, 1978) prior to separation
by acrylamide gel electrophoresis in SDS (10% gel) and
immunolocation of products with ^{125}I-labeled 6A1 or 8A4
(Fig. 5). If the products were located by Coomassie
blue staining, it was clear that endoglycosidase H
digestion caused LP2 to behave electrophoretically like
LP3 (i.e. it increased in mobility), whereas no effect
of the exoglycosidases plus endoglycosidase D was noted
(results not shown). The 6A1 and 8A4 antibodies both
reacted with the product of exoglycosidases plus
endoglycosidase D (Fig. 5, lanes 2). In fact this
combination of enzymes may have had little or no effect
on LP2. 6A1 binds to a trace of LP2 remaining after
endoglycosidase H digestion, but not to the LP3 product
(Fig. 5, lane 3). Presumably a high mannose chain
required for 6A1 binding to rat chondrosarcoma LP is
removed by this treatment and is responsible for the
diminished molecular weight of the product, LP3. The
removal of the high mannose chain does not affect the
binding of 8A4 (Fig. 5, lane 3). Therefore, 8A4 may be
a monoclonal antibody that recognizes a peptide
determinant, which would be useful in studies of
cell-free biosynthesis of link protein.

Fig. 5 <u>Immunolocation of rat chondrosarcoma link protein following digestion with glycosidases</u>

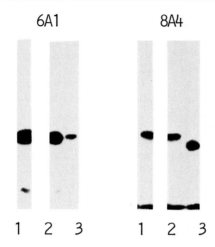

Wells

1   rat chondrosarcoma LP2
2   rat chondrosarcoma LP2 digested with
     neuraminidase, β-galactosidase, β-N-
     acetylglucosamindase and endoglycosidase D
3   rat chondrosarcoma LP2 digested with
     endoglycosidase H

     Also of importance are antibodies specific for LP
which can still bind to and recognize LP in its native
state (i.e. within the proteoglycan aggregate) and can
therefore be used for the immunohistological location of
LP. The best chance for this is with a monoclonal
antibody like 6A1 that recognizes a small
proteolytically-removable fragment of LP, which is not
involved in binding to the proteoglycan.

Acknowledgements

     We wish to thank Drs. William Butler and Ajit
Bhown, University of Alabama in Birmingham for
assistance in the amino acid sequence work. We are
grateful to Ms. Margaret Moore for skilled technical
assistance and to Ms. Sherri Nielson for typing the
manuscript.

This work was supported by NIH grants DE 02670 and AM 27791.

REFERENCES

Baker JR, Caterson B (1979). The isolation and characterization of the link proteins from proteoglycan aggregates of bovine nasal cartilage. J Biol Chem 254:2387.

Baker JR, Caterson B, Christner JE (1982). The specificity of a rabbit antiserum to link proteins from bovine nasal cartilage. In manuscript.

Burnette WN (1981). "Western blotting": Electrophoretic transfer of proteins from sodium dodecyl sulfate-polyacrylamide gels to unmodified nitrocellulose and radiographic detection with antibody and radioiodinated protein A. Anal Biochem 112:195.

Caterson B, Baker JR, Christner JE, Lee Y, Lentz M (1982). Characterization of monoclonal antibodies directed against rat chondrosarcoma link protein. Submitted to J Biol Chem.

Hascall VC (1981). Proteoglycans: Structure and function. In Ginsburg V (ed): "Biology of Carbohydrates, Vol. 1," New York: John Wiley and Sons, Inc.

Muramatsu T, Koide N, Maeyama K (1978). Further studies on endo-β-N-acetylgluconidase D. J Biochem 83:363.

Périn J-P, Bonnet F, Pizon V, Jolles J, Jolles P (1980). Structural data concerning the link proteins from bovine nasal cartilage proteoglycan complex. FEBS Lett 119:333.

Limb Development and Regeneration
Part B, pages 25–33
Published 1983 by Alan R. Liss, Inc., 150 Fifth Avenue, New York, NY 10011

THE ROLE OF CHONDRONECTIN AND CARTILAGE PROTEOGLYCAN IN
THE ATTACHMENT OF CHONDROCYTES TO COLLAGEN

A. Tyl Hewitt, Hugh H. Varner
Michael H. Silver* and George R. Martin

Laboratory of Developmental Biology and Anomalies
National Institute of Dental Research
National Institutes of Health
Bethesda, Maryland 20205

*Department of Laboratory Medicine and Pathology
University of Minnesota Medical Center
Minneapolis, Minnesota 55455

CHRONDROCYTE ATTACHMENT TO COLLAGEN

In cartilage, chondrocytes are surrounded almost entirely by a homogeneous matrix of type II collagen, chondroitin sulfate proteoglycan and other materials. Some
progress has been made in defining the molecular
mechanism used by chondrocytes to bind to their matrix.
Previous studies on fibroblasts in vitro have shown that
their attachment to collagen is stimulated by fibronectin
(Klebe 1974; Pearlstein 1976), a large glycoprotein
produced by many types of cells and present in serum and
various tissues (Stenman, Vaheri 1978; Yamada, Olden
1978). However, fibronectin does not stimulate the
attachment of isolated chondrocytes (Hewitt et al 1980)
and normal cartilage does not contain fibronectin as
judged by immunohistology (Dessau et al 1978; Stenman,
Vaheri 1978). For these reasons we speculated that a
different attachment protein was involved in binding
chondrocytes to their matrix. In our studies the
attachment to collagen of embryonic chick sternal
chondrocytes was compared with the attachment of Chinese
hamster ovary (CHO) cells, a cell line that is known to

Table 1 - Comparison of Cell Attachment

	Chondrocytes	CHO cells
Collagen Preference	II > I > III, IV	I=II=III=IV
Rate of Attachment (1/2 maximal attachment)	30 min.	15 min.
Serum required for maximal attachment	5%	1%

    Chondrocytes were isolated from 13-17 day chick
embryo sterna with 0.4% collagenase (Worthington, CLS
II). CHO cells from nearly confluent cultures were
removed from tissue culture dishes with 0.1% trypsin/0.1%
ethylenediamine tetraacetate (EDTA). Cells (1.5×10^5 in
0.1ml Ham's F12 culture medium) were added to collagen-
coated bacteriologic plastic dishes (10 μg collagen/35mm
dish) that had been preincubated with Ham's F12 medium
(1ml) containing the material to be tested for attachment
activity. After incubating for 90 min., unattached cells
were rinsed off with phosphate-buffered saline and the
attached cells removed with 0.1% trypsin/0.1% EDTA. The
numbers of attached and unattached cells were determined
using an electronic cell counter (Coulter).

require fibronectin as an adhesive factor. (See legends
to Table 1 and Figure 1 for experimental details).

    Table 1 shows some attachment characteristics of CHO
cells and chondrocytes in the presence of serum. The
attachment of both CHO cells and chondrocytes was
stimulated by serum, but there were differences between
the two cell types with regard to preference for collagen
type (Table 1). While CHO cells attached equally well to
all collagen types, chondrocytes showed a preference for
type II (cartilage-specific) collagen. Also,
chondrocytes required higher levels of serum for maximal
attachment and attached at a slower rate than CHO cells
(Table I).

    Because chondrocytes required serum for attachment,
we next studied the involvement of fibronectin in

chondrocyte attachment. Fibronectin was isolated from
serum by affinity chromatography on a column of collagen
covalently bound to Sepharose (Hopper, et al 1976;
Engvall, Rouslahti 1977). Figure 1 shows the results of
attachment studies on CHO cells and chondrocytes using
serum, fibronectin and fibronectin-depleted serum. While

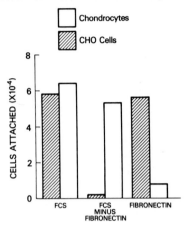

Figure 1. The effect of serum, fibronectin-free serum
and fibronectin on CHO cell and chondrocyte attachment.
Plastic petri dishes were coated with 10 μg of type II
collagen and allowed to air dry. The dishes were then
preincubated for 60 min. at 37° with Ham's F12 culture
medium (1ml) containing the material to be tested for
attachment activity. Cells (1.5 x 10^5) were then added
and incubated for an additional 90 min. At the end of
this time, the unattached and the attached cells were
removed from the dish and were counted using an
electronic cell counter (Coulter). Figure reprinted
from Kleinman, et al (1979). J Supramol Struct 11:69.

serum supported the attachment of both cell types,
fibronectin-free serum stimulated the attachment of
chondrocytes but not that of CHO cells. Fibronectin did
not stimulate chondrocyte attachment but it did stimulate
CHO cell attachment. These data suggested that chondrocytes use a serum factor other than fibronectin to
mediate their interaction with collagen.

CHARACTERIZATION OF CHONDRONECTIN--THE CHONDROCYTE
ATTACHMENT FACTOR

The chondrocyte and CHO cell attachment activities
in serum were separated from one another by DEAE-
cellulose, ion exchange column chromatography (Hewitt et
al 1980). The chondrocyte attachment protein was
isolated by chromatography on Cibacron blue-agarose and
WGA-Sepharose (Hewitt, et al 1982), and found to be a
glycoprotein with a M_r=175,000. Some of its properties

Table 2 - Characteristics of Chondronectin and
Fibronectin

	Chondronectin	Fibronectin
Molecular Weight (M_r)		
native	175,000	440,000
reduced	60,000	220,000
Heat Lability ($T_{1/2}$)	52°	57°
Amount in Serum	20 µg/ml	300 µg/ml
Tissue Distribution	cartilage, vitreous humor	dermis, bone, etc.
Amount required for cell attachment	10-100ng	1-10 µg

as well as those of fibronectin are listed in Table 2.
Chrondronectin is smaller, somewhat more heat labile and
is active in cell attachment at much lower levels than
fibronectin. While both are present in serum, chondronectin appears to be less abundant. Chondronectin also
has a different tissue distribution, being found only in
cartilage and the vitreous while fibronectin is not
present in these sites. Fibronectin is also very
sensitive to trypsin while chondronectin resists
degradation indicating that chondronectin has a compact
structure. When tested by the enzyme-linked,
immunoabsorbant assay (ELISA) (Rennard, et al 1980)

chondronectin was found to be immunologically distinct
from fibronectin (Hewitt, et al 1982).

As expected, antibodies to chondronectin block
attachment mediated by exogenously added chondronectin.
In addition, they inhibit the spontaneous attachment of
chondrocytes that occurs over extended periods of time in
the absence of either serum or chondronectin (Hewitt, et
al 1982). This indicates that chondrocytes produce an
attachment factor immunologically and functionally
similar to the serum-derived chondronectin. Immunofluorescence studies on chick limb cartilage
condensations (Figure 2) and frozen sections of embryonic
chick sternum (Hewitt, et al 1982) localize chondronectin
to a pericellular region in cartilage, a position
consistent with its postulated role of binding
chondrocytes to their matrix.

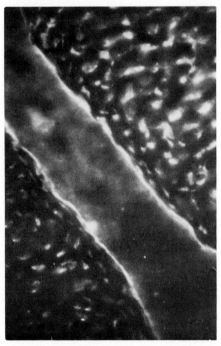

Figure 2. Immunofluorescence localization of chondronectin
in the cartilage nodules of embryonic chick limb bud. Note
that it is localized pericellularly around the chondrocytes
and not throughout the matrix.

MECHANISM OF CHONDROCYTE ATTACHMENT TO COLLAGEN

Previous studies indicate that chondronectin does
not bind readily to collagen (Hewitt et al 1980). This
conclusion was reached by preincubating collagen-coated
dishes with culture medium containing either serum or
purified attachment factors for 60 min., rinsing the
dishes with phosphate-buffered saline to remove unbound
material, and then testing for cell attachment with cells
in unsupplemented medium. When experiments were carried
out with fibronectin, the fibronectin was found to bind
to the matrix and to stimulate CHO cell attachment (not
shown). However, few chondrocytes attached under the
same conditions and little chondrocyte attachment was
observed when dishes preincubated with chondronectin were
rinsed prior to adding the cells (Figure 3; see also
Hewitt, et al 1980) Such results could indicate that
chondronectin requires an additional cartilage
macromolecule to function as an attachment protein.
    Our studies indicate that chondronectin binds to
glycosaminoglycans and that binding to the glycosaminoglycan portion of intact proteoglycan is required for
chondrocyte attachment. First, chondronectin binds well
to collagen substrates in the presence of cartilage

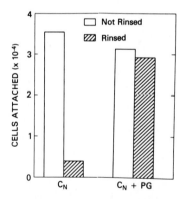

Figure 3. The requirement for proteoglycan in
chondrocyte attachement. Assays were performed as
described in Figure 1 except that some of the dishes were
rinsed and the preincubation mixture replaced with
culture medium alone before adding the chondrocytes.

proteoglycan (Figure 3). In contrast to the inability of
rinsed, chondronectin-treated collagen to support
chondrocyte attachment, collagen-coated substrates
preincubated with chondronectin and cartilage
proteoglycan monomer will stimulate attachment even when
rinsed. Proteoglycan alone does not significantly
enhance chondrocyte attachment (not shown). This
suggests that the proteoglycan stabilizes the interaction
between collagen and chondronectin. In addition,
xylosides, which interfere with the formation of normal
proteoglycans, inhibit the attachment of chondrocytes to
collagen. Further, the addition of chondronectin-4-
sulfate or chondroitin-6-sulfate to the attachment assay
inhibits chondrocyte attachment while keratan sulfate and
hyaluronic acid have no effect. However, this inhibition
is overcome by the addition of increasing levels of
intact proteoglycan monomer (not shown). These findings
suggest that chondronectin can bind to glycosaminoglycans
but that the intact proteoglycan monomer is required for
attachment.

SUMMARY

The interaction of cells with collagen has
previously been demonstrated to be of importance in the
growth and differentiation of various cells (reviewed by
Hay 1981; Kleinman, et al 1981). Chondronectin has been
demonstrated to be involved in mediating the interaction
of chondrocytes with their matrix. Unlike fibronectin,
which binds directly to collagen before the cell can
interact with this macromolecular complex, chondronectin
must interact with cartilage proteoglycan monomer in
order to function efficiently as an attachment factor.

Figure 4 is a schematic representation of the
interaction between chondrocytes and their matrix
components. We suggest that if these specific
interactions are perturbed by the introduction of some
outside factor, the phenotypic expression of the cells
could be altered. For example, chondrocytes are not
normally associated with fibronectin (Dessau, et al 1978;
Stenman, Vaheri 1978). However, if chondrocytes are
grown in the presence of fibronectin they will become
more fibroblastic and stop making the cartilage-specific
molecules of type II collagen and cartilage proteoglycan

(Pennypacker, et al 1979; West et al 1979).
Consequently, it is possible that the specific
interactions seen between chondrocytes, chondronectin,
type II collagen, and proteoglycan may be essential for
the maintenance of the chondrocyte phenotype.

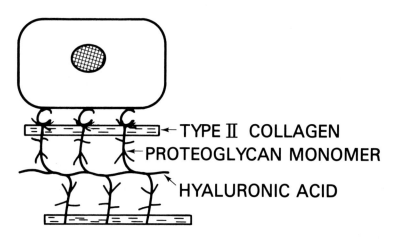

TYPE II COLLAGEN
PROTEOGLYCAN MONOMER
HYALURONIC ACID

Figure 4. Schematic representation of the interactions
between chondronectin (C), chondrocytes, type II
collagen, and cartilage proteoglycan.

REFERENCES

Dessau W, Sasse J, Timpl R, Jilek F, von der Mark K
    (1978). Synthesis and extracellular deposition of
    fibronectin in chondrocyte cultures. Response to
    the removal of extracellular cartilage matrix. J
    Cell Biol 79:342.
Engvall E, Rouslahti E (1977). Binding of a soluble form
    of fibroblast surface protein, fibronectin, to
    collagen. Int J Cancer 20:1.

Hay ED (1981). Extraceullular matrix. J Cell Biol 91
    (Part 2):205s.

Hewitt AT, Kleinman HK, Pennypacker JP, Martin GR
    (1980). Identification of an adhesion factor for
    chondrocytes. Proc Natl Acad Sci USA 77:385.

Hewitt AT, Varner HH, Silver MH, Dessau W, Wilkes CM, Martin GR (1982). The isolation and partial characterization of chondronectin, an attachment factor for chondrocytes. J Biol Chem 257:2330.

Hopper DE, Adelman BC, Gentner G, Gay S (1976). Recognition by guinea pig peritoneal exudate cells of conformationally different states of the collagen molecule. Immunology 30:249.

Klebe RJ (1974). Isolation of a collagen-dependent cell attachment factor. Nature (London) 250:248.

Kleinman HK, Klebe RJ, Martin GR (1981). Role of collagenous matrices in the adhesion and growth of cells. J Cell Biol 88:473.

Pearlstein E (1976). Plasma membrane glycoprotein which mediates adhesion of fibroblasts to collagen. Nature (London) 262:497.

Pennypacker JP, Hassell JR, Yamada KM, Pratt RM (1979). The influence of an adhesive cell surface protein on chondrogenic expression in vitro. Exp Cell Res 121:411.

Rennard SI, Berg R, Martin GR, Foidart J-M, Gehron Robey P (1980). Enzyme-linked immunoassay (ELISA) for connective tissue components. Anal Biochem 104:205.

Stenman S, Vaheri A (1978). Distribution of a major connective tissue protein, fibronectin in normal human tissues. J Exp Med 147:1054.

West CM, Lanza R, Rosenbloom J, Lowe M, Holtzer H, Avdalovic N (1979). Fibronectin alters the phenotype properties of cultured chick embyo chondroblasts. Cell 17:491.

Yamada KM, Olden K (1978). Fibronectin--adhesive glycoproteins of cell surface and blood. Nature (London) 275:177.

Limb Development and Regeneration
Part B, pages 35–43
© **1983 Alan R. Liss, Inc., 150 Fifth Avenue, New York, NY 10011**

PROTEOGLYCANS AND MATRIX PROTEINS IN CARTILAGE

Dick Heinegård, Mats Paulsson and Yngve Sommarin

Department of Physiological Chemistry
University of Lund, P.O. Box 750,
S-220 07 Lund, Sweden

Cartilage matrix contains comparatively few types of
macromolecules. Its extremely high content of fixed charged
groups (Maroudas 1979) is contributed by the proteoglycans,
being the major constituent essential for the resilience
of the tissue. Another major component is collagen, providing tensile strength and preventing changes of tissue
volume. Cartilage also contains a number of specific matrix
proteins, synthesized by the chondrocyte. The content of
plasma proteins, however, is remarkably low (Paulsson,
Heinegård 1979) probably a result of low permeability to
macromolecules. The aim of the present work is to briefly
describe the different classes of cartilage matrix molecules and to discuss molecules specific for certain types
of cartilage. The presentation will primarily deal with
proteoglycans and with non-collagenous matrix proteins.

PROTEOGLYCANS

Cartilage contains proteoglycans capable of specific
interaction with hyaluronate (Hardingham, Muir 1972; Hardingham, Muir 1973; Hascall, Heinegård 1974), Fig. 1, i.e.
aggregating proteoglycans, as well as others not capable of
interaction, i.e. non-aggregating proteoglycans.

The general structure of proteoglycans is known mainly
from studies of the aggregating cartilage proteoglycans,
Fig. 1.. Their core protein is divided into three regions of
different composition (Heinegård, Axelsson 1977). At one end
the protein forms a structure essential for the specific

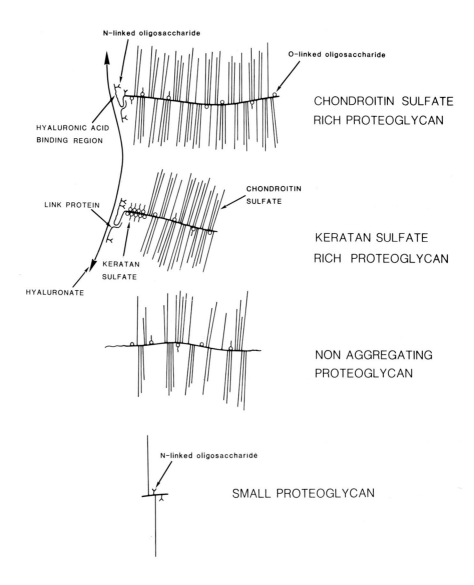

Fig. 1. Tentative structures of four populations of cartilage proteoglycans.

interaction of the proteoglycan with hyaluronic acid
(Heinegård, Hascall 1974a). This hyaluronic acid binding
region contains N-glycosidically linked oligosaccharides
(Lohmander et al 1980), but contains no glycosaminoglycan
side chains (Heinegård, Hascall 1974a). Next, the keratan
sulfate rich region contains primarily keratan sulfate
chains and a few O-glycosidically linked oligosaccharides
having a structure similar to that of the keratan sulfate
linkage region to protein (Lohmander et al 1980). The
major amino acids in this region are glutamic acid/gluta-
mine and proline, together constituting almost 50 % of the
total (Heinegård, Axelsson 1977). The chondroitin sulfate
rich region at the other end of the proteoglycan core
protein contains the chondroitin sulfate chains bound to
the protein in clusters, each containing on the average
four chains (Heinegård, Hascall 1974b). This region in addition contains a few keratan sulfate chains as well as O-
-glycosidically linked oligosaccharides. Serine, glycine
and glutamic acid/glutamine are the major amino acids
(Heinegård, Axelsson 1977).

Two types of aggregating proteoglycans have been
identified (Heinegård, Wieslander, Sheehan, unpublished).
Their structures differ and they can be identified by their
different antigenicities, as well as by different mobilities
in electrophoresis on agarose-polyacrylamide composite gels.
The larger aggregating proteoglycan, mol. weight 3.5×10^6,
contains predominantly chondroitin sulfate side chains as
well as some keratan sulfate chains. The smaller aggregating proteoglycan, mol. weight 1.3×10^6, contains a high
proportion of keratan sulfate chains and has a somewhat
lower number of chondroitin sulfate chains, Fig. 1. Other
tissues like sclera and aorta also contain proteoglycans
capable of forming aggregates with hyaluronate. These proteoglycans, however, have a different structure and notably
lack keratan sulfate.

A population of high molecular weight, non-aggregating cartilage proteoglycans, Fig. 1, represent some 10 %
of the total (Heinegård, Hascall 1979). They have a low
relative keratan sulfate content. The protein content on
the other hand is higher (12 % of dwt) than for the aggregating proteoglycans, although the clustering of chondroitin sulfate chains along the protein core is similar. The
amino acid composition of the three high molecular weight

proteoglycans is best recognized by high contents of
serine, glycine and glutamic acid.

A low molecular weight, non-aggregating proteoglycan,
representing less than 5 % of the total, is structurally
quite different, Fig. 1 (Heinegård et al 1981). Its molecular weight is 76 000. Its two chondroitin sulfate side
chains are larger than those of other cartilage proteoglycans. The protein content is high (about 25 %) and the
amino acid composition is different from other cartilage
proteoglycans, leucine (about 14 %) and aspartic acid/aspa-
ragine (about 13 %) being most prominent. Similar, antigenically related low molecular weight proteoglycans have
been identified in sclera, aorta and cornea (Heinegård,
Cöster, Malmström, Gardell, unpublished). It is possible
that this type of proteoglycan is ubiquitous in connective
tissue.

The proportions of the proteoglycans change with
age. Young cartilage contains predominantly the chondroitin sulfate rich aggregating proteoglycan, while old cartilage contains predominantly the keratan sulfate rich
aggregating proteoglycan (Inerot, Heinegård, unpublished).
The proteoglycans appear structurally very similar from
one cartilage to another, although differences in proportions of subpopulations can be demonstrated. Also within
one tissue the distribution of proteoglycans may be different between locations, i.e. the surface layer of articular
cartilage contains primarily small proteoglycans, while
deeper layers contain large proteoglycans (Franzén et al
1981a).

MATRIX PROTEINS

Cartilage contains a number of matrix proteins. The
link proteins, which are part of the proteoglycan aggregates are abundant in all cartilages (Paulsson, Heinegård
1982), Fig. 2. The apparent molecular weights of the two
major link proteins are 46 000 and 40 500, respectively,
(Baker, Caterson 1979). They interact with hyaluronate,
with the same specificity for a decasaccharide as does the
proteoglycan (Tengblad 1981; Wieslander, Heinegård, unpublished). They also interact with the hyaluronic acid
binding region of the proteoglycan monomer (Franzén et al
1981b). Although all steps of the aggregation process are

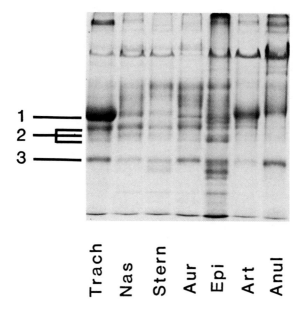

Fig. 2. SDS-polyacrylamide gel electrophoresis patterns
of proteins recovered in top fractions of direct dissociative CsCl density gradient centrifugation of extracts of
bovine cartilages. Gel concentration 8 %. Trach – tracheal,
Nas – nasal, Stern – sternal, Aur – auricular, Epi – epiphyseal, Art – articular and Anul – anulus fibrosus cartilage. (1) Position of the "cartilage matrix protein" subunits, (2) positions of the link proteins and (3) position
of the 36K protein. The prominent bands seen in samples
from articular cartilage and the anulus fibrosus, with mobilities similar to that of "cartilage matrix protein" subunits, are monomeric proteins with no antigenic relationships to this protein.

not known, it is well established that aggregates are
formed extracellularly and that an early event is the formation of a link protein-proteoglycan monomer precursor
(Kimura et al 1980; Björnsson, Heinegård 1981).

Another matrix protein, which is present at high
concentration in all cartilages, is the 36K protein,

Fig. 2. It has an apparent molecular weight of 36 000. Its amino acid composition is very similar to that of the low molecular weight cartilage proteoglycan, i.e. high contents of leucine and aspartic acid/asparagine (Paulsson, Sommarin, Heinegård, unpublished). Further support for structural similarities is that the 36K protein reacts with antibodies to the proteoglycan and vice versa. It is a glycoprotein containing primarily xylose, galactose and mannose in short oligosaccharide side chains. The 36K protein is not a degradation product of proteoglycans, since it appears rapidly as a biosynthetic product in tissue culture experiments (Paulsson, Sommarin, Heinegård, unpublished). The function of this protein is not known. It is, however, quite tightly anchored in the cartilage matrix, since high speed homogenization does not extract the protein, although about 50 % of the proteoglycans and a large proportion of the "cartilage matrix protein" discussed below are extracted. The 36K protein is best extracted with 4 M guanidine hydrochloride. Its distribution among non-cartilage tissues is presently not known.

The "cartilage matrix protein" has been purified from bovine tracheal cartilage, where it is the major non-collagenous protein. It has a molecular weight of about 150 000 and contains three subunits, each having a molecular weight of 52 000 (Paulsson, Heinegård 1981). The protein contains a comparatively large number of cysteine residues and disulphide bridges are essential for maintaining an intact trimeric structure. Its content of carbohydrate is about 4 %. More than one type of oligosaccharide has been identified, the dominating one probably being N-glycosidically linked and of high mannose type. The protein does not contain sialic acid (Paulsson, Heinegård 1981).

Turnover experiments, using tissue culture, show that the "cartilage matrix protein" as such is synthesized by the chondrocyte and that it has a very slow elimination, with a half life of the same order as the proteoglycans (Paulsson, Sommarin, Heinegård, unpublished). The function of the "cartilage matrix protein" remains to be identified. It appears, however, that the protein can bind to proteoglycans, primarily based on experiments where the matrix protein cofractionate with cartilage proteoglycans (Paulsson, Heinegård 1979). Whether or not the interaction is specific has not been shown.

The distribution among tissues of the "cartilage matrix protein" has been studied using radioimmunoassay (Paulsson, Heinegård 1982). It is not found in extracts of non-cartilage tissues. Furthermore, various types of cartilage have extremely different contents of the protein. Tracheal cartilage (1.4 % of wet weight) has the highest content followed by nasal (0.1 %), xiphosternal (0.04 %), Auricular (0.004 %) and epiphyseal (0.0003 %) cartilage, Fig. 3 (Paulsson, Heinegård 1982). Surprisingly, extracts

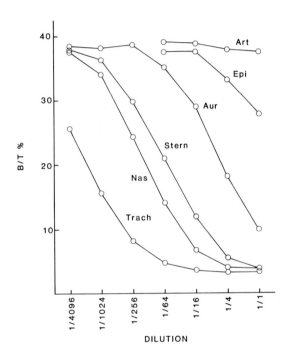

Fig. 3. Radioimmunoassay of "cartilage matrix protein" in extracts of bovine cartilages. Competition in the assay expressed as percent of labeled cartilage matrix protein precipitated over the total amount added versus the dilution of the extract. Trach – tracheal, Nas – nasal, Stern – sternal, Aur – auricular, Epi – epiphyseal and Art – articular cartilage.

of articular cartilage and of the intervertebral disc do
not contain the protein. These tissues, however, appear
to contain a unique protein having a single polypeptide
chain of an apparent molecular weight of about 60 000,
Fig. 2, (Paulsson, Heinegård 1982).

The chondrocytes, then, appear to express different
phenotypes in different cartilages. The differences cannot
be related only to the gross appearance of the tissue as
hyaline, fibrous or elastic cartilage, as shown by the
variable contents of matrix proteins. These proteins may
therefore provide important markers for studies of the
differentiation process. Other potential markers for
differentiation are the presence of particular subpopulations
of proteoglycans and/or variations in their relative
quantities (DeLuca et al 1977).

Acknowledgements: Supported by the Swedish Medical Research
Council (5668 and 5739), Kocks Stiftelser, Konung Gustaf
V:s 80-årsfond, Österlunds Stiftelse and the Medical Faculty,
University of Lund.

REFERENCES

Baker J, Caterson B (1979). J Biol Chem 254:2387
Björnsson S, Heinegård D (1981). Biochem J 199:17
DeLuca S, Heinegård D, Hascall V, Kimura J, Caplan A.
  (1977). J Biol Chem 252:6600
Franzén A, Inerot S, Hejderup S-O, Heinegård D (1981a)
  Biochem J 195:535
Franzén A, Björnsson S, Heinegård D (1981b). Biochem J
  197:669
Hardingham T, Muir H (1972). Biochem Biophys Acta 279:401
Hardingham T, Muir H (1973). Biochem J 135:905
Hascall V, Heinegård D (1974). J Biol Chem 249:4242
Heinegård D, Hascall V (1974a). J Biol Chem 249:4250
Heinegård D, Hascall V (1974b). Arch Biochem Biophys
  165:427
Heinegård D, Axelsson I (1977). J Biol Chem 252:1971
Heinegård D, Hascall V (1979). J Biol Chem 254:927
Heinegård D, Paulsson M, Inerot S, Carlström C (1981).
  Biochem J 197:355
Kimura J, Hardingham T, Hascall V (1980). J Biol Chem
  255:7134

Lohmander S, DeLuca S, Nilsson B, Hascall V, Caputo C,
  Kimura J, Heinegård D (1980). J Biol Chem 255:6084
Maroudas A (1979). In Freeman MAR (ed): "Adult Articular
  Cartilage", Tunbridge Wells: Pitman Medical p 215
Paulsson M, Heinegård D (1979). Biochem J 183:539
Paulsson M, Heinegård D (1981). Biochem J 197:367
Paulsson M, Heinegård D (1982). Biochem J, In Press
Tengblad A (1981). Biochem J 199:297

Limb Development and Regeneration
Part B, pages 45–54
© **1983 Alan R. Liss, Inc., 150 Fifth Avenue, New York, NY 10011**

DIFFERENTIAL ACCUMULATION AND AGGREGATION OF PROTEOGLYCANS
IN LIMB DEVELOPMENT

N.S. Vasan, D.V.M., Ph.D. and Karen M. Lamb, B.S.

Department of Anatomy, Cell Biology Division
New Jersey Medical School 100 Bergen Street
Newark, New Jersey 07103

Each stage in the morphogenesis and differentiation
of a particular tissue type during embryogenesis or adult
tissue remodeling (regeneration) is characterized not only
by unique cell types but also by a unique extracellular
matrix. The major macromolecular components of these matrices appear to be important both structurally and as environmental signals which influence cell activity. The mechanisms
controlling the continuously repeated sequence of differentiative events seen during embryogenesis are not clearly known.
However, collagen and proteoglycans, which accumulate in the
extracellular matrix, are known to play an important role in
the control and regulation of some developmental processes
(Nevo, Dorfman 1972; Fitton Jackson 1975; Meier, Hay 1975;
Lash, Vasan 1978). It has been clearly shown that there is
a temporal and spatial transition in collagen and proteoglycan types during embryonic limb development (Linsenmayer,
Toole, Trelstad 1973; Goetinck, Pennypacker, Royal 1974;
Vasan, Lash 1977; 1979). In this report, we extend our studies to analyze the accumulation of proteoglycans at different
stages of limb development.

SULFATED PROTEOGLYCAN SYNTHESIS DURING LIMB DEVELOPMENT

Several reports suggest that in the developing chick
limb, biosynthesis of chondroitin sulfate precedes cartilage
formation. The first obvious histologically detectable metachromasia in the limb bud is found at stage 25-26. However,
at stage 22-23, the central core of chondrogenic cells takes
up sulfate-^{35}S differentially and provides the first evi-

dence for the appearance of prospective cartilage (Searls
1965). The radioactive sulfate is incorporated by the limb
long before stage 23, and between stages 15-22 the label is
uniformly distributed throughout the limb (Searls 1965). A
mucopolysaccharide with the same electrophoretic mobility as
chondroitin sulfate (Franco-Browder, DeRydt, Dorfman 1963)
partly digested by chondroitinase ABC has been identified in
stages 12-22 limb buds (Vasan, Lash 1979).

Cartilage proteoglycans are complex heteropolysaccha-
rides in which a large number of chondroitin sulfate and
keratan sulfate chains are covalently linked to a core protein to form a proteoglycan monomer. Many such monomers
interact with a hyaluronic acid backbone to form a large
aggregate (Hascall 1977). Proteoglycan synthesis, which
is lower in precartilagenous limb tissue and increases
tremendously in differentiated wings and limbs (Goetinck,
Pennypacker, Royal 1974; Vasan, Lash 1979), is an indicator
of chondrogenic expression. Increasing amounts of radioactive sulfate are incorporated into the chondroitin sulfate
molecules which are digested by chondroitin lyase ABC (Table
1). Stage 18-22 wing bud (anterior limb) seems to incorporate more radioactive sulfate than does limb bud (posterior
limb) of the same stage. In subsequent stages this trend
reverses in favor of posterior limb (Table 1).

The distribution of Chondroitin 4 and Chondroitin 6-
sulfated molecules during the development of anterior and
posterior limb was determined with chondroitinase ABC
digestion (Saito, Yamagata, Suzuki 1968). Ch 4-sulfate
synthesis increases and Ch 6-sulfate synthesis decreases
in limb bud chondrocyte culture (DeLuca, Heinegard, Hascall,
Kimura, Caplan 1977). In vivo study (Vasan, Lash 1979)
also showed that Ch 4-sulfate increases during embryronic
wing and limb development. Up to stage 25 synthesis of
Ch 4-sulfate in wing bud increases when compared to that of
limb bud, but in subsequent stages limb bud shows a marked
increase in Ch 4-sulfate (Table 1). This observation suggests that during this period of development more 4-sulfo-
transferase than 6-sulfotransferase was present (Honda,
Murota, Mori 1982). However, Meezan, Davidson (1967) proposed an alternate hypothesis: they suggest that the nature
of the sulfate acceptor may alter the location of the sulfate. The results in Table 1, where anterior limb shows an
advanced phase of differentiation compared to posterior limb
in early development, suggest a possible cranio-caudal direc-

Table 1. Total and relative amounts of ^{35}S-sulfate incorporated into sulfated glycosaminoglycans.

	^{35}S CPM/μg DNA	ChS*	Other Sulfated GAGs**	Ch4-S	Ch6-S	Ch4-S/Ch6-S
Stage 18						
wing bud	1452	76.3	23.7	29.7	70.3	0.43
limb bud	1030	64.4	35.6	21.2	78.8	0.27
Stage 22						
wing bud	1827	81.1	18.9	41.6	58.4	0.72
limb bud	1192	69.8	30.2	28.2	71.8	0.39
Stage 25						
wing bud	2038	80.4	19.6	43.0	57.0	0.75
limb bud	2641	77.8	22.2	41.9	58.1	0.72
Stage 35						
wing bud	2647	70.6	29.4	41.7	58.3	0.72
limb bud	9611	79.7	20.3	58.6	41.1	1.43

Proteoglycan monomer (D1) isolated from the 4.0M guanidinium chloride/CsCl gradient centrifugation was used in this analysis. The enzyme resistant material was subsequently identified by cellulose acetate electrophoresis containing mostly heparan sulfate in the earlier stages and also keratan sulfate in the later stages.

* Chondroitinase ABC sensitive materials

** Enzyme resistant materials.

The sulfated materials are expressed as percent of the total.

tion in the expression of chondrogenesis.

Fully expressed chondrocytes in culture synthesize proteoglycans distinctly different from the prechondrogenic or
early undifferentiated embryonic cells (Goetinck, Pennypacker,
Royal 1974; DeLuca, Heinegard, Hascall, Kimura, Caplan 1977).
We made similar observations of stage 12-22 limb buds in
which proteoglycans are different in physical and chemical
properties from those in stage 35 limb buds (Vasan, Lash
1977; 1979). The proteoglycans (ubiquitous) made by prechondrogenic tissues were smaller, contain a higher ratio of
Ch 6-sulfate to Ch 4-sulfate, and do not interact to any
appreciable extent with hyaluronic acid. Large proteoglycans
synthesized by the fully expressed chondrocytes are described
as 'cartilage specific' molecules (Goetinck, Pennypacker,
Royal 1974; Vasan, Lash 1979), and smaller proteoglycans
synthesized by non-cartilagenous tissue and pre-chondrogenic
cells are defined as 'ubiquitous proteoglycans' (Vasan 1982).

The anterior limb at stages 12-24 synthesizes predominantly small size proteoglycans (Table 2) with a kav. 0.8;
however, by stage 35 the number of large size molecules
(kav. 0.8-1.1) gradually increases. Interestingly, the
accumulation of large size proteoglycans between stages 15-24
in posterior limb is much slower than in anterior limb
(Table 2). This again suggests possible cranio-caudal expression of chondrogenesis. The increase in the size of the
large proteoglycan molecule at the later stages of limb development was attributed to the presence of two glycoprotein
link factors (necessary for stable aggregate formation,
Hascall 1977) as opposed to only one link factor in early limb
buds (Vasan, Lash 1977), where the predominant proteoglycan
was smaller in size. Subsequent studies also showed that
proteoglycans synthesized by limb buds at later stages of development (Vasan, Lash 1979), and fully expressed chondrocytes
in culture (DeLuca, Heinegard, Hascall, Kimura, Caplan 1977)
interact with hyaluronic acid much more than the proteoglycans
of early stage limb buds or prechondrogenic cells.

The proteoglycan monomer which forms one single unit of
a large aggregate also increases in size as the limb buds
start to attain chondrogenic expression (Fig. 1). The Sepha-
rose-2B elution patterns of major proteoglycan synthesized
by the developing limb buds (present study) are similar to
those found when chondrogenesis is examined in high density
cell cultures derived from limb buds (DeLuca, Heinegard,

Hascall, Kimura, Caplan 1977) and the increase in sedimentation rate of the dissociated monomer on sucrose gradient
centrifugation (Royal, Sparks, Goetinck 1980). The increase

Table 2. Controlled pore glass (CPG-2500) chromatography of
^{35}S-labeled proteoglycans extracted from embryonic tissues.

Tissue source	kav.	Percent of total
Stage 15 Ant. limb	0.09	12
	0.85	88
Post. limb	0.08	7
	0.84	93
Stage 18 Ant. limb	0.08	15
	0.84	85
Post. limb	0.08	9
	0.92	91
Stage 24 Ant. limb	0.08	23
	0.79	77
Post. limb	0.09	18
	0.92	82
Stage 35 Ant. limb	0.11	49
	0.81	51
Post. limb	0.08	80
	0.53	20

The limb buds from various stages were separated and
exposed to F12 medium containing 25-50µCi/ml of radioactive
sulfate. The 4.0M GuHCl extract was subjected to chromatography after dialysis against elution buffer. The proteoglycans in all cases eluted as two peaks with distinct kav.

in the size of the proteoglycan monomer could be due to one
of four causes, it could be caused by increases in either
first, the length of the protein core; or second, the number
of chondroitin sulfate chains attached to the protein core;
or third, the length of the glycosaminoglycan chains. Fourth,
it could be caused by large variations in the length of the
portion of protein core to which the chondroitin sulfate chains
are attached (Hascall, Heinegard 1975; Hardinghan, Ewins, Muir
1976; Heinegard 1977; Vasan 1982). However, polyacrylamide
gel electrophoretic analysis of the protein core of stages
24, 30 and 35 did not reveal any appreciable difference, suggesting the length of the core remains unaltered. On the

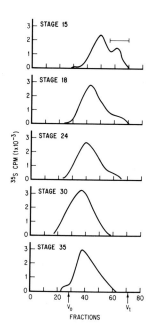

Fig. 1. Sephorose-2B elution profile of ^{35}S-labeled monomer
proteoglycans isolated from stage 15-35 limb buds. The monomers (D1) were isolated by subjecting the 4.0M extract to
a direct dissociative CsCl gradient. The D1 fraction was
collected from the lower half of the gradient for stages
15, 18, and 24, and the lower two-fifths of the gradient
for stages 30 and 35. The column was eluted with 0.5M sodium
acetate pH 6.5 buffer. The bar represents proteoglycan of
ubiquitous type.

other hand, the uronic acid/protein ratio (6.1, 7.2, 11.7
and 13.5 respectively for stages 18, 24, 30 and 35) and glycosaminoglycan chain length analysis (Fig. 2) show that during limb development there is an increase in both the glycosylation of the protein core (more chondroitin sulfate chains
are attached to the protein) and elongation of glycosaminoglycan chains.

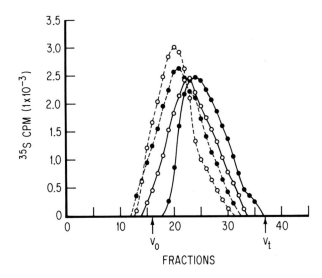

Fig. 2. Sephadex G-200 elution pattern of the single chondroitin sulfate chains isolated with papain of the ^{35}S-labeled monomer proteoglycans (D1) from stage 18 (●—●), 24 (o—o), 30 (●--●) and 35 (o--o). The column was eluted with 0.2M sodium acetate pH 6.5 buffer.

HYALURONIC ACID SYNTHESIS DURING LIMB DEVELOPMENT

Toole (1982) has suggested that hyaluronate production accompanies the initial phase of mesenchymal cell movement, accumulation and aggregation, followed by its removal during the differentiation of the cells to chondroblasts. This removal of hyaluronate is due to the increase of hyaluronidase between stages 22-30, implying that cartilage formation must await removal of hyaluronate by hyaluronidase (Toole 1972), and that exogenous addition of hyaluronate inhibits chondrogenic expression (Toole 1982). The formation of large proteoglycan aggregates during chondrogenesis requires the presence of much hyaluronic acid and the hyaluronic acid binding region in the protein core of the monomers. Earlier reports showed that prechondrogenic cells synthesize proteoglycans having less affinity for large molecular weight cockscomb hyaluronic acid, suggesting the possible absence of a hyaluronic acid binding region (Hascall, Oegema, Brown, Caplan

1976; Vasan, Lash 1979) in the monomers. The heterogeneity in the size of the proteoglycan aggregates (A1 fraction) resolved on molecular seive chromatography during limb development was attributed to the possible presence of hyaluronate of different sizes; some are small enough that they cannot form huge aggregates that exclude from the column (Vasan 1982). This possibility is further demonstrated in the present study (Fig. 3); the ^{3}H-glucosamine labeled hyaluronate increased in sedimentation rate as the cells in the developing limb ceased to migrate and started to differentiate. Thus it seems during chondrogenesis, the excess of hyaluronate is removed by the hyaluronidase, and at the same time an increase in the size of the hyaluronate molecules facilitates the attachment of a greater number of monomers to form large aggregates.

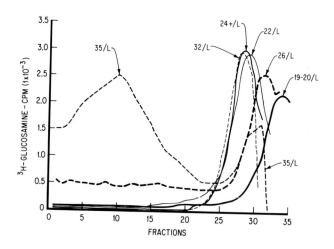

Fig. 3. Profiles of ^{3}H-glucosamine labeled proteoglycans sedimented on 5-20% linear sucrose gradients under dissociated conditions (4.0M guanidinium chloride). The proteoglycans were obtained from intact limb buds of various stages.

The data presented here indicate that the proteoglycans undergo physical and chemical changes in their properties characteristic of the relative state of development of the

cells. The significance of alteration in the composition of
extracellular proteoglycans in development is still unclear.
However, studies on various experimental model systems and
genetic mutants suggest that these changes are important for
structural integrity of future bone and provide biomechanical
stability.

This investigation was supported by a grant from
NICHHD 1 R01 HD 16147. The author wishes to thank Dianne
Prime for typing the manuscript.

DeLuca S, Heinegard D, Hascall VC, Kimura JH, Caplan AI
(1977). Chemical and physical changes in proteoglycans
during development of chick limb bud chondrocytes grown
in vitro. J Biol Chem 252:6600.
Fitton Jackson S (1975). The influence of tissue interactions and extracellular macromolecules on the control
of phenotypic expression and synthetic capacity of bone
and cartilage. In Slavkin HC, Greulich RC (eds):
"Extracellular Matrix Influences on Gene Expression",
New York: Academic Press, p 489.
Franco-Browder S, DeRydt J, Dorfman A (1963). The identification of a sulfated mucopolysaccharide in chick
embryos, stages 11-23. Proc Natl Acad Sci USA 49:643.
Goetinck PF, Pennypacker JP, Royal PD (1974). Proteochondroitin sulfate synthesis and chondrogenic expression.
Exptl Cell Res 87:241.
Hardingham TE, Ewins RJF, Muir H (1976). Cartilage
proteoglycans: Structure and heterogeneity of the
protein core and the effects of protein modifications
on the binding of hyaluronate. Biochem J 157:127.
Hascall VC, Heinegard D (1975). The structure of cartilage proteoglycans. In Slavkin HC, Gurulich RC (eds):
"Extracellular Matrix Influences on Gene Expression",
New York: Academic Press, p 423.
Hascall VC, Oegema TR, Brown M, Caplan AI (1976). Isolation and characterization of proteoglycans from chick
limb bud chondrocytes grown in vitro. J Biol Chem
251:3511.
Hascall VC (1977). Interaction of cartilage proteoglycans
with hyaluronic acid. J Supramole Struct 7:101.
Heinegard D (1977). Polydispersity of cartilage proteoglycans. Structural variations with size and buoyant
density of the molecules. J Biol Chem 252:2908.
Honda A, Murota S, Mori Y (1982). Comparative study on
glycosaminoglycan sulfotransferases in rat costal

cartilage and chick embryo cartilage. Comp Biochem Physiol 71B:41.

Lash JW, Vasan NS (1978). Somite chondrogenesis in vitro: Stimulation by exogenous extracellular matrix components. Develop Biol 66:151.

Linsenmayer TF, Toole BP, Trelstad RL (1973). Temporal and spatial transitions in collagen types during embryonic chick limb development. Develop Biol 35:232.

Meezan E, Davidson EA (1967). Mucopolysaccharide sulfation in chick embryo cartilage. 1. Properties of sulfation system. 2. Characterization of endogenous acceptor J Biol Chem 242:1685;4956.

Meier S, Hay E (1975). Stimulation of corneal differentiation by interaction between cell surface and extracellular matrix. 1. Morphometric analysis of transfilter 'induction'. J Cell Biol 66:275.

Nevo Z, Dorfman A (1972). Stimulation of chondromucoprotein synthesis in chondrocytes by extracellular chondromucoprotein. Proc Natl Acad Sci USA 69:2069.

Royal PD, Sparks KJ, Goetinck PF (1980). Physical and immunochemical characterization of proteoglycans synthesized during chondrogenesis in the chick embryo. J Biol Chem 255:9870.

Saito H, Yagamata T, Suzuki S (1968). Enzymatic methods for the determination of small quantities of isomeric chondroitin sulfates. J Biol Chem 243:1536.

Searls RL (1965). An autoradiographic study of the uptake of ^{35}S-sulfate during the differentiation of limb bud cartilage. Develop Biol 11:155.

Toole BP (1972). Hyaluronate turnover during chondrogenesis in the developing chick limb and axial skeleton. Develop Biol 29:321.

Toole BP (1981). Glycosaminoglycans in morphogenesis. In Hay (ed): "Cell Biology of Extracellular Matrix", New York: Plenum Press, p 259.

Vasan NS, Lash JW (1977). Heterogeneity of proteoglycans in developing chick limb cartilage. Biochem J 164:179.

Vasan NS, Lash JW (1979). Monomeric and aggregate proteoglycans in the chondrogenic differentiation of embryonic chick limb buds. J Embryol Exp Morp 49:47.

Vasan NS (1982). Analysis of the intermediate size proteoglycans from the developing chick limb buds. J. Embryol Exp Morphol 69 (in press).

Limb Development and Regeneration
Part B, pages 55–65
© **1983 Alan R. Liss, Inc., 150 Fifth Avenue, New York, NY 10011**

MODULATION OF PROTEOGLYCAN SYNTHESIS IN MATURE CARTILAGE

J.D. Sandy and Helen Muir

Kennedy Institute of Rheumatology

London W6 7DW, U.K.

INTRODUCTION

Recent studies on a number of cell culture systems have shown that the rate of synthesis and the structure of newly formed proteoglycans may be influenced by the differentiated state of the cell, by interaction of the cell with components of the extracellular matrix or by chemical disturbance of the biosynthetic process. In cultures of mesenchymal cells established from embryonic chick limb buds, there is a reproducible chondrogenic process, each stage of which is characterised by the synthesis of a particular proteoglycan structure at a defined rate (see Caplan and Hascall, 1980). While addition of hyaluronate to mesenchymal cells inhibits chondrogenesis in these cultures (Toole et al., 1972), it appears that the chondrocyte maturation process, once begun, is not markedly influenced by altering the extracellular matrix by xyloside treatment or enzyme digestion (see Caplan, 1981).

In chondrocyte cultures derived from the epiphyseal cartilage of 13-day old chick embryos inhibition of proteo-glycan synthesis, by addition of proteoglycan monomer or omission of glutamine, did not markedly influence the structure of proteoglycan formed (Handley et al., 1980) whereas incubation with β-D-xylosides resulted in the synthesis of a poorly glycosylated proteoglycan monomer with abnormally short chondroitin sulphate (CS) chains (Schwartz, 1978). Further, in cultures of rat chondrosarcoma cells, the structure of proteoglycan formed can be readily altered by short-term treatments of cells; thus, decreasing the

rate of proteoglycan synthesis with cycloheximide or low
incubation temperature results in the synthesis of long CS
chains, whereas increasing the rate of synthesis by addition
of β-D-xylosides results in the synthesis of abnormally short
chains (Mitchell and Hardingham, 1982).

Conversely, the results to be presented in this paper
show that in two separate experimental systems, where an
increase in the rate of proteoglycan synthesis has been
observed in mature articular cartilage, it also has been accompanied by the synthesis of proteoglycans with abnormally
long CS chains. It is suggested that such modulation of
proteoglycan synthesis in mature cartilage might be explained
by a change in extracellular matrix properties inducing the
expression of an altered, apparently immature, chondrocyte
type.

MATERIALS AND METHODS

Tissue Culture of Articular Cartilage

Bovine articular cartilage was obtained from the meta-
carpal-phalangeal joints of 2-3 year-old steers within 1h of
slaughter at a local abbattoir. The cartilage was dissected
free of perichondrium and sub-chondral bone, sliced into
small pieces (approx. 1mm x 1mm x 2mm), washed in phosphate
buffered saline pH 7.4 and maintained in Dulbecco's modified
Eagles basal medium containing 20% foetal calf serum, as
previously described (Sandy et al., 1980).

Induction of Osteoarthritis and Cartilage Preparation

The dogs used were female beagles, aged 6-8 years, that
had been used solely for breeding. Osteoarthritis was
induced in the right hind knee by severing the anterior
cruciate ligament by a stab incision, the unoperated left
knee serving as a control (Pond and Nuki, 1973). The surgery
and post-operative care was carried out by Dr. M.E.J.
Billingham, I.C.I. Ltd, Alderley Park, Cheshire, UK. The
dogs were killed by intravenous injection of sodium pento-
barbitone 3 weeks after severance of the ligament and
cartilage was taken separately from the lateral and medial
tibial plateaux, the lateral and medial femoral condyles
and the patellar groove. Cartilage was sliced into small

pieces, washed in phosphate buffered saline pH 7.4 and distributed into vials containing 50-100 mg wet wt/2 ml of Dulbecco's modified Eagles basal medium, supplemented with 4 mM glutamine.

Determination of the Rate of PG Synthesis

The methods for incubation of bovine cartilage with ^{35}S-sulphate and isolation of labelled glycosaminoglycans have been described (Sandy et al., 1980). Canine cartilage samples were incubated at 37°C under 5% CO_2/95% air and after 30 min ^{35}S-sulphate (50-100 µCi) was added and incubation was continued for 2h in the presence of isotope followed by 4h in the absence of isotope. Cartilage was then washed twice in 0.15M sodium acetate pH 6.8 (containing protease inhibitors) and extracted with 4M guanidine HCl/ 0.5M sodium acetate pH 6.8 (containing protease inhibitors) for 48h at 4°C. The cartilage residue was washed with water and digested with papain. Portions of the guanidine extracts and papain digests were dialysed exhaustively against water before radioactive counting and hydroxyproline assay (Stegeman, 1967).

Chromatographic Methods

^{35}S-labelled proteoglycans were prepared from bovine and canine cartilage by extraction in 4M guanidine HCl/0.05M sodium acetate pH 6.8 (containing protease inhibitors) for 48h at 4°C and applied to columns of Sepharose CL-2B eluted with 2M guanidine hydrochloride/0.5M sodium acetate pH 6.8. ^{35}S-labelled glycosaminoglycans were prepared from guanidine extracts by β-elimination in 0.5M NaOH or by papain digestion of cartilage; samples were applied to columns of Sepharose CL-6B eluted with 2M guanidine hydrochloride/0.5M sodium acetate pH 6.8.

RESULTS

Tissue Culture of Mature Articular Cartilage

When bovine articular cartilage slices were maintained in culture and the rate of ^{35}S-sulphate incorporation was determined during 1h labelling periods after various times in culture, a rapid increase in proteoglycan synthesis was observed (Fig.1). A significant activation was observed after only 10h of culture and maximum activity was reached after 24-48h.

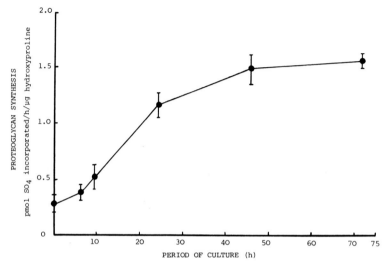

FIG.1 Activation of proteoglycan synthesis during culture of bovine articular cartilage.

The ^{35}S-proteoglycans synthesised by fresh cartilage and cartilage after 72h of culture were extracted (see methods) and fractionated on Sepharose CL-2B under dissociative conditions (Fig.2). It was found that the ^{35}S-proteoglycans extracted from the high activity cultured cartilage contained more high molecular weight proteoglycan than did similar extracts of fresh cartilage. This difference was seen as an enrichment in ^{35}S-radioactivity on the high molecular weight side of the proteoglycan monomer profile.

This increase in the average size of newly synthesised proteoglycan monomer was at least partly due to an increase in the size of the substituent ^{35}S-glycosaminoglycans. Thus, when papain digests of 4h labelled cartilage samples were fractionated on CL-6B (Fig.3), the average molecular weight of the free ^{35}S-glycosaminoglycans was found to be about 16,000 in fresh cartilage and 28,000 in cultured cartilage.

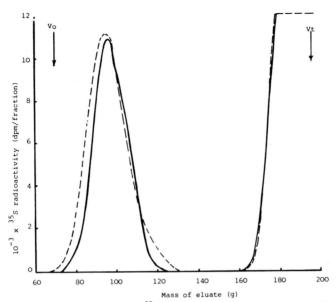

FIG.2 Chromatography of ^{35}S-proteoglycans on a column (1.9 cm x 65 cm) of Sepharose CL-2B in 2M guanidine hydrochloride/0.5M sodium acetate pH 6.8 (——) Proteoglycans formed in fresh cartilage; (----) proteoglycans formed in cartilage previously cultured for 72h.

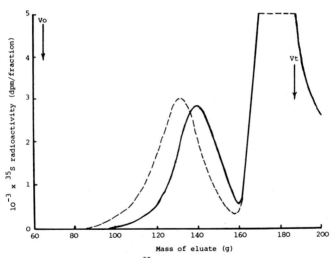

FIG.3 Chromatography of ^{35}S-glycosaminoglycans on a column (1.5 cm x 80 cm) of Sepharose CL-6B.
(——) ^{35}S-glycosaminoglycans formed in fresh cartilage;
(----) ^{35}S-glycosaminoglycans formed in cartilage previously cultured for 72h.

Proteoglycan Synthesis in Articular Cartilage in Experimental Osteoarthritis

The rate of proteoglycan synthesis determined in incubations of articular cartilage (see methods) was generally 1.5-2.5 fold higher in arthritic cartilage than in controls (Table 1).

Table 1. Proteoglycan synthesis in articular cartilage in experimental osteoarthritis

| Cartilage Source | Rate of proteoglycan synthesis ($pmol\ SO_4/h/\mu g$ hydroxyproline) | | | |
| | Control Joint Cartilage | | Arthritic Joint Cartilage | |
	Dog 1	Dog 2	Dog 1	Dog 2
Lateral Tibial Plateau	0.98	0.55	1.54	1.13
Medial Tibial Plateau	0.83	0.97	2.37	2.12
Lateral Femoral Condyle	0.85	0.92	2.53	1.86
Medial Femoral Condyle	0.99	0.55	1.71	1.27
Patellar Groove	1.36	0.42	1.57	2.22

Newly synthesised ^{35}S-proteoglycans were extracted (see methods) and fractionated on Sepharose CL-2B under dissociative conditions (Fig.4). In all cases about 50% of the newly synthesised ^{35}S-proteoglycan was extracted from the tissue. Elution profiles are shown for a cartilage area showing a mild (1.57-fold) increase in activity (Fig.4a) and one showing a marked (5.29-fold) increase in activity (Fig.4b).

In both cases the proteoglycan monomers extracted from osteoarthritic cartilage were of a greater average size than those extracted from control cartilage. Similar fractionation of ^{35}S-proteoglycans extracted from all other areas of cartilage (see Table 1) showed that the ^{35}S-proteoglycans formed at a higher rate in arthritic cartilage were consistently of a larger average hydrodynamic size than in controls.

This increase in average molecular size of the ^{35}S-proteoglycan in arthritic cartilage was also at least

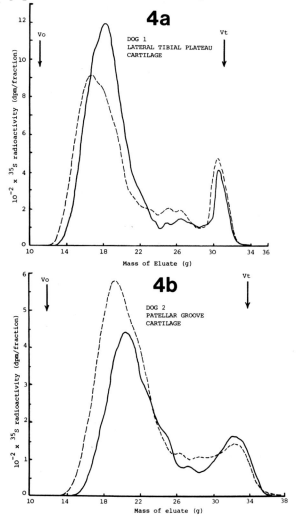

Fig.4 Chromatography of ^{35}S-proteoglycans on columns
(0.6 cm x 100 cm) of Sepharose CL-2B
(———) Proteoglycan formed in control cartilages
(————) Proteoglycan formed in osteoarthritic cartilage

partly due to the synthesis of abnormally long ^{35}S-glycosa-
minoglycans. This can be seen from the elution profiles
on Sepharose CL-6B (Fig.5) of the glycosaminoglycans prepared
from the proteoglycan preparations described in Fig.4.

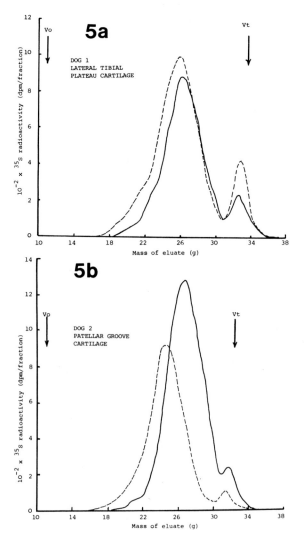

Fig.5 Chromatography of ^{35}S-glycosaminoglycans on a column
(0.6cm x 100cm) of Sepharose CL-6B. (———)Glycosaminoglycans
formed in control cartilages. (----) Glycosaminoglycans
formed in osteoarthritic cartilages.

The difference in size of the ^{35}S-proteoglycans present in the extracts of arthritic cartilages (Fig.4) was not due to the selective extraction of high molecular weight products from arthritic cartilage, since when the cartilage residues left after guanidine extraction were digested with papain and the ^{35}S-glycosaminoglycans were fractionated on Sepharose CL-6B, the arthritic samples again showed the presence of markedly larger chains.

DISCUSSION

In previous work on the modulation of proteoglycan synthesis in articular cartilage, incubation of mature rabbit cartilage in a serum-rich medium resulted in extensive loss of proteoglycan into the medium (Sandy et al., 1978) and this was followed by an increase in the rate of proteoglycan synthesis (Sandy et al., 1980). It was suggested that the stimulation of chondrocyte activity might be a cellular response to the loss of extracellular proteoglycan. On the other hand, similar incubation of bovine articular cartilage resulted in marked increases in chondrocyte activity in the first 24h of culture even though very little proteoglycan was lost from the tissue. Further, this increase could be partially inhibited by omitting serum (30% inhibition) or by maintaining a high concentration of proteoglycan in the culture medium (50% inhibition) suggesting that serum factors and extracellular proteoglycan concentration might both influence the cellular response. Moreover, recent studies with bovine articular cartilage in explant culture (Hascall et al., 1981) have indicated that serum components markedly influence chondrocyte activity in these cultures.

The activation of proteoglycan synthesis seen here in bovine cartilage over 72h in culture (Fig.1) was accompanied by the synthesis of abnormally long CS chains (Fig.3). However to what extent these two changes are closely linked or fortuitous is not known.

It is of interest that a similar change in the rate of proteoglycan synthesis and the structure of the proteoglycan formed has now also been observed in experimental osteoarthritis (Table 1, Figs. 4 and 5) although no serum was present during the labelling period. These apparently innate differences between control and osteoarthritic cartilage suggest that other, as yet unknown mechanisms, may be

responsible for modulation of proteoglycan synthesis in
mature articular cartilage.

One possibility is that in the early stages of this
model arthritis there is considerable chondrocyte mitosis;
histological sections taken at 2 weeks after joint surgery
have indicated an increase in cell density and some chondrocyte clones in osteoarthritic cartilage (McDevitt et al.,
1977). Moreover, chondrocyte mitosis has been observed in
^{3}H-thymidine labelled autoradiographs of articular cartilage
following short periods of traumatic experimental osteoarthritis in rabbits (Telhag, 1972). These newly divided
cells may then contribute to the high rates of proteoglycan
synthesis and be responsible for the synthesis of the immature
type CS-rich proteoglycan observed. Alternatively, changes
in the organisation of the cartilage matrix, following
hydration of the tissue, loosening of the intercellular
collagen network, and disorganisation of the peri-lacunar
collagen (Altman et al., 1981) may induce ultrastructural
changes in chondrocytes. Alterations in the mode of
attachment of cells to their lacunae may disturb the cyto-
skeleton sufficiently to interfere with the normal mechanisms
controlling the transport of proteoglycan through the endoplasmic reticulum and so influence chondroitin sulphate
chain length. Indeed, electron microscopy of chondrocytes
in the deep zones of osteoarthritic human cartilage indicate
that matrix disorganisation is accompanied by marked changes
in cell ultrastructure (Weiss, 1973). Such mechanisms may
also be operating during the changes in proteoglycan synthesis
seen in culture of mature articular cartilages (Sandy et al.,
1980) and might also partly determine the type and quantity
of proteoglycan formed in the different zones of mature
articular cartilage in vivo.

REFERENCES

Altman RD, Tenenbaum J, Pardo V, Blanco LN, Howell DS (1981).
   Morphological changes and swelling properties of osteoarthritic dog cartilage. Sem Arth Rheum XI (1):39.
Caplan AI, Hascall VC (1980). Structure and developmental
   changes in proteoglycans. In Naftolin F, Stubblefield PG
   (eds): "Dilatation of the Uterine Cervix", New York:
   Raven Press, p 79.
Caplan AI (1981). The molecular control of muscle and
   cartilage development. In "Levels of Genetic Control in

Development", Alan R. Liss, New York, p 37.

Handley CJ, Speight G, Robinson HC, Lowther DA (1980). Molecular size distribution of proteoglycan sub-units and glycosaminoglycans synthesised by chondrocytes under conditions of reduced proteoglycan synthesis. Biochim Biophys Acta 631:124.

Hascall VC, Handley CJ, Robinson HC, Lowther DA (1981). Biosynthesis and turnover of proteoglycans in explant cultures of articular cartilage. Proc Aust Biochem Soc 14:2.

McDevitt CA, Gilbertson E, Muir H (1977) An experimental model of osteoarthritis; early morphological and biochemical changes. J Bone Joint Surg 58B:24.

Mitchell D, Hardingham T (1982). The control of chondroitin sulphate biosynthesis and its influence on the structure of cartilage proteoglycans. Biochem J 202:387.

Pond MJ, Nuki G (1973). Experimentally induced osteoarthritis in the dog. Ann Rheum 32:387.

Sandy JD, Brown HLG, Lowther DA (1978). Degradation of proteoglycan in articular cartilage. Biochim Biophys Acta 543:536.

Sandy JD, Brown HLG, Lowther DA (1980). Control of proteoglycan synthesis. Studies on the activation of synthesis observed during culture of articular cartilage. Biochem J 188:119.

Sandy JD, Adams ME, Billingham MEJ, Plaas HAK, Muir H (1982). Biochemical studies in early experimental canine osteoarthritis. Submitted Arth & Rheum.

Schwartz NB (1979). Synthesis and secretion of an altered chondroitin sulphate proteoglycan. J Biol Chem 254:2271.

Stegeman H, Stalder K (1967). Determination of hydroxyproline. Clin Chim Acta 18:267.

Telhag H (1972). Mitosis of chondrocytes in experimental osteoarthritis in rabbits. Clin Orthop and Rel Res 86:224.

Toole BP, Jackson G, Gross J (1972) Hyaluronate in morphogenesis: Inhibition of chondrogenesis in vitro. Proc Nat Acad Sci USA, 69:1384.

Weiss C (1973) Ultrastructural characteristics of osteoarthritis. Fed Proc 32:1459.

Limb Development and Regeneration
Part B, pages 67–84
© **1983 Alan R. Liss, Inc., 150 Fifth Avenue, New York, NY 10011**

ISOLATION, CHARACTERIZATION AND IMMUNOHISTOCHEMICAL LOCALIZATION OF A DERMATAN SULFATE-CONTAINING PROTEOGLYCAN FROM BOVINE FETAL EPIPHYSEAL CARTILAGE

L. Rosenberg,[*] L. Tang,[*] H. Choi,[*] S. Pal,[*] T. Johnson,[*]
A.R. Poole,[†] P. Roughley,[†] A. Reiner,[†] and I. Pidoux[†]

[*]Montefiore Hospital and Medical Center
111 East 210th Street, Bronx, NY 10467
[†]Shriners Hospital for Crippled Children
Montreal, Quebec, CANADA H3G 1A6

Prior to the advent of chondrogenesis, the developing limb bud consists of a core of mesenchyme covered by ecto-derm. The mesenchymal cells are closely spaced and sepa-rated by relatively small amounts of extracellular matrix. Before chondrogenesis begins, the extracellular matrix of the mesenchymal core contains mainly a mesenchymal proteo-glycan and much smaller amounts of cartilage-specific pro-teoglycan. With the advent of chondrogenesis, synthesis of the cartilage-specific proteoglycan suddenly increases, and synthesis of the mesenchymal proteoglycan is relatively decreased (Goetinck *et al.*, 1974; Pennypacker, Goetinck, 1976; Goetinck, Pennypacker, 1977; Royal, Goetinck, 1977; McKeown, Goetinck, 1979).

Cartilage-specific proteoglycans are readily isolated in amounts sufficient for their detailed characterization, and much is known about their chemical, physical and immuno-logic properties (Hascall, 1981; Hascall, 1982). The car-tilage-specific proteoglycan monomer is composed of chondroi-tin sulfate and keratan sulfate chains covalently bound to a protein core. Cartilage-specific proteoglycan monomers also contain oligosaccharides whose structure resembles that of the keratan sulfate linkage region, and mannose-containing oligosaccharides (Lohmander *et al.*, 1980; Hascall, 1982). Cartilage-specific monomers are polydisperse in size and chemical composition (Rosenberg, *et al.*, 1976; Buckwalter, Rosenberg, 1982). However, most of the cartilage-specific proteoglycan monomers isolated from bovine fetal epiphyseal cartilage (Pal *et al.*, 1981) or bovine nasal cartilage (Tang *et al.*, 1979; Reihanian *et al.*, 1979; Reihanian *et al.*, 1981) are of relatively large size with sedimentation coefficients of approximately 22 S and molecular weights of approximately 3×10^6 daltons. In native cartilages, most of the cartilage-specific proteoglycans are aggregates formed by the non-covalent association of proteoglycan monomer with hyaluronic acid and link protein. In a proteoglycan aggregate from bovine fetal epiphyseal cartilage, over one hundred proteoglycan monomers may be bound to a single hyaluronic acid filament to form an aggregate with a sedimentation coefficient of 150 S (Pal *et al.*, 1981) and a molecular weight of over 3×10^8 daltons. Link protein is centrally located where proteoglycan monomer binds to hyaluronic acid. Link protein appears to bind simultaneously to the hyaluronic acid binding region of proteoglycan monomer and to hyaluronic acid. It stabilizes the binding of proteoglycan monomer to hyaluronate (Tang *et al.*, 1979; Hardingham, 1979).

Antisera have been prepared against cartilage-specific proteoglycan monomers and link proteins (Poole *et al.*, 1980a). Using immunofluorescence microscopy (Poole *et al.*, 1980b; Poole *et al.*, 1982a) and immunoelectron microscopy (Poole *et al.*, 1982b) we have studied the localization and organization of these molecules in mature and developing cartilages.

Much less is known about the structure and localization of the mesenchymal proteoglycan monomer, because of the difficulty in isolating this species in amounts sufficient for its detailed characterization. Most of the present knowledge of the properties of the mesenchymal proteoglycan have come from *in vitro* studies of cultured limb bud mesenchymal cells.

Goetinck and his co-workers have shown that [^{35}S] sulfate-labelled mesenchymal proteoglycan monomer synthesized by chick limb bud mesenchymal cells is smaller in size than cartilage-specific proteoglycan monomer, possesses an immunologically different core protein, but like cartilage monomer, is capable of binding to hyaluronic acid to form proteoglycan aggregates (Goetinck *et al.*, 1974; Pennypacker, Goetinck, 1976; Goetinck, Pennypacker, 1977; Royal *et al.*, 1980).

Royal and Goetinck (1977) have also studied the changes in proteoglycan synthesis by mouse limb bud mesenchymal cells

during *in vitro* chondrogenesis. Limb bud cells were dissociated from 11 day old embryos and cultured at high density.
In 24 hour cultures, cells with the morphology of mesenchymal
cells synthesized mainly a mesenchymal proteoglycan monomer
of relatively small size. After 5 days, with the advent of
chondrogenesis, there was an increase in the synthesis of
cartilage-specific proteoglycan and a relative decrease in
the amount of mesenchymal proteoglycan. Whereas cartilage-
specific proteoglycan contained 80% chondrotin 4-sulfate,
20% chondroitin 6-sulfate and negligible amounts of dermatan
sulfate, the mesenchymal proteoglycan contained 55% chondroitin 4-sulfate, 4% chondroitin 6-sulfate and 20% dermatan
sulfate, indicating that the mesenchymal proteoglycan from
mouse limb bud mesenchymal cells is a dermatan sulfate-
containing proteoglycan.

In the studies described above, the mesenchymal proteoglycan monomer was still detectable shortly after the onset
of chondrogenesis in limb bud cell cultures. We wondered
whether biosynthesis of the dermatan sulfate-containing monomer continues long after the advent of chondrogenesis, and
whether a dermatan sulfate-containing proteoglycan monomer
is present in developing and even mature cartilages in amounts
sufficient to allow its isolation and characterization.
Using chondroitinase assays, we detected a dermatan sulfate-
containing monomer in associative density gradient fractions
from bovine fetal epiphyseal cartilage, which we have isolated, partially characterized, and localized in the extracellular matrix of hyaline cartilages by immunohistochemical
methods. In this paper we briefly describe our new findings.

EXPERIMENTAL PROCEDURES

The analytical methods used were uronate (Bitter, Muir,
1962), hexose (Yemm, Willis, 1954), sialate (Warren, 1959),
glucosamine and galactosamine (Rosenberg *et al.*, 1976) and
protein (Lowry *et al.*, 1951). Chondroitin sulfate and dermatan sulfate contents were determined using chondroitinases
AC and ABC. Chondroitinase digestions were in 0.1 M sodium
acetate, 0.1 M Tris, pH 7.3 for 2 hours at 37° using 0.5
unit of enzyme per mg of sample. The amount of unsaturated
disaccharide released by the chondroitinases was determined
after periodate degradation by the thiobarbituric acid assay
(Hascall *et al.*, 1972). Neutral sugars, glucosamine and
galactosamine were determined as the alditol acetates by

gas liquid chromatography.

Isolation of Proteoglycans:

Bovine fetuses obtained immediately after the slaughter of pregnant cows were submerged in ice and removed to a cold room. The femora, tibiae and humeri were removed by dissection. Samples of epiphyseal cartilage were taken at least 4 mm from the growth plate, or from the secondary center of ossification. The samples were immediately added to ice-cold 4 M GdmCl, 0.15 M sodium acetate, 0.05 M EDTA, pH 7, containing 0.005 M benzamidine hydrochloride, 0.005 M iodoacetamide and 0.005 M phenylmethylsulfonyl fluoride, and extracted for 24 hours at 5°. The extract was filtered, then dialyzed at 5° for 16 hours against 20 volumes of 0.15 M sodium acetate, 0.05 M EDTA, pH 7 containing 0.001 M iodo-acetamide, 0.001 M benzamidine hydrochloride, and 0.001 M phenylmethylsulfonyl fluoride. The extract was then frac-tionated by equilibrium density gradient centrifugation under associative conditions in 3.5 M CsCl at 5° for 60 hours at 40,000 rpm. The gradient was divided into six equal frac-tions called A1 through A6. Aliquots of each fraction were dialyzed against 0.15 M sodium acetate, pH 6.3, precipitated with ethanol, collected by centrifugation, washed with ethanol and ether and dried in a vacuum.

The chondroitin sulfate and dermatan sulfate (DS) con-tents of the fractions from the associative gradient were determined using chondroitinase AC and ABC. Fraction A1 which contained little or no dermatan sulfate was further fractiona-ted by equilibrium density gradient centrifugation under dissociative conditions in 4 M GdmCl, 3 M CsCl, 0.05 M EDTA, 0.15 M sodium acetate, pH 7 at 40,000 rpm for 60 hours at 5° to prepare cartilage-specific proteoglycan monomer (A1D1D1). Fraction A2 which contained dermatan sulfate was further fractionated by dissociative density gradient centrifugation to prepare the mesenchymal proteoglycan monomer. Most of the dermatan sulfate in fraction A2 was recovered in frac-tions D5 and D6 in the top one-third of the dissociative gradient. Fraction A2D5D6 from the dissociative gradient was subjected to chromatography on DEAE-Sephacel in 6 M urea, 0.025 M Tris, pH 6.5. The column was eluted with a 0 to 1 M NaCl gradient. Fractions were analyzed for uronate and pooled.

The dermatan sulfate-containing proteoglycan monomer was

again identified by analyzing the pooled fractions with chondroitinases AC and ABC. It was then further purified by gel
chromatography on Sephacryl S-200 in 4 M GdmCl, 0.01 M sodium
acetate, pH 6.5.

Immunologic Methods:

Antibodies to the purified dermatan sulfate containing
proteoglycan were prepared in rabbits using standard immuni-
zation protocols (Poole *et al.*, 1980a). Antibodies were
concentrated by precipitation of the sera with ammonium
sulfate and F(ab')2 antibody subunits were prepared as described earlier (Poole *et al.*, 1980b). Antibody reactivity
was studied by enzyme linked immunosorbent assay (Poole *et
al.*, 1982c). These molecules were localized by immunofluorescence microscopy (Poole *et al.*, 1980b, 1982a, 1982c) and
by immunoelectron microscopy (Poole *et al.*, 1982b) using
published methods as indicated.

Agarose/Polyacrylamide Gel Electrophoresis:

Proteoglycan fractions were subjected to electrophoresis
in gels composed of 0.6% agarose and 1.2% polyacrylamide.
The sample (20 μl) consisted of the proteoglycan (2 mg/ml
in water and 50% sucrose containing 0.05% bromophenol blue
in a ratio of 1:1. The gels were stained with toluidine
blue.

RESULTS

Proteoglycans were extracted from bovine fetal epiphyseal
cartilage in 4 M GdmCl containing protease inhibitors, then
fractionated by equilibrium density gradient centrifugation
under associative conditions. The yields and chemical composition of fractions A1 through A6 from a representative
associative density gradient are shown in Table 1. Fractions A1 through A6 were examined for the presence of dermatan sulfate using chondroitinases AC and ABC. Dermatan sulfate was usually not detectable in fraction A1. Little or
no dermatan sulfate was found in fractions A5 and A6. The
amount of dermatan sulfate increased in fractions A3 and A4.
Fraction A2 usually contained the largest amount of dermatan
sulfate. As shown at the bottom of Table 1, dermatan sulfate

accounted for 5-7% of the dry weight of fraction A2, and represented ~13% of the total glycosaminoglycan in this fraction.

TABLE 1

Yields and chemical composition of fractions isolated from
bovine fetal epiphyseal cartilage by equilibrium density
gradient centrifugation under associative conditions.
Analytical data are shown as percent of dry weight.

Fraction	A1	A2	A3	A4	A5	A6
Yield, g/g	0.683	0.007	0.005	0.091	0.053	0.160
Uronate, %	25.0	13.1	9.6	8.0	4.2	1.0
Hexose	8.2	8.0	4.6	4.4	3.2	2.0
Protein	7.0	31.0	41.0	48.5	54.2	-
CS	86.2	35.7	22.7	18.6	6.2	0.2
DS	0	5.4	4.2	1.5	0.2	0
(DS/DS+CS) x 100	0	13.1	15.6	7.5	3.1	0

Sedimentation velocity studies of fraction A2 in the
analytical ultracentrifuge showed that fraction A2 contained
three species: proteoglycan aggregate and proteoglycan monomer $(s_{20}^0 = 22\ S)$ of high molecular weight, presumed to represent the cartilage-specific proteoglycans, and a proteoglycan monomer $(s_{20}^0 = 5\ S)$ of much lower molecular weight,
presumed to represent the dermatan sulfate-containing mesenchymal proteoglycan monomer. The dermatan sulfate-containing
proteoglycan monomer in fraction A2 was separated from the
cartilage-specific proteoglycan monomer by equilibrium density gradient centrifugation under dissociative conditions
in 4 M GdmCl and 3 M CsCl. The yields and chemical compositions of fractions A2D1 through A2D6 from a representative dissociative gradient are shown in Table 2.

Because of its high protein content, the dermatan sulfate-containing proteoglycan monomer is concentrated mainly
in fractions A2D5 and A2D6. The cartilage-specific proteoglycan monomer is concentrated mainly in A2D1 and A2D2. This

is shown by the fact that 40% of the total uronate contained
in all fractions, but none of the dermatan sulfate, is recovered in A2D1 and A2D2. Approximately 31% of the total
uronate, and 81% of the total dermatan sulfate contained in
all fractions is recovered in fractions A2D5 and A2D6, reflect-
ing the separation of the dermatan sulfate-containing proteoglycan monomer into these low density fractions. There was
also a significant amount of the dermatan sulfate-containing
proteoglycan monomer in fraction A2D3.

TABLE 2

Yields and chemical composition of fractions isolated from
fraction A2 by dissociative density gradient centrifugation.
Analytical data are shown as percent of dry weight.

Fraction	D1	D2	D3	D4	D5	D6
Yield, g/g	0.251	0.039	0.088	0.134	0.166	0.321
Uronate, %	25.3	18.7	11.8	13.9	13.1	7.3
Hexose	10.5	9.8	5.9	5.6	5.6	4.9
Protein	8.9	-	-	-	40.0	51.6
CS	80.7	46.4	31.0	23.9	20.7	9.6
DS	0	0	5.9	1.0	9.4	3.8
(DS/DS+CS) × 100	0	0	16.0	4.0	31.2	28.4

The dermatan sulfate-containing proteoglycan monomer
present in fraction A2D5D6 was further purified by chromatography on DEAE-Sephacel in 6 M urea, then by gel chromatography on Sephacryl S-200 in 4 M GdmCl. Fig. 1 shows the
effect of chromatography of fraction A2D5D6 on DEAE-Sephacel
in 6 M urea, using a 0 to 0.5 M NaCl gradient followed by
stepwise elution with 1 M NaCl, then with 2 M NaCl. The
first small peak contained no chondroitin sulfate or dermatan
sulfate on analyses with the chondroitinases. The second
large peak, eluted at 0.5 M NaCl, contained almost all of
the dermatan sulfate proteoglycan. A third small fraction
eluted at 1 M NaCl contained small amounts of dermatan sulfate. Peak 2 from DEAE-Sephacel was further purified by
gel chromatography on Sephacryl S-200 in 4 M GdmCl with the
result shown in Fig. 2.

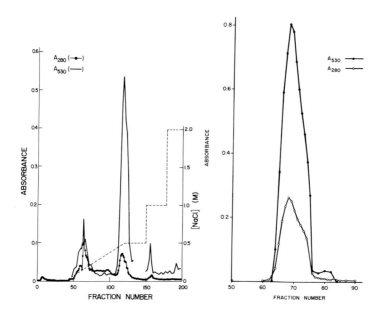

Fig. 1 (top, left). Chromatography of fraction A2D5D6 on DEAE-Sephacel in 6 M urea, the second large peak eluted at 0.5 M NaCl contained almost all of the dermatan sulfate proteoglycan.

Fig. 2 (top, right). Gel chromatography of the dermatan sulfate proteoglycan on Sephacryl S-200 in 4 M GdmCl.

Aliquots of the same twelve fractions from the main peak shown in Fig. 2 were subjected to electrophoresis on agarose/polyacrylamide gels, with the result shown in Fig. 3. Because of its relatively small size, the mobility of the dermatan sulfate-containing monomer is much faster than that of the cartilage-specific proteoglycan monomer.

Table 3 gives the chemical composition of the dermatan sulfate-containing proteoglycan monomer. In the purified monomer, dermatan sulfate represents 39.9% of the total glycosaminoglycan. The dermatan sulfate-containing proteoglycan monomer has a much higher protein content than the cartilage-specific monomer. It contains 4% glucosamine,

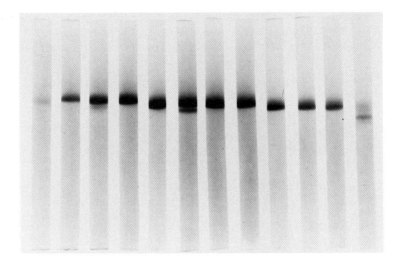

Fig. 3. Agarose/polyacrylamide gel electrophoresis of the twelve fractions from the main peak shown in Fig. 2.

TABLE 3

Chemical composition of the dermatan sulfate-containing proteoglycan monomer from bovine fetal epiphyseal cartilage. Values are shown as percent of dry weight.

Protein	35.50%
Hexuronate	11.51
Galactosamine	13.25
Glucosamine	4.13
Mannose	1.03
Galactose	0.88
Xylose	0.45
Fucose	0.21
Glucose	0.16
Sialate	1.00
Chondroitin sulfate	15.8
Dermatan sulfate	10.5
(DS/CS+DS) x 100	39.9

most of which cannot be associated with keratan sulfate because of the low galactose content. Taken together, the
glucosamine, mannose and sialate values suggest that the
dermatan sulfate-containing proteoglycan monomer contains
a substantial number of mannose-rich oligosaccharides.

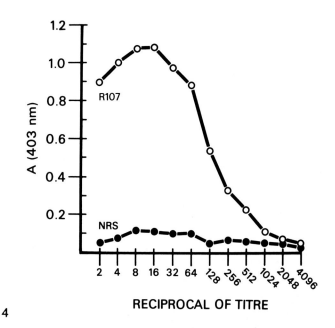

4

Fig. 4. Enzyme (peroxidase) linked immunosorbent assay of
rabbit antibodies to dermatan sulfate proteoglycans (R107,
o-o-o). The antigen bound to the microtitre wells was dermatan sulfate proteoglycan. The reaction with non-immune rabbit
serum is also shown (●-●-●). For further details of method
see Poole *et al.*, 1982c.

     Rabbit antibodies to dermatan sulfate proteoglycan
reacted well with these molecules as demonstrated by enzyme
linked immunosorbent assay (Fig. 4). They showed no reacti⁻
vity with cartilage proteoglycans of high buoyant density
(A1D1D1) but did react with dermatan sulfate proteoglycan
isolated from bovine sclera (Poole *et al.*, 1982c). Antibodies to A1D1D1 proteoglycans did not cross-react with
dermatan sulfate proteoglycan from sclera (Poole *et al.*, 1982c)

or cartilage. These observations clearly confirmed the chemical dissimilarity of mesenchymal proteoglycans and revealed that they and immunochemically related molecules, could be detected within tissues and in tissue extracts using these antibodies.

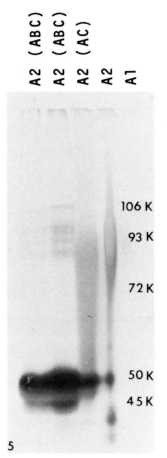

Dermatan sulfate proteoglycans were detected immunologically in tissue extracts and fractions thereof using a combination of sodium dodecyl sulfate polyacrylamide gel electrophoresis followed by electrophoretic transfer from the gel onto an overlay of nitrocellulose. This method permits the identification and sizing of small proteoglycans with small impure samples. Using this approach, we have determined the molecular size of these molecules to be within the range 72K to 106K (Fig. 5). We have also confirmed that they are most concentrated in fraction A2 isolated from fetal epiphyseal cartilage, although they are also present in fractions A3, A4 and A5. Differential treatment before gel electrophoresis with chondroitinase ABC and AC also permits the demonstration of dermatan sulfate since only chondroitinase ABC produces maximal reduction in molecular size, while chondroitinase AC produces molecular

Fig. 5. Western electrophoretic blot on nitrocellulose paper of fraction A2 isolated from bovine fetal epiphyseal cartilage and electrophoresed in polyacrylamide with sodium dodecyl sulfate. The blot was stained first with antibody to dermatan sulfate proteoglycan followed by peroxidase-labeled pig antibody to rabbit immunoglobulin G. Fractions A1 and A2 were examined before and after digestion with chondroitinases ABC and AC. Molecular sizes are indicated. For further details see Poole *et al.*, 1982c.

species of intermediate molecular size.

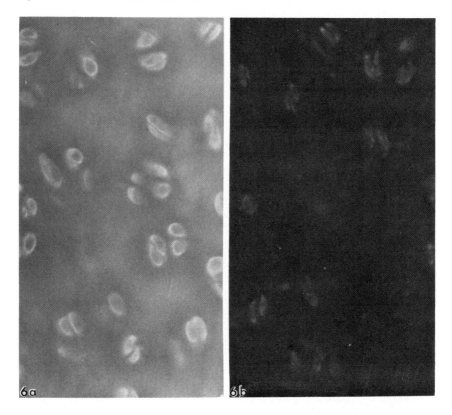

Fig. 6. Immunofluorescent localization of dermatan sulfate proteoglycan in mature bovine articular cartilage matrix using the indirect method, (a) test and (b) control. For further details of methods see Poole *et al.*, 1980b.

Using immunofluorescence microscopy, dermatan sulfate proteoglycans have been demonstrated in the extracellular matrix of non-mineralizing fetal epiphyseal cartilage and of adult bovine articular cartilage: in the latter case these molecules are most concentrated in the deeper layers, removed from the articular surface (Fig. 6). These or re-lated molecules are also present within the media of bovine aorta and in the stratum germinativum and the dermis of skin. Their presence in the sclera of the eye can also be demon-strated, confirming the earlier biochemical studies of

Coster and Fransson (1980).

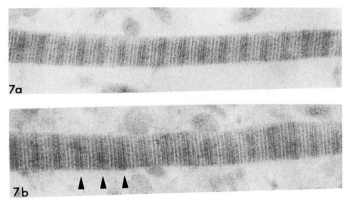

Fig. 7. The presence of dermatan sulfate proteoglycan (arrow heads) on collagen fibrils in adult bovine articular cartilage demonstrated with immunoelectron microscopy, (a) control and (b) test. For further details of immunoelectron microscopy see Poole *et al.*, 1982b. Magnification x 120,000.

An examination of the ultrastructural organization of dermatan sulfate proteoglycans in mature articular cartilage has identified at least two organizational patterns involving those molecules. The majority of these molecules seem to be associated with collagen fibrils (Fig. 7) where they are regularly arranged across the width of the type II collagen fibril, in a manner very similar to that described for dermatan sulfate proteoglycans arranged on type I collagen fibrils in rat tail tendon (Scott, Orford, 1981). In addition, dermatan sulfate proteoglycans were sometimes detected between collagen fibrils in a lattice work (Fig. 8) reminiscent of that produced in interterritorial sites in the deep zone by the major proteoglycan population (examined with antibodies to proteoglycan fractions A1D1 or A1D1D1). Unlike these more common proteoglycans, dermatan sulfate proteoglycans seem to be differently arranged in that the core proteins appear to be more closely packed (Fig. 9 and Table 4). Whereas the organization of A1D1 proteoglycans is thought to be based on hyaluronic acid (Poole *et al.*, 1982b) that of interfibrillar dermatan sulfate proteoglycans may result from dermatan sulfate chain interactions (Fig. 9). Previously, Fransson *et al.* (1979) showed that iduronic acid-rich dermatan sulfate

chains can interact directly leading to proteoglycan aggregation which is independent of hyaluronic acid.

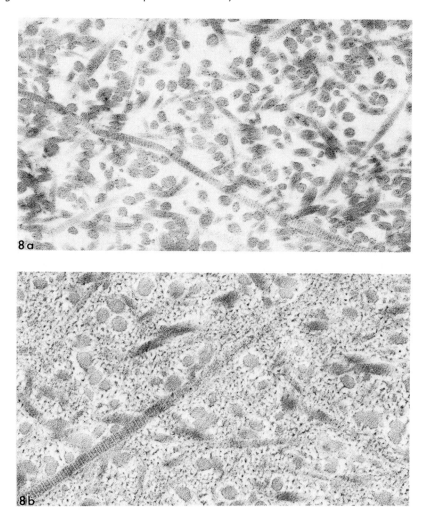

Fig. 8. Dermatan sulfate proteoglycan present between collagen fibrils in adult bovine articular cartilage (b). The
control is (a). For further details of the method see Poole
et al., 1982b. Magnification x 50,000.

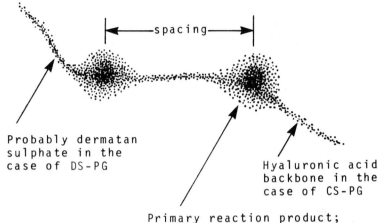

|←———spacing———→|

Probably dermatan
sulphate in the
case of DS-PG

Hyaluronic acid
backbone in the
case of CS-PG

Primary reaction product;
represents collapsed CS-PG
(Poole et al., 1982b) or
core protein of DS-PG

9

Fig. 9. Diagrammatic representation of the observed organization of interfibrillar proteoglycans in the interterritorial deep zone of mature bovine articular cartilage as
seen with immunoelectron microscopy. The spacings between
primary reaction products are shown in Table 4.

TABLE 4

The sizes and spacings of proteoglycan reaction product in
the interterritorial deep zone of mature bovine articular
cartilage.

Primary Reaction Product	Proteoglycan (PG)	
	DS-PG	CS-PG (A1D1D1)
Spacing (nm)	32	85^1
Dimensions (nm)	14×9	27×21^2

n = 20-30
DS represents dermatan sulfate; CS represents chondroitin
    sulfate
[1,2] Data from Poole *et al.*, 1982b.

DISCUSSION

The studies described in this report demonstrate that a dermatan sulfate-containing proteoglycan monomer is present in the extracellular matrix of developing epiphyseal cartilage and even mature articular cartilage. The results reported here, together with the observations of Royal and Goetinck (1977, 1980) indicate that the mesenchymal proteoglycan monomer is a dermatan sulfate-containing proteoglycan monomer, whose core protein is immunologically different from that of the cartilage-specific proteoglycan monomer. Since the dermatan sulfate proteoglycan is present in fetal epiphyseal and mature cartilage long after the onset of chondrogenesis, the biosynthesis of this proteoglycan must continue in cartilage postnatally and even after maturation.

Admittedly, only a very small amount of dermatan sulfate proteoglycan is present in fetal epiphyseal cartilage. From 250 g of fresh wet fetal epiphyseal cartilage (32 g dry weight), the yield of isolated, purified dermatan sulfate proteoglycan monomer is only 15 mg, while the yield of cartilage-specific proteoglycan monomer (A1D1D1) is 8 g. Thus, approximately 500 times as much cartilage-specific proteoglycan monomer as dermatan sulfate proteoglycan monomer can be isolated from bovine fetal epiphyseal cartilage. Nevertheless, the dermatan sulfate-containing proteoglycan monomer can be isolated from bovine fetal epiphyseal cartilage in amounts sufficient for its chemical characterization, and to prepare a mono-specific antiserum. Using the antisera to both the dermatan sulfate-containing proteoglycan monomer and the cartilage-specific proteoglycan monomer in immunohistochemical studies, it will be fascinating to examine simultaneously the changes in the ultrastructure, localization and distribution of these two species which occur in the developing limb bud during chondrogenesis.

REFERENCES

Bitter T, Muir H (1962). A modified uronic acid carbazole reaction. Anal Biochem 4:330.
Buckwalter JA, Rosenberg L (1982). Electron microscopic studies of cartilage proteoglycans. Direct evidence for the variable length of the chondroitin sulfate rich region of proteoglycan subunit core protein. J Biol Chem In press.
Coster L, Fransson L-A (1980). Isolation and characterization of dermatan sulphate proteoglycans from bovine sclera. Biochem J 193:143.

Fransson L-A, Nieduszynski IA, Phelps CF, Sheehan JK (1979). Interactions between dermatan sulphate chains. III. Light scattering and viscometry studies of self-association. Biochim Biophys Acta 586:179.

Goetinck PF, Pennypacker JP, Royal PD (1974). Proteochondroitin sulfate synthesis and chondrogenic expression. Exptl Cell Res 87:241.

Goetinck PF, Pennypacker JP (1977). Controls in the acquisition and maintenance of chondrogenic expression. In Ede DA, Hinchliffe JR, Balls M (eds): "Vertebrate Limb and Somite Morphogenesis," Cambridge: Cambridge University Press.

Hardingham TE (1979). The role of link-protein in the structure of cartilage proteoglycan aggregates. Biochem J 177:237.

Hascall VC (1981). Proteoglycans: Structure and function. In Ginsburg V, Robinson P (eds): "The Biology of Carbohydrates, Volume I," New York: John Wiley, p 1.

Hascall VC (1982). Structure and biosynthesis of proteoglycans with keratan sulfate. In "Proceedings of the Third International Conference on Limb Development and Regeneration," (These Proceedings).

Hascall VC, Riolo R, Hayward J, Reynolds CC (1972). Treatment of bovine nasal cartilage proteoglycan with chondroitinases from *Flavobacterium heparinum* and *Proteus vulgaris*. J Biol Chem 247:4521.

Lohmander LS, De Luca S, Nilsson B, Hascall VC, Caputo CB, Kimura JH, Heinegard D (1980). Oligosaccharides on proteoglycans from the Swarm rat chondrosarcoma. J Biol Chem 255:6084.

Lowry OH, Rosebrough NJ, Farr AL, Randall RJ (1951). Protein measurement with the folin phenol reagent. J Biol Chem 193:265.

McKeown PJ, Goetinck PF (1979). A comparison of the proteoglycans synthesized in Meckel's and sternal cartilage from normal and nanomelic chick embryos. Develop Biol 71:203.

Pal S, Tang L, Choi H, Habermann E, Rosenberg L, Roughley P, Poole AR (1981). Structural changes during development in bovine fetal epiphyseal cartilage. Coll Res 1:151.

Pennypacker JP, Goetinck, P (1976). Biochemical and ultrastructural studies of collagen and proteochondroitin sulfate in normal and nanomelic cartilage. Develop Biol 50:35.

Poole AR, Reiner A, Tang L-H, Rosenberg L (1980a) Proteoglycans from bovine nasal cartilage. Immunochemical studies of link protein. J Biol Chem 255:9295.

Poole AR, Pidoux I, Reiner A, Tang L-H, Choi H, Rosenberg L (1980b). Localization of proteoglycan monomer and link protein in the matrix of bovine articular cartilage: An immunohistochemical study. J Histochem Cytochem 28:621.

Poole AR, Pidoux I, Rosenberg L (1982a). Role of proteoglycans in endochondral ossification: Immunofluorescent localization of link protein and proteoglycan monomer in bovine fetal epiphyseal growth plate. J Cell Biol 92:249.

Poole AR, Pidoux I, Reiner A, Rosenberg L (1982b) An immunoelectron microscopic study of the organization of proteoglycan monomer, link protein and collagen in the matrix of articular cartilage. J Cell Biol In press.

Poole AR, Pidoux I, Reiner A, Coster L, Hassell J (1982c) Mammalian eyes and associated tissues contain molecules which are immunologically related to cartilage proteoglycan and link protein. J Cell Biol In press.

Reihanian H, Jamieson AM, Tang L-H, Rosenberg L (1979). Hydrodynamic properties of proteoglycan subunit from bovine nasal cartilage. Self-association behavior and interaction with hyaluronate studied by laser light scattering. Biopolymers 18:1727.

Reihanian H, Jamieson AM, Blackwell J, Tang L-H, Rosenberg L (1981). Structural characterization of proteoglycan subunit from nasal septum by laser light scattering. In Brant DA (ed): "Solution Properties of Polysaccharides," Washington DC: American Chemical Society, p 201.

Rosenberg L, Wolfenstein-Todel C, Margolis R, Pal S, Strider W (1976). Proteoglycans from bovine proximal humeral articular cartilage. Structural basis for the polydispersity of proteoglycan subunit. J Biol Chem 251:6439.

Royal PD, Goetinck PF (1977). *In vitro* chondrogenesis in mouse limb mesenchymal cells: Changes in ultrastructure and proteoglycan. J Embryol Exp Morph 39:79.

Royal PD, Sparks KJ, Goetinck PF (1980). Physical and Immunochemical characterization of proteoglycans synthesized during chondrogenesis in the chick embryo. J Biol Chem 255:9870.

Scott JE, Orford CR (1981). Dermatan sulphate-rich proteoglycan associates with rat tail-tendon collagen at the d band in the gap region. Biochem J 197:213.

Tang L-H, Rosenberg L, Reiner A, Poole AR (1979). Proteoglycans from bovine nasal cartilage. Properties of a soluble form of link protein. J Biol Chem 254:10523.

Yemm EW, Willis AJ (1954). Estimation of carbohydrates in plant extracts by anthrone. Biochem J 57:508.

Warren L (1959). The thiobarbituric acid assay of sialic acids. J Biol Chem 234:1971.

Limb Development and Regeneration
Part B, pages 85–95
© **1983 Alan R. Liss, Inc., 150 Fifth Avenue, New York, NY 10011**

PROTEOGLYCANS IN THE CARTILAGE OF THE TURKEY CHONDRODYSTROPHY
MUTANT

Ann Dannenberg,[1] E. G. Buss,[2] and P. F. Goetinck[1]

Department of Animal Genetics,[1] University of
Connecticut, Storrs, CT 06268 and Department of
Poultry Science,[2] Pennsylvania State University,
University Park, PA 16802

INTRODUCTION

The synthesis of sulfated proteoglycans can be viewed
as taking place in three steps. These consist of the synthesis of the core protein, the glycosylation of that core
protein, and finally, the sulfation of appropriate carbohydrate residues. The major carbohydrate components of
cartilage proteoglycan are the two glycosaminoglycans,
chondroitin sulfate and keratan sulfate, as well as a number
of N- and O-linked oligosaccharides. Sulfated proteoglycans
from cartilage are capable of forming aggregates by interacting with hyaluronic acid through a specialized region of
the core protein. This aggregation is stabilized by the
action of a link protein which interacts both with the core
protein of proteoglycan and with hyaluronic acid (Hascall
and Hascall, 1981).

The proteoglycans of cartilage and those of the mesenchymal precursor cells differ both in their carbohydrate
and their core protein. Chemical, immunochemical, genetic,
and developmental evidence indicates that the synthesis of
core protein is a developmentally regulated event which
results in the synthesis of a cartilage-specific proteoglycan (Goetinck, 1982).

A number of inherited disorders of cartilage proteoglycans have been described in the chicken and in the mouse
(see Goetinck, 1983, for review). These disorders have been
shown to result from mutations which affect the synthesis of
core protein (nanomelia in the chicken, cartilage matrix

deficiency in the mouse, and the sulfation of the glycosaminoglycans (brachymorphy in the mouse). In the present
report we give the results of an analysis of a mutation
which affects the cartilage proteoglycans of the turkey.
The results suggest that this mutation may affect the glycosylation of the proteoglycans.

MATERIALS AND METHODS

Embryos

Mutant and normal embryos were obtained from matings
between parents heterozygous for the recessive lethal gene,
chondrodystrophy. These birds are being maintained by the
Poultry Science Department at Pennsylvania State University.

Sucrose Density Gradient Centrifugation

Sucrose density gradients (5-20%) were prepared in 4 M
guanidine hydrochloride, 0.05 M sodium acetate, pH 5.8,
according to the method of Okayama *et al.* (1976). Centrifugation was for 18.5 hours at 36,500 rpm in a Beckman SW40
rotor.

Antiserum

The antiserum against chicken cartilage proteoglycan
monomer has been described (Sparks *et al.*, 1980).

Column Chromatography

Molecular sieve chromatography on controlled pore glass
was described in Lever and Goetinck (1976) and McKeown and
Goetinck (1979).

Polyacrylamide Gel Electrophoresis

The electrophoretic analyses were done as described in
Argraves *et al.* (1981).

RESULTS AND DISCUSSION

The reduced content of galactosamine in the mutant cartilage (Leach and Buss, 1977; Table 1) suggests that the
quantity of chondroitin sulfate is reduced in that tissue.
Since the measurements obtained by these authors reflect the
quantity of material accumulated throughout the development
of the cartilaginous organs, the reduced value could represent either a decrease in the synthesis or an increase in
the degradation of the affected molecules. In order to distinguish between these two possibilities the synthesis of
the sulfated proteoglycans was monitored in both organ and
cell cultures (Table 1). After an 18-hour exposure to [^{35}S]-
sulfate the mutant sterna incorporated the isotope at levels
which were 28.7% of normal. Similar tests with chondrocytes
cultures in suspension reveal that the mutant chondrocytes
incorporate the isotope at 26.2% of normal levels. The
levels of isotope incorporation are in close agreement with
the galactosamine content of the cartilage and these results
are taken to mean that the reduction in chondroitin sulfate
is the result of reduced synthesis and not of excessive
degradation.

Table 1. Turkey Cartilage Proteoglycan

	Galactosamine*	[^{35}S] sulfate incorporation	
		DPM/sternum	DPM/10^5 cells
Normal	1.80	27,139	5,994
ch/ch	0.58	7,788	1,570
$\frac{\text{Mutant}}{\text{Normal}}$ X 100	32.2	28.7	26.2

*Mg/gm of dry cartilage (Leach and Buss, 1977).

When proteoglycans were extracted from normal and mutant
cartilage and subjected to molecular sieve chromatography on
controlled pore glass, no difference could be observed between the two genotypes (Table 2). On CPG-1400, 79% of the
radioactivity of the normal and 88% of the mutant chromatographed in the void volume. Both monomer and aggregates of
cartilage proteoglycan chromatograph in the void volume of

such columns. In contrast, on CPG-2500 columns proteoglycan aggregates chromatograph in the void volume and the monomers are included. When CPG-1400 V_o material is chromatographed on CPG-2500, 70.6% of the normal and 75.7% of the mutant material chromatographs as aggregates. These results indicate that the mutant proteoglycans possess a hyaluronic acid binding region, and therefore that this part of the core protein is normal. Further tests using antibodies directed against chicken cartilage proteoglycan monomer indicate that there are no detectable immunochemical differences between normal and mutant either when the measurements are made on the total extract or on CPG-1400 void volume material. The corrected specific binding of the turkey proteoglycan by the antiserum was similar for both genotypes for each of the antigenic preparations (Table 3). Antibody

Table 2. Distribution of Proteoglycans upon Chromatography on Controlled Pore Glass

	Extract on CPG-1400		CPG-1400 V_o on CPG-2500	
	V_o	Inc	V_o	Inc
Normal	79.0%	21.0%	70.6%	29.4%
Mutant	88.2%	11.8%	75.7%	24.3%

Table 3. Corrected Specific Binding of Turkey Proteoglycans by Antiserum to Chicken Cartilage Proteoglycan Monomer

	Extract	CPG-1400 V_o
Normal	86.6%	83.0%
Mutant	84.3%	77.2%

dilution experiments with monomer as the antigen gave nearly identical dilution curves for the two phenotypes.

When chicken proteoglycan aggregates are mixed with

antiserum elicited against chicken cartilage proteoglycan
monomer and then sedimented on a 5-20% sucrose gradient in
the presence of 4 M guanidine hydrochloride, more than 30%
of the proteoglycan is recovered in the bottom fraction
(Sparks and Goetinck, 1979; Goetinck, 1982). Because of the
sharpness of the peak in the profile it has been referred to
as a "spike". The spiking phenomenon requires proteoglycan
aggregates since it is not observed when monomers are used,
and it is interpreted to reflect the stabilization of the
aggregates by the antiserum even in the presence of 4 M
guanidine hydrochloride when   aggregates are usually dissociated. When this test was carried out with CPG-1400 void
volume material of both normal and mutant extracts, the
spiking phenomenon could be demonstrated in both instances
(Fig. 1). Since aggregation takes place through the hyaluronic acid binding region of the core protein, these results
support further that the mutant proteoglycan possesses a
functional hyaluronic acid binding region.

A very striking difference is observed between normal
and mutant in the sedimentation rate of the cartilage proteoglycan monomer on 5-20% sucrose density gradients in 4 M
guanidine hydrochloride. In Fig. 2 are shown the results of
sedimenting whole extracts or that fraction of the extract
which chromatographs in the void volume of a CPG-1400 column
on sucrose density gradients. The sedimentation rate of the
mutant cartilage proteoglycan is much slower than that of
its normal counterpart and resembles that of the major proteoglycan of a number of non-cartilaginous tissues such as
chicken skin fibroblasts (Goetinck, 1982) and stage 24 limb
buds (Royal et al., 1980). The mutant cartilage proteoglycan,
however, differs from these two non-cartilaginous proteoglycans in its immunochemical properties. The former cross
reacts with an antiserum to chicken cartilage proteoglycan
monomer and is, in these tests, immunologically indistinguishable from normal turkey cartilage proteoglycan, whereas the
latter do not cross react with the antiserum. The information presented up to this point then indicates that the
mutant chondrocytes synthesize a small cartilage-specific
proteoglycan which contains approximately one-third the
chondroitin sulfate found in the larger proteoglycan synthesized by normal turkey chondrocytes.

Possible structural models for explaining the mutant
proteoglycans are represented diagramatically in Fig. 3, A-
C, and compared to normal. In this oversimplified diagram

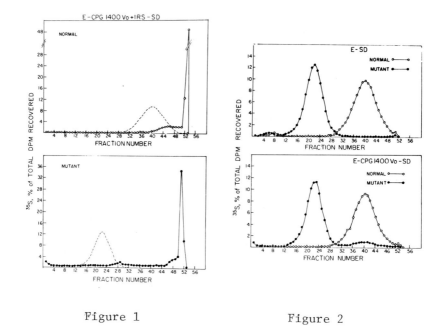

Figure 1   Figure 2

Fig. 1. Cartilage extract which chromatographs in the void volume of a CPG-1400 column was mixed with immune rabbit serum against proteoglycan monomer and sedimented on a dissociative 5-20% sucrose density gradient. The dotted lines represent the sedimentation profiles of the proteoglycans when mixed with normal rabbit serum or in the absence of serum. Direction of sedimentation is from left to right.

Fig. 2. Sedimentation profiles of normal and mutant extract (top) or extract which chromatographs in the void volume of a CPG-1400 column (bottom) on dissociative 5-20% sucrose density gradients. Direction of sedimentation is from left to right.

the only type of carbohydrate side chain indicated represents chondroitin sulfate. Three models are presented. In Model A, the length of the chondroitin sulfate side chain is one-third the length of normal. In Model B, there is only one-third the number of chondroitin sulfate side chains of normal length as a result of a shorter core protein, and in Model C there is a third of the number of chondroitin sulfate side chains of normal length on a core protein of normal length.

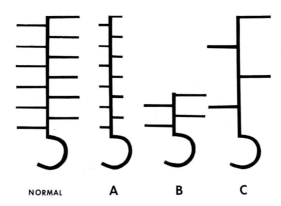

NORMAL   A   B   C

Fig. 3. Diagramatic representation of a normal proteoglycan monomer with only chondroitin sulfate side chains attached to the core protein. The curved bottom of the core protein represents the hyaluronic acid binding region. Diagrams A, B, and C represent structural models for the mutant proteoglycan monomer.

Model A (Fig. 3) was tested in two manners: First, the capacity of the chondrocytes to synthesize chondroitin sulfate was tested by the addition of para-nitrophenyl-β-D xyloside to the culture medium. Xylosides or xylose act as initiators for the synthesis of free chondroitin sulfate by substituting for xylosylated protein backbone, and can be used to measure indirectly the activity of the glycosyltransferases distal to xylosyltransferase. If [^{35}S]-sulfate is used to measure the synthesis, the tests also measure indirectly the enzymes in the sulfation pathway. The results shown in Fig. 4 clearly indicate that the [^{35}S]-sulfate incorporation can be increased in the mutant by a xyloside to the same level as in similarly treated normal cells. We conclude from this that the mutation does not affect the synthesis of chondroitin sulfate. Support for this conclusion is also obtained from the analysis of the sulfated glycosaminoglycans obtained by alkaline hydrolysis from the intact proteoglycans. When analyzed by molecular sieve chromatography on CPG-500 the size of the sulfated glycosaminoglycans is similar to those from normal proteoglycans (Fig. 5).

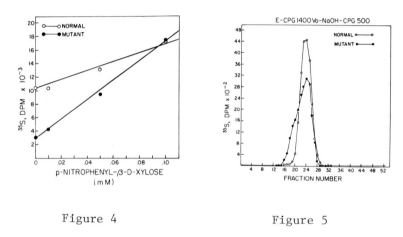

Figure 4   Figure 5

Fig. 4. Stimulation of chondroitin sulfate synthesis by normal and chondrodystrophic turkey chondrocytes by paranitrophenyl-β-D xylose.

Fig. 5. Molecular sieve chromatography on controlled pore glass (CPG-500) of [^{35}S]-sulfated glycosaminoglycans isolated by alkaline hydrolysis from cartilage extracts which chromatograph in the void volume of a CPG-1400 column.

Model B (Fig. 3), in which a shorter core protein is proposed, was tested by labeling chondrocytes with [^{35}S]-cysteine and by electrophoresing the intracellular proteins on polyacrylamide gels (Fig. 6). Based on a previous study

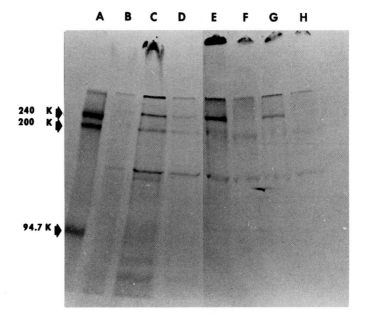

Fig. 6. Fluorogram of [^{35}S] cysteine labeled proteins of normal (Lane C) and chondrodystrophic (Lane D) turkey chondrocytes. For comparison, labeled proteins from chicken chondrocytes were electrophoresed in Lane B. Indirect antibody immunoprecipitation of the radiolabeled intracellular proteins utilizing antiserum elicited against chicken cartilage proteoglycan monomer (E, normal; G, chondrodystrophic) or normal rabbit serum (F, normal; H, chondrodystrophic). Lane A contains molecular weight markers.

with chicken chondrocytes one can detect in such extracts a 246,000 dalton intracellular protein which is the core protein of cartilage proteoglycan (Argraves *et al.*, 1981).

In the present study both normal and mutant turkey chondrocyte extracts contain a protein with a molecular weight of approximately 246,000 daltons (Fig. 6, Lanes C and D). These molecules can be precipitated by antiserum to chicken cartilage proteoglycan monomer (Fig. 6, Lanes E and G) but not by non-immune rabbit serum (Fig. 6, Lanes F and

H). Some smaller molecular weight material was precipitated by both the antiserum and non-immune rabbit serum in both the normal and the mutant, and therefore, does not represent material related to core protein. The core protein of the turkey proteoglycans is identical in molecular weight to that found in chicken chondrocyte extracts (Fig. 6, Lane B). The mutant turkey chondrocytes, therefore, synthesize a core protein with a molecular weight identical to that synthesized by normal chondrocytes.

CONCLUSION

Taken together, the above results eliminate both models A and B (Fig. 3) and suggest that the mutant proteoglycan has a core protein of normal length to which fewer chondroitin sulfate chains of normal length are attached (Model C, Fig. 3). Whether these chondroitin sulfate chains are evenly distributed along the core protein or clustered is not known. The present working hypothesis to explain Model C is that the mutation may affect either the synthesis of UDP-xylose or the xylosyltransferase reaction.

REFERENCES

Argraves WS, McKeown-Longo PJ, Goetinck PF (1981). Absence of proteoglycan core protein in the cartilage mutant nanomelia. FEBS Letters 131:265-268.

Goetinck PF (1982). Proteoglycans in developing embryonic cartilage. In Horowitz MI (ed): "The Glycoconjugates," New York: Academic Press, Vol. 3, pp 197-229.

Goetinck PF (1983). Mutations affecting limb cartilage. In Hall BK (ed): "Cartilage," New York: Academic Press, in press.

Hascall VC, Hascall GK (1981). Proteoglycans. In Hay ED (ed): "Cell Biology of Extracellular Matrix", New York: Plenum Publishing Corporation, pp 39-63.

Leach RM, Jr, Buss EG (1977). The effect of inherited chondystrophy on the hexosamine content of cartilage from turkey embryos. Poultry Sci 56:1043-1045.

Lever PL, Goetinck PF (1976). Molecular sieve chromatography of proteoglycans. A comparative analysis. Anal Biochem 75:67-76.

McKeown PJ, Goetinck PF (1979). A comparison of the proteoglycans synthesized in Meckel's and sternal cartilage from normal and nanomelic chick embryos. Dev Biol 71:203-215.

Okayama M, Pacifici M, Holtzer H (1976). Differences among sulfated proteoglycans synthesized in non-chondrogenic cells, presumptive chondroblasts and chondroblasts. Proc Natl Acad Sci, USA 73:3224-3228.

Royal PD, Sparks KJ, Goetinck PF (1980). Physical and immuno-chemical characterization of proteoglycans synthesized during chondrogenesis in the chick embryo. J Biol Chem 255:9870-9878.

Sparks KJ, Goetinck PF (1979). Detection of an antigenic determinant unique for avian cartilage proteoglycan. J Cell Biol 74:466a.

Sparks KJ, Lever PL, Goetinck PF (1980). Antibody binding of cartilage-specific proteoglycans. Arch Biochem Bio-phys 199:579-590.

ACKNOWLEDGEMENTS

This work was supported by Grant HD-09175 from the NICHHD to P.F.G., and by Pennsylvania Agricultural Experiment Station Project 2552 to E.G.B.

This paper is Scientific Contribution No. 975, of the Storrs Agricultural Experiment Station, University of Connecticut, Storrs, Connecticut 06268.

Limb Development and Regeneration
Part B, pages 97–103
© **1983 Alan R. Liss, Inc., 150 Fifth Avenue, New York, NY 10011**

DEFECT IN PROTEOGLYCAN SYNTHESIS IN BRACHYMORPHIC MICE

Nancy B. Schwartz, Ph.D.

Departments of Pediatrics and Biochemistry
University of Chicago
Chicago, IL 60637

The proteoglycan from cartilage, which is considered the prototype of these kind of molecules, consists of a protein core to which are covalently linked oligo and poly-saccharide chains. An average proteoglycan molecule is approximately 2.5×10^6 daltons, with a protein core of about 200,000 daltons. Glycosaminoglycan chains are asymmetrically distributed along the core protein with one end, the hyaluronic acid binding region, relatively free of carbohydrate. An interior portion is enriched in keratan sulfate and glycoprotein-type oligosaccharide components, and the remainder of the molecule is highly substituted with chondroitin sulfate chains. Some variations and heterogeneity have been observed in the general features of this structure of proteoglycans isolated from different cartilage sources.

We are particularly interested in synthesis of proteoglycans from a regulatory point of view (Schwartz 1982). These are large molecules of extremely complex structure, requiring a highly ordered, well-controlled system to coordinate the more than 10,000 reactions necessary for synthesizing an individual proteoglycan molecule. Furthermore, synthesis has been shown to change in certain instances; most importantly a significant increase in synthesis of proteoglycans is observed during differentiation of chondrocytes from mesenchyme cells. Thus, we are especially concerned with those aspects of biosynthesis which may be involved in the overall regulation of this dynamic process.

Certainly one of the most powerful tools for elucidating

structure, function or control mechanisms, is to study systems exhibiting defects in the process. In order to elucidate possible control mechanisms in proteoglycan synthesis, as well as the role of proteoglycans in tissue organization and development of cartilage, we initiated studies on mutant mouse systems exhibiting defects in production of proteoglycans and manifesting growth disorders. This discussion will be limited to the brachymorphic mutant, a homozygous-recessive non-lethal mutation in which is produced a proteoglycan molecule that is undersulfated 50-65%.

Sulfation of proteoglycan or any naturally occurring sulfated compound requires the components shown in Fig. 1. Those include the acceptor molecule to be sulfated, a sulfate donor, which universally is the high energy compound phosphoadenosylphosphosulfate (PAPS), and one or more sulfotransferases. By a series of experiments which will not be reviewed (Schwartz et al. 1978; Sugahara, Schwartz 1979; 1982a), the possibilities of an abnormal substrate which cannot accept sulfate, abnormal sulfotransferases which are unable to transfer sulfate, as well as increased degradation of any of the products, have been eliminated. Thus, the defect in brachymorphic mice was postulated to be related to the availability of the sulfate donor, PAPS. Furthermore, when sulfation of proteoglycan was initiated in vitro by addition of PAPS, comparable levels of sulfate were incorporated by normal C57B16J and bm/bm cartilage; whereas when sulfation was initiated with ATP and SO_4^{2-}, a 75% reduction in incorporation was observed by mutant system, suggesting a possible defect in synthesis of PAPS (Schwartz et al. 1978).

Since two steps are involved in synthesis of PAPS (Fig. 1), the following approaches were taken to elucidate the defect. Intermediates were quantitated following incubation of cartilage extracts with ATP and $^{35}SO_4^{2-}$ and separated by high voltage paper electrophoresis. Routinely, synthesis of macromolecular products were substantially reduced. PAPS was reduced to about 8% of control; whereas APS levels were comparable or actually higher in incubations of mutant tissues (Sugahara, Schwartz 1979). This was the first indication that the defect was at the step of conversion of APS to PAPS.

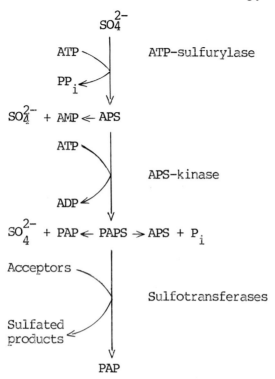

Fig. 1. Sulfate Activation Pathway.

More directly, the two enzymes, ATP-sulfurylase and
APS-kinase, constituting the PAPS-synthetic pathway were
measured individually by two different assay procedures.
By both assay procedures, reductions of 85-90% of APS-kinase
activity from mutant tissue were observed (Table 1). Less
pronounced reductions in ATP-sulfurylase activity to approximately 50% of control, have also been consistently observed
(Table 1). Several possible explanations for the partial
reduction in ATP-sulfurylase activity, centering on it being
secondary to the profound decrease in APS-kinase activity
have been explored and eliminated. Rather, it is believed
that ATP-sulfurylase is defective and that both enzymes
are affected by the mutation in brachymorphic mice. Three
lines of evidence lead to this conclusion.

TABLE 1

PERCENT OF CONTROL SPECIFIC ACTIVITIES AND SULFATED INTERMEDIATES IN
VARIOUS TISSUES OF NORMAL AND bm/bm MICE

TISSUE	APS	PAPS	SULFATED PRODUCTS	ATP-SULFURYLASE	APS-KINASE	RESIDUAL ATP-SULFURYLASE AT 50°
	%	%	%	%	%	%
Cartilage	300	8	10	53	7	33
Liver	80	7	14	31	5	36
Kidney	90	45	63	66	37	40
Skin	94	99	102	110	100	100
Brain	110	105	102	107	100	100

First, ATP-sulfurylase from bm/bm cartilage is more labile. Thermal inactivation studies have shown that ATP-sulfurylase from mutant tissue decays with a $t_{\frac{1}{2}}$ of about 10 minutes, while the enzyme from normal tissue is stable more than 2 hours at 50°. The mutant enzyme is also inactivated to a greater extent (61%) by repeated freezing and thawing compared to normal (25%) (Sugahara, Schwartz 1982a).

Additional evidence that both enzymes of the PAPS synthesizing pathway in brachymorphic mice are defective was obtained from a study of the tissue distribution of the defect. This study was prompted by a concern that if the defect limited the availability of PAPS, then other sulfated molecules in other tissues should be affected as well; and perhaps the manifestations of the defect might be expected to be more severe. Thus, a study of the distribution of the defect in tissues which should have a requirement for a sulfation pathway was instituted with respect to four criteria: levels of all intermediates, ATP-sulfurylase and APS-kinase activities, and thermostability of ATP-sulfurylase. The results of this series of experiments (Table 1) indicate that the tissues examined fell into two categories; those that exhibited the defect (cartilage and liver) and those that did not (skin and brain) (Sugahara, Schwartz 1982b). Kidney consistently showed intermediate levels with respect to all four parameters. Furthermore, the sulfation of specific products by these tissues was examined and a similar pattern obtained. Proteoglycan was undersulfated in mutant cartilage, and detoxification, which occurs predominantly by sulfation in normal liver switched to glucuronidation in mutant liver (unpublished). In contrast to both of these systems which were apparently affected by the defect in bm/bm mice, synthesis of sulfated glycolipids in mutant and normal brain and kidney were comparable (unpublished). These results have led to the conclusion that both enzymes are defective, and to hypothesize that since they share two substrates in the coupled reactions, ATP and APS, that they may share a common subunit that is affected by the mutation in brachymorphic mice.

Evidence for this hypothesis must await the availability of purified enzymes to examine their structural relatedness. Toward this objective, we have been purifying ATP-sulfurylase and APS-kinase from rat chondrosarcoma. In preliminary experiments, both activities co-elute through gel filtration on Sephadex G-200 and ion exchange chromatography, on DEAE-

Sephadex and hydroxylapatite, illustrating the intimate association of these enzymes through these purification steps. Most recently, an additional 10-fold purification by affinity chromatography on ATP-Sepharose has been achieved, yielding preparations now purified > 2000 fold (unpublished).

These results all suggest that both enzymes may be affected in the brachymorphic mutant and may be due to a common component of the enzyme complex. Furthermore, these results demonstrate a multiple, but not universal distribution of defect, and may suggest distinct forms of the enzymes exist in different tissues. These findings also explain why sulfation in skin and brain are normal and therefore why the manifestations of the mutation are not more severe and widespread. The defect predominantly affects cartilage, where there is a great demand for sulfation of proteoglycan and the primary lesion is expressed as a growth disorder.

These findings also indicate, with respect to proteoglycan chemistry, that a large portion of the sulfate and hence polyanionic nature of proteoglycans can be eliminated and this is not of lethal consequence. However, it is necessary for proper development and maintenance of the integrity of the extracellular matrix. From the early work of Orkin et al. (1976; 1977), reduction in the net ionic charge of proteoglycan resulted in a change in conformation of proteoglycan aggregates from large polygonal granules to thin filamentous forms. The effect on tissue growth was such that a progressive reduction in size of the columnar and hypertrophic zones from day 5 to 16 of age was observed. The overall size of the epiphyseal plate in mutant mice was less than half of that in normals, illustrating the profound effect on cartilage development caused by this alteration in the proteoglycans.

In conclusion, the bm/bm mouse is an interesting metabolic mutant, resulting from a possible double enzyme defect. Furthermore, it represents a model system in which one characteristic of proteoglycans as a major component of extracellular matrix is altered, thus allowing the examination of the consequence of this alteration to the development and organization of the connective tissue.

Orkin RW, Pratt RM, Martin GR (1976). Undersulfated chondroitin sulfate in the cartilage matrix of brachymorphic mice. Dev Biol 50:82.

Orkin RW, Williams BR, Cranley RE, Poppke DC, Brown KS (1977). Defects in the cartilagenous growth plates of brachymorphic mice. J Cell Biol 73:287.

Schwartz NB (1982). Regulatory mechanisms in proteoglycan biosynthesis. In Varma R (ed) "Glycosaminoglycans and proteoglycans in physiological and pathological processes of body systems", Basel: S. Karger AG., p 41.

Schwartz NB, Ostrowski V, Brown KS, Pratt RM (1978). Defective PAPS-synthesis in epiphyseal cartilage from brachymorphic mice. Biochem Biophys Res Commun 82:173

Sugahara K, Schwartz NB (1979). Defect in 3'-phosphoadenosine 5'-phosphosulfate formation in brachymorphic mice. Proc Natl Acad Sci USA 76:6615.

Sugahara K, Schwartz NB (1982a). Defect in 3'-phosphoadenosine 5'-phosphosulfate synthesis in brachymorphic mice. I. Characterization of the defect. Arch Biochem Biophys 214:589.

Sugahara K, Schwartz NB (1982b). II. Tissue distribution of the defect. Arch Biochem Biophys 214:602.

Limb Development and Regeneration
Part B, pages 105–111
Published 1983 by Alan R. Liss, Inc., 150 Fifth Avenue, New York, NY 10011

ISOLATION OF A PROTEIN FACTOR WHICH SUPPRESSES THE
CHONDROCYTE PHENOTYPE

Hugh H. Varner, A. Tyl Hewitt, John R. Hassell
Janice A. Jerdan, Elizabeth A. Horigan and
George R. Martin

Laboratory of Developmental Biology and Anomalies
National Institute of Dental Research
National Institutes of Health
Bethesda, Maryland 20205

It is well known that chondrocytes in culture have an
unstable phenotype and, over a period of 3-20 days,
convert to fibroblastic cells (Holtzer, et al., 1960;
Abbott, Holtzer, 1966). This conversion can be monitored
by changes in the morphology of the cells and by changes
in their biosynthetic activities (Schiltz, et al., 1973;
Muller, et al., 1977). Cultured chondrocytes produce
certain cartilage-specific macromolecules including type
II collagen and cartilage proteoglycan. The
"dedifferentiated" cells produce type I collagen instead
and do not synthesize cartilage proteoglycan.

Dedifferentiation of the chondrocytes occurs under
standard culture conditions but is accelerated in the
presence of embryo extract (Coon, Cahn, 1966) and extracts
of a variety of tissues (Mayne et al., 1976), vitamin A
(Solursh, Meier, 1973; Vasan, Lash, 1975; Shapiro, Poon,
1976), 5-bromo-2'-deoxyuridine (Abbott, Holtzer, 1968;
Mayne, et al., 1975) and fibronectin (Pennypacker, et al.,
1979; West, et al., 1979). The molecular mechanisms
involved in this phenomenon are not understood.

In testing the ability of various enzymes to induce
the dedifferentation of chondrocytes in culture,
Pennypacker and Goetinck (1979) noted that exposure of
isolated chick chondrocytes to certain preparations of
bovine testicular hyaluronidase caused a rapid loss of the

cartilage phenotype. Since other more purified
preparations of the enzyme did not have the same effect,
it seemed possible that a molecule other than
hyaluronidase was responsible for the change in the
chondrocytes.

We have extended these studies and have partially
purified the testicular factor. It appears to be a
protein (M_r=150,000) which lacks hyaluronidase activity.
The protein has been found to alter the differentiated
properties of other cells and could play a significant
role in maintaining germ cells and other stem cells in an
undifferentiated state.

EFFECTS ON CHONDROGENIC EXPRESSION

In these studies, we have assayed the activity of the
testicular factor on chick embryo chondrocytes isolated
from 14-17 day embryonic sterna by standard methods. The
chondrocytes were plated in 35mm tissue culture dishes in
Ham's F-12 media containing 10% fetal calf serum, bovine
serum albumin, HEPES buffer and gentamycin. The presence
of 20-50 μg/ml of the partially purified testicular
preparation caused the cells to assume a flattened
elongated shape while control cells retained the polygonal
or rounded morphology characteristic of cultured chondrocytes. Further, when cells incubated with the testicular
factor were labeled with ^{35}S-methionine, we found that
they produced type I collagen. In contrast, control
cultures produced primarily type II collagen. These
observations confirm those of Pennypacker and Goetinck
with the crude hyaluronidase showing that both
preparations suppressed chondrogenic expression by
cultured chondroctyes.

Similar studies were carried out with micromass
cultures of limb bud mesenchymal cells. These cells
differentiate into chondrocytes and deposit considerable
amount of matrix (Goetinck, et al., 1974; Lewis, et al.,
1978). The amount of cartilage matrix produced can be
estimated by the uptake of Alcian blue dye which stains
the matrix proteoglycans at acid pH. Inhibitors of
chondrogenesis such as vitamin A and 5-bromo-2'-
deoxyuridine prevent the accumulation of proteoglycan.
Similarly, the addition of the testicular protein also
blocked chondrogenesis in a dose-dependent manner (Figure

1). Differentiation of the limb bud cells into
chondrocytes was inhibited at even lower concentrations
than required to cause the dedifferentiation of sternal
chondrocytes. The cultures of limb bud cells were used to
assay for activity of the factor during purification.

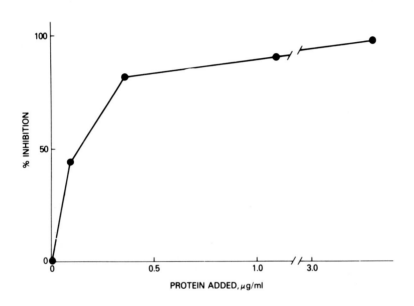

Figure 1. Inhibition of chondrogenesis in micromass
cultures of limb bud mesenchyme. Factor was added on day
3 of culture. Cultures were stained on day 6 with Alcian
blue at acid pH. Dye was extracted with 4M guanidine-HCl
and quantitated by measuring absorption at 600 nm.

PURIFICATION OF TESTICULAR FACTOR

The testicular factor was partially purified by
chromatofocusing, a chromatographic procedure which
separates proteins on the basis of their isoelectric
points. Using a pH gradient of 9-7, an enrichment of
activity was observed in pooled fractions eluted between

pH 8.5 and 8.0. An apparent purification of 29-fold was obtained with a 50% recovery of activity. The activity of the crude hyaluronidase may not accurately reflect the activity under consideration since hyaluronidase itself could break down portions of the matrix without necessarily affecting the phenotypic expression of the cells (Pennypacker, Goetinck, 1979). Thus, the estimate of a 29-fold purification is probably conservative. Gel filtration gives an estimate of the native molecular weight of 150,000. Periodic acid-Schiff staining of the suspected active band on SDS-PAGE gives evidence that the factor is a glycoprotein. The activity is also heat labile and protease-sensitive.

EFFECT ON OTHER CULTURED CELLS

In addition to its effect on chondrogenesis, preliminary studies on non-chondrogenic cell types suggest that the testicular factor may affect a broad range of cells (Table I). For example, addition of the factor to cultured embryonic chick cranial neural crest cells caused a stimulation of pigment synthesis as judged by ^{14}C-tyrosine incorporation into melanin. However, when the

Table 1: EFFECT OF TESTICULAR FACTOR ON CULTURED CELLS

Chondrocytes	Assume fibroblastic appearance Stops synthesis of proteoglycan Switch from type II to type I collagen
Limb bud mesenchyme	Inhibits chondrogenesis
Cranial neural crest B16/C3 mouse melanoma	Stimulates melanogenesis Inhibits melanogenesis
PYS-2 (Epithelial cell line)	Decreases cell-associated proteoglycan
Osteosarcoma cells	Increase basal and PTH-stimulated cAMP levels

factor was added to cultures of mouse melanoma B16/C3 cells, pigment synthesis was inhibited. Previously, we have reported the partial purification from calf serum of a protein which promotes pigmentation in both neural crest and melanoma cells (Jerdan, et al., 1980). The differential effects on these two cell types suggest that the testicular factor acts by a different mechanism than that of the serum-derived pigmentation factor.

DISCUSSION

A protein from certain crude commercial preparations of bovine testicular hyaluronidase has been partially purified by chromatofocusing and appears to be a glycoprotein (M_r=150,000) without hyaluronidase activity. The protein suppresses the chondrocyte phenotype in the same manner as reported for the crude hyaluronidase. Chondrogenesis in limb bud mesenchyme spot cultures is also inhibited by the factor, but at much lower concentration.

While these studies are interesting in terms of providing yet another factor with which to perturb the chondrocyte phenotype, an obvious relationship between the regulation of chondrogenic expression and a testicular protein is not readily apparent. However, preliminary immunological studies show that an immunologically cross-reactive protein is present in serum. The observed effect on the phenotypic expression of several cell types suggests that this factor may be involved in the regulation of differentiated function for a variety of cells.

REFERENCES

Abbott J, Holtzer H (1966). The loss of phenotypic traits by differentiated cells. IV. The reversible behavior of chondrocytes in primary cultures. J Cell Biol 28:473.

Abbott J, Holtzer H (1968). 5-Bromodeoxyuridine: Effect on cell surfaces of chondrocytes in vitro. Proc Natl Acad Sci USA 59:1144.

Coon HG, Cahn R (1966). Differentiation in vitro: effects
of Sephadex fractions of chick embryo extract. Science
153:1116.

Goetinck PF, Pennypacker JP, Royal PD (1974). Proteochondroitin sulfate synthesis and chondrogenic
expression. Exp Cell Res 87:241.

Holtzer H, Abbott J, Lash J, Holtzer S (1960). The loss
of phenotypic traits by differentiated cells in vitro.
I. Dedifferentiation of cartilage cells. Proc Natl
Acad Sci USA 46:1533.

Jerdan JA, Greenberg JH, Varner HH (1980). Partial
purification of a serum factor which promotes
pigmentation in cultured neural crest cells. J Cell
Biol 87:19a.

Lewis CA, Pratt RM, Pennypacker JP, Hassell JR (1978).
Inhibition of limb chondrogenesis in vitro by vitamin A:
alterations in cell surface characteristics. Develop
Biol 64:31.

Mayne R, Vail MS, Miller EJ (1975). Analysis of changes
in collagen biosynthesis that occur when chick
chondroctyes are grown in 5-bromo-2'-deoxyuridine. Proc
Natl Acad Sci USA 72:4511.

Mayne R, Vail MS, Miller EJ (1976). The effect of embryo
extract on types of collagen synthesized by cultured
chick chondrocytes. Develop Biol 54:230.

Muller PK, Lemmen C, Gay S, Gauss V, Kuhn K (1977).
Immunochemical and biochemical study of collagen
synthesis by chondrocytes in culture. Exp Cell Res
108:47.

Pennypacker JP, Goetinck PF (1979). Reversible inhibition
of chondrogenic expression by certain hyaluronidase
preparations. J Embryol exp Morph 53:91.

Pennypacker JP, Hassell JR, Yamada KM, Pratt RM (1979).
The influence of an adhesive cell surface protein on
chondrogenic expression in vitro. Exp Cell Res 121:411.

Schiltz JR, Mayne R, Holtzer H (1973). The synthesis of
collagen and glycosaminoglycans by dedifferentiated
chondroblasts in culture. Differentiation 1:97.

Shapiro SS, Poon JP (1976). Effect of retinoic acid on
chondrocyte glycosaminoglycan biosynthesis. Arch
Biochem Biophys 174:74.

Solursh M, Ahrens PB, Reiter RS (1978). A tissue culture
analysis of the steps in limb chondrogenesis. In vitro
14:51.

Solursh M, Meier S (1973). The inhibition of
mucopolysaccharide synthesis by vitamin A treatment of
cultured chick embryo chondrocytes. Calcif Tissue Res
13:131.
Vasan N, Lash JW (1975). Chondrocyte metabolism as
affected by vitamin A. Calcif Tissue Res 19:99.
West CM, Lanza R, Rosenbloom J, Lowe M, Holtzer H,
Avdalovic N (1979). Fibronectin alters the phenotypic
properties of cultured chick embryo chondroblasts. Cell
17:491.

Limb Development and Regeneration
Part B, pages 113–123
© **1983 Alan R. Liss, Inc., 150 Fifth Avenue, New York, NY 10011**

CARTILAGE PROTEOGLYCAN INTERACTIONS WITH C1q AND THE
LOCALIZATION OF AGGREGATED IgG COMPLEXES IN CARTILAGE

Kenneth Sparks, Paul Goetinck, and Mark Ballow[*]

Department of Animal Genetics, University of
Connecticut, Storrs, CT, and Department of Pediatrics, U-Conn Health Center, Farmington, CT (*)

INTRODUCTION

C1, the first component of the complement system, is a
trimolecular complex held together by calcium ions. The
three components of this complex are designated C1q, C1r, and
C1s. C1q serves to recognize and bind immune complexes which
then activates C1r and C1s resulting in the initiation of the
complement cascade. A serum inhibitor (C1qI) of C1q has been
described (Conradie et al., 1975) and found to be a chond-
roitin-4-sulfate proteoglycan (Silvestri et al., 1981).
Further, the binding of C1q by C1qI has been shown to be
through the chondroitin-4-sulfate on the latter. Although
the C1qI was shown to inhibit the hemolytic activity of C1q,
it is not known whether it interferes with activation of C1r
and C1s, or whether it prevents C1q binding of immune complexes. In this report we describe a simple ELISA system to
demonstrate C1q-proteoglycan interaction and show that C1q
is capable of simultaneously binding aggregated IgG and
proteoglycan (PG).

MATERIALS

Rabbit antiserum to human C1q (RA-C1q) was purchased
from Behring Diagnostics (Somerville, NJ). Rabbit antiserum
to human IgG (RA-HIgG), peroxidase-linked goat anti-rabbit
IgG (Per-GAR), and peroxidase-linked goat anti-human IgG
(Per-GAH) were purchased from Cappel Laboratories (Cochran-
ville, PA). Human C1q was obtained from the Center for Blood
Research (Boston, MA) and human IgG (Gamastan) was purchased
from Cutter Laboratories (Berkeley, CA). The peroxidase

substrate 2,2'-Azino-Di-(3-Ethyl Benzthiazoline sulfonic acid) (ABTS) was obtained from Sigma Chemical Company (St. Louis, MO). Normal human serum and serum from a patient with systemic lupus erythematosus (SLE) containing 45 µg/ml immune complexes, were gifts from Dr. Thomas Kennedy (University of Connecticut Health Center, Farmington, CT).

METHODS

A_1D_5 Preparation

Bovine A_1D_5 was prepared from bovine nasal septum cartilage. Cartilage was extracted in 4 M GuHCl containing 0.05 M sodium acetate, pH 5.8. The extraction solution contained protease inhibitors (Oegema et al., 1975) 6-amino caproic acid (0.1 M), benzamidine-HCl (0.005 M), and sodium EDTA (0.05 M). The extract was then adjusted to 0.4 M GuHCl by dialysis and CsCl equilibrium density centrifugation was performed according to the method of Sajdera and Hascall (1969).

Heat Aggregation of IgG

A solution of human IgG in phosphate buffered saline (PBS), pH 7.4 (1 mg/ml) was heated at 63°C for 20 min. The aggregated material (HA-IgG) was centrifuged, and protein measured by absorbance at 280 nm.

ELISA for Clq Binding

The assay was performed essentially by the method of Bullock and Walls (1977). Bovine A_1D_5 was diluted to 10 µl/ml with 0.05 M carbonate-bicarbonate buffer, pH 9.6, and 50 µl of this solution was added to the wells of polystyrene microtiter plates (Flow Laboratories, McLean, VA). The plates were incubated at 37°C for 1 hr, washed 3 times with PBS + 0.05% Tween-20 (PBS-T). Fifty microliters of human Clq in PBS-T were added and the plates returned to 37°C for 30 min. The plates were again washed 3 times and 50 µl of a 1:50 dilution of RA-Clq in PBS-T were added, followed by further incubation at 37°C for 30 min. The plates were washed again in PBS-T and 50 µl of Per-GAR (1:200 dilution in PBS-T were added. After a final incubation at 37°C for 30 min, the plates were washed as above and 50 µl of the peroxidase substrate were added. The substrate consisted of 0.012% H_2O_2 and 0.67% ABTS in pH 4.0, 0.05 M citrate buffer. The enzyme substrate reaction was carried out at room temperature for 10 min. The absorbances were then read at 414 nm in a Titertek Multiscan ELISA plate reader (Flow Laboratories).

ELISA for Aggregated IgG Binding

This assay was essentially the same as above except HA-IgA was substituted for RA-C1q and Per-GAH was used in place of Per-GAR. Controls for the above ELISAs included substituting NRS or RA-HIgG for RA-C1q. RA-HIgG controls were necessary because the C1q contained trace amounts of human IgG. Serum was also completely omitted and wells were just treated with PBS-T. The above controls were also performed when no C1q was added to the wells and was replaced by PBS-T. Finally, all the above groups were repeated on wells which were not coated with A_1D_5 but only treated with carbonate coating buffer. The results given in the text have all been corrected for the control values.

In Situ ELISA

A 1 cm diameter plug was punched out of a piece of bovine nasal septum cartilage with a cork borer. The plug was then sliced into 0.2 mm slices with a gel slicer (Joyce Loebl, Inc., England). Each slice was placed in a separate tube and washed 3 times with PBS-T. From this point the slices were treated in the same manner as the C1q ELISA above. Controls for this experiment were identical to those for the plate ELISA except that no controls for the absence of PG were done.

RESULTS

Interaction of A_1D_5 and C1q

The ELISA is dependent upon a reagent build-up on a solid phase surface. Following the adsorption of the first reagent to the solid phase, each succeeding reagent specifically binds to only the preceding one. For the initial attempt to show PG binding of C1q, either the A_1D_1 or A_1D_5 fraction of bovine nasal cartilage extract was adsorbed to the wells of micro-

titer plates. Subsequent reagent addition followed the
sequence: human Clq; RA-Clq; Per-GAR. The results of the
A_1D_5 ELISA can be seen in Figure 1. Addition of increasing
concentrations of Clq produces a dose-dependent increase in
activity when RA-Clq is added but essentially no response

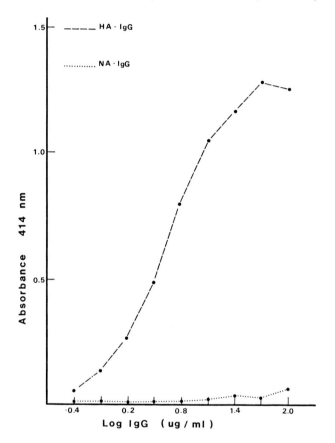

FIGURE 1. ELISA Dose Response Curve for Clq. Fifty microliters of a bovine A_1D_5 (10 µg/ml) were added to the wells
of polystyrene microtiter plates and allowed to passively
adsorb to the wells. The adsorption was followed by the
sequential addition of human Clq (10 µg/ml), RA-Clq, and
Per-GAR. Each reagent was followed by 3 washes with PBS-T.
Controls included the substitution for RA-Clq with NRS, RA-
HIgG, or PBS-T. The above substitutions were done both with
Clq addition and with PBS-T substituted for Clq.

when NRS is added. Since the C1q preparation contained small amounts of human IgG, RA-HIgG was also tested and produced identical results as the NRS. When C1q is omitted no activity is seen regardless of whether RA-C1q, RA-HIgG, or NRS is used. Likewise, if nothing is adsorbed to the cells no activity is seen. This result clearly indicates that C1q binds to the solid phase A_1D_5. The results of the ELISA using A_1D_1 adsorbed to the plate are not shown. Although C1q binding to the A_1D_1 can be shown, high concentrations of the A_1D_1 (1000 µg/ml) are required to obtain sufficient adsorption of this material to the plate.

A_1D_5 Binding of Serum C1q

Since C1q concentration is relatively high (180 µg/ml) in normal human serum, it was felt that this ELISA might be used to quantify serum C1q. The ELISA for C1q binding was repeated, testing both purified C1q and normal human serum. The test was performed with and without 0.01 M EDTA. EDTA disrupts the C1qrs complex in serum. The results can be seen in Table 1. There is only minimal activity in normal

A_1D_5 BINDING OF SERUM C1q

Test Material	Absorbance 414 nm	
	with EDTA	without EDTA
C1q 50 ug/ml	1.205	1.240
12.5	0.516	0.592
3.125	0.150	0.106
NHS 1/2	0.125	0.104
1/8	0.094	0.071
1/32	0.037	0.031

TABLE 1. A_1D_5 Binding of Serum C1q. The ELISA for C1q was repeated, substituting dilutions of normal human serum for purified C1q. Serum was tested in the presence of 0.01 M EDTA to disrupt the C1qrs complex or in the absence of EDTA. Similarly, purified C1q was tested with and without EDTA.

human serum even at the 1:2 dilution. The presence or
absence of EDTA apparently made little difference in either
purified Clq or the normal serum. Since the test could
measure as little as 3.125 µg/ml of purified Clq, the Clq in
the serum dilutions was well within the detectable range of
the assay. The data suggest the presence of a substance
preventing serum Clq from binding the A_1D_5.

Serum Inhibitor of A_1D_5-Clq Binding

To test the possibility of a serum inhibitor of A_1D_5-
Clq binding, purified Clq was added alone or with normal
human serum. The results in Table 2 show a considerable
inhibition of binding. Concentrations of Clq from 50 µg/ml
to 6.25 µg/ml are inhibited between 65.0 to 71.5% by a 1:2
dilution of serum.

SERUM INHIBITION OF A_1D_5 BINDING

Clq (ug/ml)	Absorbance 414 nm		% Inhibition
	Clq alone	Clq + NHS	
50.00	1.346	0.409	69.6
25.00	0.821	0.287	65.0
12.50	0.479	0.158	67.0
6.25	0.253	0.072	71.5

TABLE 2. Serum Inhibitor of Clq-A_1D_5 Binding. To test the
possibility of a serum inhibitor of Clq binding by A_1D_5, the
Clq ELISA was done by adding Clq alone or with normal human
serum.

Simultaneous Binding of A_1D_5 and Aggregated IgG by Clq

An important function of Clq is the binding of aggregated
IgG. It was desirable to know if the Clq bound to A_1D_5 could
in turn bind aggregated IgG. This possibility was tested by
a modification of the above ELISA. A_1D_5 was adsorbed to the

plate followed by the sequential binding of purified Clq, HA-IgG, and Per-GAH. The results of this experiment can be seen in Figure 2. HA-IgG binds A_1D_5-Clq in a dose-dependent fashion, while little or no binding is seen with non-aggregated IgG (NA-IgG).

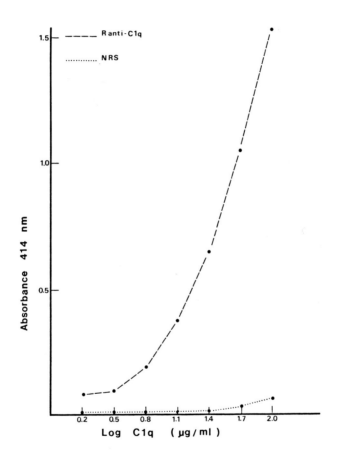

FIGURE 2. Simultaneous Binding of A_1D_5 and Aggregated IgG by Clq. A_1D_5 was adsorbed to the plate followed by the sequential binding of purified Clq, HA-IgG, and Per-GAH. Similar concentrations of NA-IgG were used as controls.

It was important to determine whether the binding seen above was restricted to HA-IgG or whether *in vivo* formed immune complexes could also be bound by A_1D_5 C1q. Using the above ELISA, NHS and a serum from a patient with SLE containing 45 µg/ml IgG immune complexes were tested. The values in Table 3 show good agreement for SLE serum at dilutions from 1:5 to 1:125, the average of all three measurements being 153.15 µg/ml. The normal human serum value falls outside the linear portion of the standard curve and the value 3.8 µg/ml is an overestimate.

A_1D_5 - C1q BINDING OF IN VIVO FORMED IC

Serum Dilutions	Absorbance 414 nm	IgG Complexes[1] ug / ml
SLE 1 / 5	0.277	144.20
1 / 25	0.173	165.25
1 / 125	0.055	150.00
NHS 1 / 5	0.022	3.80

[1]Calculated on the basis of a standard curve with HA-IgG.

TABLE 3. A_1D_5-C1q Binding of *In Vivo* Formed Immune Complexes (IC). Using the same ELISA scheme as in Fig. 2, normal human serum and a serum from a patient with SLE containing immune complexes were substituted for NA-IgG and HA-IgG, respectively.

In Situ ELISA

It can be argued that the binding of C1q to PG is an artifact of the extraction of PG from cartilage, since PG exists in a large ordered complex in the extracellular matrix. To investigate this possibility the ELISA in Fig. 1 was repeated but the plate bound A_1D_5 was substituted with 1 cm diameter, 0.2 mm thick slices of bovine nasal septum cartilage. The results of this assay can be seen in Figure 3 and are similar to those in Fig. 1. Significant amounts of

activity can be seen from all the C1q dilutions tested when
RA-C1q is present, while minimal activity is seen for NRS.
This would indicate that C1q can still bind to PG present in
the matrix.

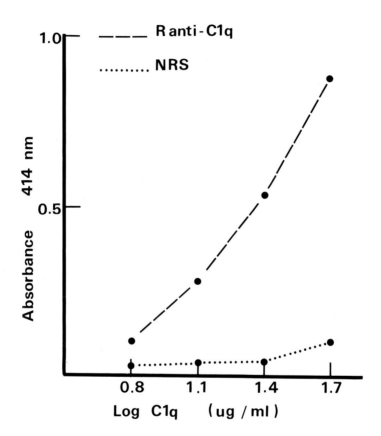

FIGURE 3. *In Situ* ELISA. Plugs (1 cm diameter) were punched
from a piece of bovine nasal septum cartilage with a cork
borer. From these plugs, 0.2 mm cross section slices were
made. Each slice was placed in a glass tube and washed 3
times with PBS-T. The subsequent reagent build-up was identical to that in Figure 1.

DISCUSSION

The ELISA system described here provides a simple rapid means of measuring C1q-PG interaction and does so with a minimum of reagent expenditure. The concentration of the A_1D_1 can be lowered to 1 µg/ml without any real loss in sensitivity and the volume of the A_1D_5 sample can be lowered to a 5 µl drop on the bottom of the well. However, when concentrations and volume are decreased it is necessary to increase the incubation time with the peroxidase substrate to obtain good color development. This ELISA has also been done with chicken cartilage A_1D_5 and both chicken and bovine A_1D_1. Although the A_1D_1 works, it requires greater concentrations of the PG and a longer time for the enzyme-substrate reaction (30-60 min). This is due to the poor binding characteristics of the A_1D_1 to the polystyrene.

It is most probable that the C1q-PG interaction seen here is identical to that described by Silvestri *et al.* (1981) and is occurring through the chondroitin-4-sulfate on the PG. This would explain the failure of serum C1q to be bound by the A_1D_5 and the inhibition of purified C1q-A_1D_5 binding by normal human serum. Further, in experiments not reported, bovine chondroitin sulfate inhibits the binding of C1q to A_1D_5.

The finding that C1q can bind both the A_1D_5 and immune complexes concurrently indicates that the binding sites on the C1q for these molecules are different. This and the results of the *in situ* ELISA suggest a possible biological role for C1q-PG interaction. Immune complexes can be shown to be associated with a variety of inflammatory joint diseases and eluates from articular tissues obtained from rheumatoid patients have been shown to contain IgG antibodies (Munthe and Natvig, 1970). Indeed, in experimental arthritis produced by the injection of antigen into the knee joints of immunized rabbits, antigen can be shown on the surface of the cartilage and to persist over long periods of time (Cooke *et al.*, 1972). Immune complexes by themselves do not bind cartilage, however, C1q may provide immune complexes with a link for the binding to cartilage surfaces.

REFERENCES

Bullock SL, Walls RW (1977). J Infect Dis 136:S279.
Conradie JD, Volanakis JE, Stroud RM (1975). Immunochemistry 12:967.
Cooke TD, Hurd ER, Ziff M, Jasin HE (1972). J Exp Med 135: 323.
Munthe E, Natvig JB (1971). Clin Exp Immunol 8:249.
Oegema TR, Hascall VC, Dziewiatkowski DD (1975). J Biol Chem 250:6151.
Sajdera SW, Hascall VC (1969). J Biol Chem 244:77.
Silvestri L, Baker JR, Rodén L, Stroud RM (1981). J Biol Chem 256:7383.

Limb Development and Regeneration
Part B, pages 125–135
© 1983 Alan R. Liss, Inc., 150 Fifth Avenue, New York, NY 10011

NEW COLLAGENS AS MARKER PROTEINS FOR THE CARTILAGE PHENOTYPE

Richard Mayne, Charles A. Reese, Cynthia C.
Williams and Pauline M. Mayne

Department of Anatomy, University of Alabama in
Birmingham, University Station, Birmingham,
Alabama 35294 U.S.A.

The major protein component of the extracellular matrix
of hyaline cartilage is type II collagen of chain composition
$[\alpha1(II)]_3$. This collagen forms a network of fine fibrils
within which the large proteoglycan aggregates are entrapped
(Miller 1976; Mayne, von der Mark 1982). Small amounts of
other collagens are also present in cartilages and three
chains have been isolated after pepsin digestion and desig-
nated 1α, 2α and 3α (Burgeson, Hollister 1979; Reese, Mayne
1981). The 1α and 2α chains appear closely related to the
$\alpha1(V)$ and $\alpha2(V)$ chains of type V collagen, sharing similar
solubility properties, amino acid compositions and migration
positions on SDS-polyacrylamide gel electrophoresis. The
native forms of 1α and 2α chains are also cleaved by a metal-
loproteinase which cleaves the triple helix of type V collagen but not of other collagens (Liotta et al 1982). However,
these chains differ in peptide mapping experiments after
cleavage with cyanogen bromide or S. aureus V8 protease, and
in elution positions on carboxymethyl-cellulose chromatography (Burgeson, Hollister 1979; Reese, Mayne 1981). The 3α
chain appears closely related to the $\alpha1(II)$ chain of type II
collagen and may be a variant of this chain which has been
subjected to extensive post-translational modification, or
could be the $\alpha1(II)$ Minor chain which sequencing studies
suggest is present in bovine hyaline cartilage (Butler et al
1977).

Several disulfide-bonded collagenous fragments have
also been isolated from the cartilagenous tissues of both
mammals (Shimokomaki et al 1980; Ayad et al 1981; Shimokomaki
et al 1981; Ayad et al 1982) and the chicken (Reese, Mayne

1981). Although the exact relationship between these fragments still remains poorly understood, it is likely that
these fragments represent the pepsin-resistant domains of
much larger proteins. Two disulfide-bonded fragments have
been isolated from chicken hyaline cartilage and called HMW
and LMW (Reese, Mayne 1981). A detailed model for the
structure of HMW has been presented which is based in part
on replicas of HMW observed in the electron microscope after
rotary shadowing (Reese et al 1982). The molecule is a typical collagen triple helix in which each chain has a molecular weight of 51,000. However, one of the chains has been
cleaved within the triple helix to give rise to peptides of
molecular weight 36,400 and 14,000. The result of this
cleavage is a recognizable kink in molecules of HMW when observed after rotary shadowing. LMW after rotary shadowing
also is a typical collagen triple helix in which each chain
has a molecular weight of 10-12,000 (Mayne, von der Mark
1982).

DISTRIBUTION OF 1α, 2α and 3α COLLAGEN CHAINS BETWEEN THE
CELL LAYER AND MEDIUM OF CHICK CHONDROCYTE CULTURES.

Several previous studies have shown that for a variety
of cells in culture, type V collagen is retained within the
cell layer and is not secreted as intact molecules into the
culture medium (Haralson et al 1980; Sage et al 1981; Tseng
et al 1981; Sasse et al 1981; Fessler et al 1981). Immunoelectron microscope studies with type V-specific antibodies
suggest that this collagen may be located in the immediate
pericellular environment of smooth muscle cells (Gay et al
1981), and may act as a bridge between the smooth muscle
basement membrane and the interstitial collagen fibrils
(Martinez-Hernandez et al 1982). It is possible that the
native form or forms of the 1α and 2α chains may serve a
similar bridging function in the pericellular matrix of
hyaline cartilage, and a series of experiments was performed
to determine the distribution of these collagen chains
between the cell layer and medium of embryonic chick chondrocyte cultures.

Embryonic chick chondrocytes were initially selected as
'floaters' and plated at a density of 5×10^4 cells/ml in
F10 medium plus supplements as described previously (Schiltz
et al 1973; Mayne et al 1975). After five days, cells were
labeled for 24 h with [^{3}H]glycine (10μC/ml) in the presence

of β-aminopropionitrile (100μg/ml) and ascorbic acid (50μg/ml). Floating cells were centrifuged from the labeling medium and combined with cells which had been scraped from the surface of the dish into cold medium. A solution of carrier collagen was added to both the labeling medium and the cold medium, and consisted of 1 mg each of the precipitates obtained from a pepsin digest of chicken sterna at 0.9 M NaCl 0.5 M HAc (Type II), 1.2 M NaCl 0.5 M HAc (native form of 1α, 2α, 3α chains) and 2.0 M NaCl 0.5 M HAc (HMW and LMW) as described in Reese and Mayne 1981. The carrier plus labeled collagens were precipitated with ammonium sulfate (35% of saturation) and, after standing overnight, the precipitates were centrifuged out, dissolved in 0.5 M HAc and incubated with pepsin (200μg/ml) for 6 h at 4°C. Subsequently, each solution was dialyzed against several changes of 0.5 M NaCl, 0.05 M Tris HCl, pH 8.0. After removal of undigested material by centrifugation, collagen was again precipitated with ammonium sulfate (35% of saturation), centrifuged out and dissolved in 0.5 M HAc. Differential salt precipitation was performed at 0.9 M NaCl, 1.2 M NaCl and 2.0 M NaCl, and each precipitate was dissolved in 0.5 M HAc, dialyzed extensively against 0.1 M HAc and lyophilized.

Figure 1 shows the separation by SDS-polyacrylamide gel electrophoresis of the fractions obtained by differential salt precipitation using a 5-10% gradient slab gel followed by fluorography as described (Reese, Mayne 1981). Lanes 1 and 2 show a single band at the location of α1(II) chains for the 0.9 M NaCl precipitate from both the cells and medium. Further characterization of this material was performed by CM-cellulose chromatography in denaturing conditions (Mayne, Zettergren, 1980), and it was estimated that 55% (range 50-58%; 5 experiments) of the type II collagen was released into the medium. In contrast, lanes 3 and 4 show that the 1α, 2α and 3α chains were predominantly retained within the cell layer. Further quantitation of the 1.2 M NaCl precipitate was performed by molecular sieve chromatography (Biogel A-5m) and the radioactivity present as α-sized material was determined for both the cells and medium. It was estimated that 86% (range 75-95%; 4 experiments) of the 1α, 2α and 3α chains were retained by the cell layer.

A band of radioactivity of lower molecular weight was, however, observed in the 1.2 M NaCl precipitate of the medium (lane 4), and considerably more of this material was present

in the 2.0 M NaCl precipitate of the medium (Lane 6). This
material has been isolated by molecular sieve chromatography
and possesses an apparent molecular weight of 43,000. It is
cleaved by highly purified bacterial collagenase (Advanced
Biofactures), and elutes from CM-cellulose at the location
of $\alpha2(I)$ chains. It does not contain interchain disulfide

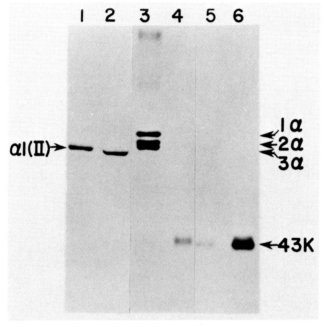

Fig. 1. Fluorogram after SDS-polyacrylamide gel electrophoresis of the collagens associated with the cells and released into the medium of chick chondrocyte cultures. Lane
1, 0.9 M NaCl ppt, cells; lane 2, 0.9 M NaCl ppt, medium;
lane 3, 1.2 M NaCl ppt, cells; lane 4, 1.2 M NaCl ppt, medium;
lane 5, 2.0 M NaCl ppt, cells; lane 6, 2.0 M NaCl ppt, medium.

bridges and therefore appears unrelated to HMW or LMW. We
have not performed a detail characterization of this material,
but it appears identical to a short chain collagenous molecule recently isolated by two other groups from chick chondrocyte cultures (Gibson et al 1981; Schmid et al 1982).

EFFECT OF BUdR ON THE SYNTHESIS OF 1α, 2α AND 3α COLLAGEN
CHAINS BY CHICK CHONDROCYTE CULTURES.

Numerous studies have shown that growth of chick chondrocytes in a variety of agents including bromodeoxyuridine
(BUdR) (Abbott, Holtzer 1968; Schiltz et al 1973; Mayne et
al 1975), embryo extract (Coon 1966; Schiltz et al 1973;
Mayne et al 1976), chicken serum (Gauss, Müller 1981) and
fibronectin (West et al 1979) results in the loss of the
typical chondrocyte morphology with the cells assuming a more
fibroblastic appearance. This is accompanied by a change in
synthesis from type II collagen to a mixture of type I collagen and type I trimer. It was therefore of interest to
determine whether or not a change in synthesis from the
cartilage-specific 1α and 2α chains to the $\alpha1(V)$ and $\alpha2(V)$
chains of type V collagen would also occur during growth of
chick chondrocytes in BUdR.

Chick chondrocytes selected as floaters were grown for
6 days in BUdR ($10\mu g/ml$) and then labeled for 24h with
[^{3}H]glycine ($10\mu C/ml$) in the presence of β-aminopropionitrile
($100\mu g/ml$) and ascorbic acid ($50\mu g/ml$). Cells were scraped
into the medium and carrier collagens added as described
earlier. Collagen was precipitated with ammonium sulfate
(35% of saturation) and the precipitate incubated at 4° with
pepsin ($200\mu g/ml$) in 0.5 M HAc for 6 h. After dialysis to
pH 8.0 and subsequent centrifugation to remove undigested
material, differential salt precipitation was performed and
precipitates obtained at 0.9 M NaCl, 0.5 M HAc and at 2.0 M
NaCl, 0.5 M HAc. Figure 2A shows SDS-polyacrylamide gel
electrophoresis and subsequent fluorography of the precipitates obtained for control (C) and BUdR (B) cultures at 0.9 M
NaCl, 0.5 M HAc. For control cultures, a single band was
present at the location of $\alpha1(II)$ chains whereas for BUdR-
treated cultures bands were present at the location of both
$\alpha1(I)$ and $\alpha2(I)$ chains. In subsequent experiments (not
presented) further characterization of the 0.9 M NaCl precipitates was performed by CM-cellulose chromatography and
analysis of the peaks by cyanogen bromide peptide mapping
using SDS-polyacrylamide gel electrophoresis and subsequent
fluorography. These results clearly demonstrated the switch
in synthesis from $\alpha1(II)$ chains to $\alpha1(I)$ and $\alpha2$ chains as
has been shown previously (Mayne et al 1975). Figure 2B
shows that for control cultures (C) synthesis of 1α, 2α, 3α
and the 43K component occurred whereas for BUdR-treated cultures (B) the synthesis of 3α chains and the 43K component

were not detectable, whereas two α-sized bands were present which could be either the 1α and 2α chains or the α1(V) and α2(V) chains.

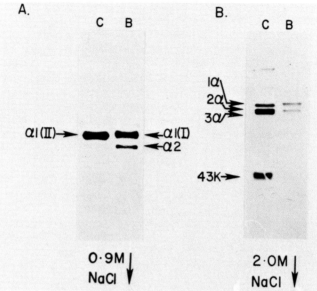

Fig. 2. Fluorogram after SDS-polyacrylamide gel electrophoresis of the collagens isolated from control and BUdR-treated cultures. Panel A. 0.9 M NaCl, 0.5 M HAc precipitates of control (C) and BUdR (B) cultures. 5-7 1/2% gradient gel. Panel B. 2.0 M NaCl, 0.5 M HAc precipitates. 5-10% gradient gel. Lanes were loaded with amounts of sample corresponding to equal DNA contents for control and BUdR-treated dishes. DNA analyses were performed fluorimetrically on duplicate dishes by the method of Puzas and Goodman 1978.

To investigate this further the precipitates obtained at 2.0 M NaCl were initially fractionated by molecular sieve chromatography (Biogel A-5m) and the α-sized peaks desalted and lyophilized. Figure 3 shows the separation of the α-sized peaks by CM-cellulose chromatography in the presence of carrier type I collagen. The upper panel shows that for control cultures three major peaks were obtained which eluted at the expected locations of the 3α, 1α and 2α chains respectively as shown previously (Reese, Mayne 1981). Each of these peaks was desalted and confirmed to be the 3α, 1α and

2α chains after cleavage with cyanogen bromide and mapping
by SDS-polyacrylamide gel electrophoresis followed by fluorography (data not presented). In contrast, the lower panel
shows that for BUdR-treated cultures only two peaks were
present which eluted between the α1(I) and α2 chains of the
carrier collagen as would be expected for the α2(V) and
α1(V) chains of type V collagen. These peaks were subsequently desalted and shown to migrate at the expected locations of the α2(V) and α1(V) chains after SDS-polyacrylamide
gel electrophoresis and fluorography (data not presented).

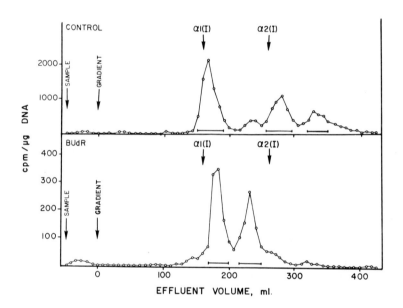

Fig. 3. CM-cellulose chromatography of the α-sized material
present in the 2.0 M NaCl precipitates from control and BUdR-
treated cultures. Initially, the 2.0 M precipitates were
fractionated by agarose gel filtration after denaturation
and the α-sized material was desalted and carrier type I collagen (5mg) added. Arrows show the location of the α1(I)
and α2(I) chains present in carrier type I collagen. The
results are expressed as cpm/fraction/μg DNA, the analyses
for DNA content having been performed on duplicate cultures.
Bars indicate the fractions which were pooled for further
analysis.

These results show that for BUdR-treated chondrocytes two chains are synthesized with very similar properties to the α2(V) and α1(V) chains of type V collagen. However, the limited incorporation of radioactivity has meant that we have been unable to confirm this identification by CNBr-peptide mapping. Also, the two chains were not obtained in a 2:1 proportion as would be expected for a molecule of chain composition [α1(V)]$_2$α2(V). This molecular organization appears to be the predominant form of type V collagen in most tissues (Kumamoto, Fessler 1980; Fessler et al 1981). The results have been expressed per DNA content of control and BUdR-treated cultures and we have been unable to demonstrate that there is a net increase in synthesis of the α1(V) and α2(V)-like chains. It is possible that our control cultures also synthesize small amounts of the α1(V) and α2(V) chains, and this may explain the isolation of α1(V) chains from some hyaline cartilages (Rhodes, Miller 1978; Gay et al (1981).

In general, the present results correspond closely to the results of Benya et al 1977 who reported that rabbit articular chondrocytes after several passages in culture synthesize two collagen chains called X and Y. These chains appear very similar in properties to the α1(V) and α2(V) chains of type V collagen.

Previous analyses of the 1α and 2α chains have suggested that these chains are very closely related to the α1(V) and α2(V) chains of type V collagen (Burgeson, Hollister 1979; Reese, Mayne 1981). The present results further support the proposal that the 1α and 2α chains be regarded as the carti-lage-equivalent of a type V family of collagens found in other tissues (Reese, Mayne 1981). The 1α and 2α chains are cell-associated and suppression of the synthesis of these chains occurs after growth in BUdR.

This research was supported by National Institutes of Health Grant AM 30481. R.M. is an Established Investigator of the American Heart Association.

Abbott J, Holtzer H (1968). The loss of phenotypic traits by differentiated cells. V. The effect of 5-bromodeoxyuri-dine on cloned chondrocytes. Proc Natl Acad Sci USA 59: 1144.

Ayad S, Abedin MZ, Grundy SM, Weiss JB (1981). Isolation and characterisation of an unusual collagen from hyaline cartilage and intervertebral disc. FEBS Lett 123:195.

Ayad S, Abedin MZ, Weiss JB, Grundy SM (1982). Characterisation of another short-chain disulphide-bonded collagen from cartilage, vitreous and intervertebral disc. FEBS Lett 139:300.

Benya PD, Padilla SR, Nimni ME (1977). The progeny of rabbit articular chondrocytes synthesize collagen types I and III and type I trimer, but not type II. Verifications by cyanogen bromide peptide analysis. Biochemistry 16:865.

Burgeson RE, Hollister DW (1979). Collagen heterogeneity in human cartilage: Identification of several new collagen chains. Biochem Biophys Res Commun 87:1124.

Butler WT, Finch JE, Miller EJ (1977). The covalent structure of cartilage collagen. Evidence for sequence heterogeneity of bovine $\alpha 1$(II) chains. J Biol Chem 252:639.

Coon HG (1966). Clonal stability and phenotypic expression of chick cartilage cells in vitro. Proc Natl Acad Sci USA 55:66.

Fessler LI, Kumamoto CA, Meis ME, Fessler JH (1981). Assembly and processing of procollagen V (AB) in chick blood vessels and other tissues. J Biol Chem 256:9640.

Fessler LI, Robinson WJ, Fessler JH (1981). Biosynthesis of procollagen [$(pro\alpha 1\ V)_2 (pro\alpha 2\ V)$] by chick tendon fibroblasts and procollagen $(pro\alpha 1\ V)_3$ by hamster lung cell cultures. J Biol Chem 256:9646.

Gauss V, Müller PK (1981). Change in the expression of collagen genes in dividing and nondividing chondrocytes. Biochim Biophys Acta 652:39.

Gay S, Martinez-Hernandez A, Rhodes RK, Miller EJ (1981). The collagenous exocytoskeleton of smooth muscle cells. Collagen Rel Res 1:377.

Gay S, Rhodes RK, Gay RE, Miller EJ (1981). Collagen molecules comprised of αI(V)-chains (B-chains): an apparent localization in the exocytoskeleton. Collagen Rel Res 1:53.

Gibson GJ, Schor SL, Grant ME (1981). Partial characterization of a low-molecular-weight collagen synthesized by chondrocytes cultured in collagen gels. Biochem Soc Transact 9:550.

Haralson MA, Mitchell WM, Rhodes RK, Kresina TF, Gay R, Miller EJ (1980). Chinese hamster lung cells synthesize and confine to the cellular domain a collagen composed solely of B chains. Proc Natl Acad Sci USA 77:5206.

Kumamoto CA, Fessler JH (1980). Biosynthesis of A,B procollagen. Proc Natl Acad Sci USA 77:6434.
Liotta LA, Kalebic T, Reese CA, Mayne R (1982). Protease
susceptibilities of HMW, 1α, 2α but not 3α cartilage collagens are similar to type V collagen. Biochem Biophys
Res Commun 104:500.
Martinez-Hernandez A, Gay S, Miller EJ (1982). Ultrastructural localization of type V collagen in rat kidney. J
Cell Biol 92:343.
Mayne R, Vail MS, Miller EJ (1975). Analysis of changes in
collagen biosynthesis that occur when chick chondrocytes
are grown in 5-bromo-2'-deoxyuridine. Proc Natl Acad Sci
USA 72:4511.
Mayne R, Vail MS, Miller EJ (1976). The effect of embryo
extract on the types of collagen synthesized by cultured
chick chondrocytes. Develop Biol 54:230.
Mayne R, Zettergren JG (1980). Type IV collagen from chicken
muscular tissues. Isolation and characterization of the
pepsin-resistant fragments. Biochemistry 19:4065.
Mayne R, von der Mark K (1982). Collagens of cartilage. In
Hall BK (ed): "Cartilage", Vol 1, San Diego: Academic
Press, in press.
Miller EJ (1976). Biochemical characteristics and biological
significance of the genetically-distinct collagens. Mol
Cell Biochem 13:165.
Puzas JE, Goodman DBP (1978). A rapid assay for cellular
deoxyribonucleic acid. Anal Biochem 86:50.
Reese CA, Mayne R (1981). Minor collagens of chicken hyaline
cartilage. Biochemistry 20:5443.
Reese CA, Wiedemann H, Kühn K, Mayne R (1982). Characterization of a highly soluble collagenous molecule isolated from
chicken hyaline cartilage. Biochemistry 21:826.
Rhodes RK, Miller EJ (1978). Physicochemical characterization and molecular organization of the collagen A and B
chains. Biochemistry 17:3442.
Sage H, Pritzl P, Bornstein P (1981). Characterization of
cell matrix associated collagens synthesized by aortic
endothelial cells in culture. Biochemistry 20:436.
Sasse J, von der Mark H, Kühl U, Dessau W, von der Mark K
(1981). Origin of collagen types I, III, and V in cultures of avian skeletal muscle. Develop Biol 83:79.
Schiltz JR, Mayne R, Holtzer H (1973). The synthesis of collagen and glycosaminoglycans by dedifferentiated chondroblasts in culture. Differentiation 1:97.

Schmid TM, Conrad HE, Bruns R, Linsenmayer TF (1982). A low molecular weight collagen synthesized by hypertrophying chondrocytes. Anat Rec 202:169a.

Shimokomaki M, Duance VC, Bailey AJ (1980). Identification of a new disulfide bonded collagen from cartilage. FEBS Lett 121:51.

Shimokomaki M, Duance VC, Bailey AJ (1981). Identification of two further collagenous fractions from articular cartilage. Biosci Rep 1:561.

Tseng SCG, Savion N, Gospodarowicz D, Stern R (1981). Characterization of collagens synthesized by cultured bovine corneal endothelial cells. J. Biol Chem 256:3361.

West CM, Lanza R, Rosenbloom J, Lowe M, Holtzer H, Avdalovic N (1979). Fibronectin alters the phenotypic properties of cultured chick embryo chondroblasts. Cell 17:491.

SECTION SIX
CARTILAGE AND BONE

Limb Development and Regeneration
Part B, pages 139–148
© **1983 Alan R. Liss, Inc., 150 Fifth Avenue, New York, NY 10011**

CELL INTERACTIONS DURING IN VITRO LIMB CHONDROGENESIS

Michael Solursh, Ph.D.

Department of Zoology
University of Iowa
Iowa City, Iowa   52242

INTERACTIONS BETWEEN MESENCHYME CELLS

One approach to studying cell interactions, whether
between mesenchymal cells or epithelia and mesenchyme, is
to separate the limb into components, dissociate and ran-
domize the constituent cells, and then examine the behavior
of cells in reconstituted cultures.   In cell cultures of
limb mesenchyme, a variety of cell types will differentiate
(Umansky 1966; Caplan 1970) with cartilage predominating
when the cells are inoculated at densities greater than
confluency.   In such cultures, small cell aggregates are
formed.   Subsequently, these aggregates become centers of
cartilage tissue formation.   Progressively, these centers
enlarge as adjacent cells also become chondrogenic (Ahrens
et al 1977) in a sequence which suggests the occurrence of
assimilative induction between mesenchyme cells (Solursh
1980).

The specificity of the interactive process has been
studied by examining the behavior of cultures prepared from
mixtures of various types of cells.   In cultures prepared
from mixtures of limb mesenchyme and mesonephric cells,
Moscona (1956) found that cartilage formation increased in
direct relation to the proportion of limb mesenchyme cells
present.   Cell mixture experiments show that the interactive capacity not only is a tissue-specific but is a stage-
specific property, which is acquired in the chick after
Hamburger and Hamilton (1951) stage 19 (Solursh, Reiter
1980) and lost before stage 26 (Solursh et al 1982a).   Furthermore, the interactive stimulus is made defective by the

brachypod mutation in the mouse (Owens, Solursh 1982). The
temporal control of this capacity might underly the process
of cartilage histogenesis (Solursh et al 1981a; Owens,
Solursh 1981).

The cellular effects of interactions between mesenchymal cells can be studied by finding manipulations which promote chondrogenesis by bypassing cell interactions (see
Solursh 1982a for review). Two such treatments are known.
Treatment of high density cultures of limb mesenchyme with
dibutyryl cyclic AMP or related compounds will promote chondrogenesis and apparently bypass required cell interactions
(Ahrens et al 1977; Solursh et al 1981a). A similar treatment will even promote chondrogenesis in clonal cultures
of limb mesenchyme (Solursh, Reiter 1975). Dibutyryl cAMP
also promotes chondrogenesis by explants of limb mesenchyme
(Kosher et al 1979). While the mechanism by which dibutyryl
cAMP promotes chondrogenesis is not known, endogenous cyclic
AMP levels do increase significantly early during chondrogenesis (Solursh et al 1979). This result supports the
hypothesis that changes in intracellular cAMP levels mediate the onset of cartilage differentiation (Solursh et al
1978; see chapter by Kosher).

Another treatment which promotes chondrogenesis by
limb mesenchymal cells is maintenance in suspension culture
(Levitt, Dorfman 1972; Solursh, Reiter 1975). Under these
conditions, DNA synthesis is blocked (Solursh, Reiter 1975)
and the cells are spherical. While it was unlikely that
cell aggregation mediated this effect, such a possibility
could not be conclusively eliminated. More recently, it
has been demonstrated that single mesenchymal cells immo-
bilized in or on type I collagen gels become chondrocytes,
as defined by the presence of immunologically detectable
type II collagen (Figure 1), or an alcian blue staining
extracellular matrix (Solursh et al 1982b). Under these
conditions, the cells are round and remain single. These
results emphasize the importance of the cell cycle and cell
shape in regulating cartilage cell differentiation. It is
noteworthy that mesenchymal cells become round in shape
early during cartilage differentiation in situ (Thorogood,
Hinchliffe 1975; Singley, Solursh 1981).

These treatments can be used to study the distribution
of potentially chondrogenic cells within the limb. A number of years ago Zwilling demonstrated the potentially

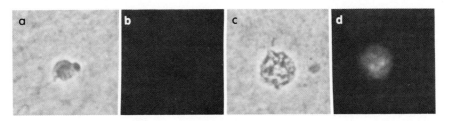

Fig. 1. Stage 23 limb mesenchymal cells were dissociated as
described previously (Ahrens et al 1977) and inoculated in
type I collagen gels (1 mg/ml in Ham's F12 medium containing
10% fetal bovine serum (Elsdale, Bard 1972)). After various
times, the cells were fixed and stained with monoclonal antibody directed against type II collagen (Linsenmayer,
Hendrix 1980) as described in detail elsewhere (Solursh et
al 1982b). a and c are phase contrast micrographs of cells
after 3 hrs and 7 days, respectively, of culture. b and d
are the corresponding immunofluorescent micrographs. While
initially type II collagen is not detected, it is clearly
present by day 7. The fluorescence is largely intracellular
due to the scorbutic culture conditions used (Meier, Solursh
1978). (600X)

chondrogenic capacity of the normally non-chondrogenic limb
periphery (Zwilling 1966) when placed in explant culture.
While there are quantitative differences in chondrogenic
capacity of early limb bud mesenchyme isolated from normally chondrogenic or non-chondrogenic regions (Ahrens et
al 1979), treatment with dibutyryl cAMP can overcome these
differences (Solursh et al 1981a). Single cells from the
peripheral region of the stage 24 chick embryo wing bud also
become chondrocytes when placed in collagen gels (unpublished observation). The fact that this potentially chondrogenic capacity in the limb periphery is not expressed in
situ is of interest in relation to the development of skeletal pattern and suggests the presence of an inhibitor of
chondrogenesis in the limb periphery. One component of the
limb bud which is lost during culture preparation is the
limb ectoderm.

EFFECTS OF LIMB ECTODERM ON MESENCHYME BEHAVIOR

The role of the apical ectodermal ridge (AER) in distal

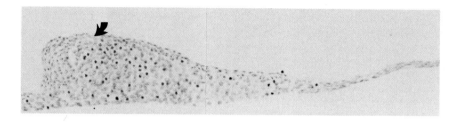

Fig. 2. Mitogenic effect of the AER. A micromass culture
was established from dissociated stage 23 chick wing buds
as described previously (Solursh et al 1981b). After 24
hrs, quail wing ectoderm containing AER was attached. Cartilage nodules form all through the culture except in the
vicinity of the ectoderm. During the last four hours of a
three day period, 0.05 µg/ml of colcemid was present to
allow the accumulation of mitotic figures (Reiter, Solursh
1982). The cultures were fixed in Kahle's fixative, stained
with Schiff's reagent (Humason 1972) and alcian blue, embedded in paraffin, and sectioned. The quail ectoderm can
be distinguished from the chick mesenchyme by the presence
of the nucleolar marker (Le Douarin 1973). Note that, unlike the adjacent mesenchyme, the mesenchyme associated with
the AER (arrow) contains numerous mitotic figures and has
formed an outgrowth. (260X)

limb outgrowth (Saunders 1948) is well established (see
chapter by Saunders). It is thought to maintain the underlying mesenchyme in a labile state (MacCabe et al 1973).
The nature of its actions is not known. By use of the test
system described by Globus and Vethamany-Globus (1976) in
which ectoderm is attached to high density mesenchyme cultures, it has been possible to examine some actions of the
AER and non-ridge ectoderm on limb mesenchyme (Solursh et
al 1981b).

The AER in this in vitro test system consistently stimulates the outgrowth of associated mesenchyme as well as
delays the differentiation of the mesenchyme (Figure 2;
Solursh et al 1981b). The presence of the AER increases
the mitotic and labeling indices in underlying mesenchyme
compared to those in mesenchyme alone or with non-ridge
ectoderm, indicating that the AER has mitogenic properties
(Reiter, Solursh 1982). It is possible that distal limb

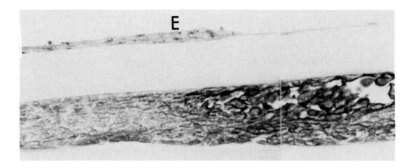

Fig. 3. Floating filter culture. A micromass culture was
established from stage 23-24 chick wing bud mesenchyme on
a poly-lysine coated Millipore filter (THWP01300 filter,
0.45 μm pore size, 25 μm thick), as described previously
(Solursh et al 1981b). The next day wing ectoderm from
stage 23-24 quail embryos was placed on the opposite side
of the filter. After four days in culture with a daily
change of medium, the cultures were fixed, stained with alcian blue at pH 1 (Lev, Spicer 1964) and hematoxylin, embedded in paraffin and sectioned. Note the increased
staining with alcian blue and extent of cartilage differentiation with distance of the mesenchyme from the ectoderm
(E). Since the culture was floating in the culture medium,
the antichondrogenic effect of the ectoderm is unrelated to
nutrient availability but appears to be mediated by a diffusible, ectoderm-derived influence. While the cells
closest to the ectoderm are still mesenchymal, the remainder of the culture consists primarily of cartilage. (485X)

outgrowth in situ is stimulated through such a mitogenic
action of the AER.

The non-ridge ectoderm in this test system also delays
differentiation, but no outgrowth is formed (Solursh et al
1981b). In contrast to cells under the AER, eventually
loose connective tissue differentiates in association with
the ectoderm, while the rest of the culture forms cartilage.
The influence of the non-ridge ectoderm can be transmitted
across filters 25 μm thick (Figure 3; Solursh et al 1981b)
but not across 150 μm thick filters. Since cell processes
are not observed to cross these filters, the ectodermal
influence is likely to be mediated by a diffusible influence.

EXTENSION OF IN VITRO RESULTS TO THE DEVELOPMENT OF SKELE-
TAL PATTERN

While considerable caution is necessary in attempting
to extend observations made in vitro to development in situ,
it is worth considering the in vitro studies described here
in relation to previous knowledge of limb development. The
Saunders-Zwilling hypothesis concerning the reciprocal interaction between the AER and the limb mesenchyme is central
to current ideas on limb morphogenesis. The AER is required
for limb outgrowth and influences its symmetry. However,
the mesenchyme determines the extent and shape of the AER
(Zwilling 1955) in a stage-specific manner (Rubin, Saunders
1972). If one adds to this concept the notions of an intrinsic chondrogenic capacity of limb mesenchyme, the antichondrogenic effects of non-ridge ectoderm and the mitogenic action of the AER, the possible origin of skeletal pattern begins to make sense (Solursh 1982b).

The non-ridge ectoderm could create a sleeve of antichondrogenic activity around the limb periphery, where fibrous connective tissue would eventually differentiate.
The formation of muscle could reflect the pattern of invasion of somite-derived, myogenic cells (Chevallier 1978;
Newman et al 1981), which might be influenced by adjacent
mesenchyme (Chiquet et al 1981). The central limb core
could form cartilage according to its inherent capacity.

The skeletal pattern could be a secondary result of
the size and shape of the limb bud as it elongates under a
mitogenic influence of the AER. The shape of the chondrogenic core would reflect the overall contour of the limb
and its ectodermal covering (Figure 4). It is noteworthy
that at least at some axial levels the entire limb core
becomes chondrogenic initially (Dawd, Hinchliffe 1971; see
chapter by Hinchliffe) and is secondarily subdivided by the
death of intervening cells. In some mutants where such
cell death is reduced in extent, fused skeletal primordia
develop (e.g., Hinchliffe, Thorogood 1979). Proximal-
distal elongation of skeletal primordia occurs after the
initial pattern is established (Lewis 1977). Thus, if one
assumes that there is a mechanism which limits the initial
cross-sectional diameter of each skeletal primordium, the
larger the diameter of the prechondrogenic core, the
greater the number of skeletal elements formed. With the
combination of reciprocal interaction between the AER and

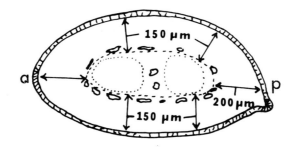

Fig. 4. A schematic representation from Solursh (1982b) of
a stage 24 chick wing bud seen in cross-section at the level
of the prospective radius and ulna (stippled outlines).
Note the asymmetries along the anterior-posterior (a,p) and
dorsal-ventral axes. As reported elsewhere (Solursh et al
1981b), the peripheral, non-chondrogenic mesenchyme forms
a uniformly thick sleeve around the chondrogenic core and
progressively accumulates a distinctive hyaluronate-rich
extracellular matrix (Singley, Solursh 1981). The thickness of the non-chondrogenic sleeve is in the range of the
distance over which the antichondrogenic activity of non-
ridge ectoderm can be detected in vitro (Solursh et al
1981b).

mesenchyme, shaping the limb outgrowth and the non-ridge
ectoderm providing a functional link between limb shape
and chondrogenesis, the origin of the skeletal pattern can
be explained in terms of activities of cells in a manner
that has not been possible previously.

REFERENCES

Ahrens PB, Solursh M, Reiter RS (1977). Stage-related
    capacity for limb chondrogenesis in cell culture. Devel-
    op Biol 60:69.
Ahrens PB, Solursh M, Reiter RS, Singley CT (1979).
    Position-related capacity for differentiation of limb
    mesenchyme in cell culture. Develop Biol 69:436.
Caplan A (1970). Effects of the nicotinamide-sensitive
    teratogen 3-acetylpyridine on chick limb bud cells in
    culture. Exp Cell Res 62:341.

Chevallier A (1978). Etude de la migration des cellules somitique dans le mésoderme somatopleural de l'ebauche de l'aile. Wilhelm Roux Arch 184:57.

Chiquet M, Eppenberger HM, Turner DC (1981). Muscle morphogenesis: evidence for an organizing function of exogenous fibronectin. Develop Biol 88:220.

Dawd DS, Hinchliffe JR (1971). Cell death in the "opaque patch" in the central mesenchyme of the developing chick limb: a cytological, cytochemical and electron microscope analysis. J Embryol Exp Morph 26:401.

Elsdale TR, Bard J (1972). Collagen substrata for studies on cell behavior. J Cell Biol 54:626.

Globus M, Vethamany-Globus S (1976). An in vitro analogue of early chick limb bud outgrowth. Differentiation 6:91.

Hamburger V, Hamilton HL (1951). A series of normal stages in the development of the chick embryo. J Morphol 88:49.

Hinchliffe JR, Thorogood PV (1974). Genetic inhibition of mesenchymal cell death and the development of form and skeletal pattern in the limbs of $talpid^3$ (ta^3) mutant chick embryos. J Embryol Exp Morph 31:747.

Humason GL (1972). "Animal Tissue Techniques," 3rd ed, San Francisco: WH Freeman and Co.

Kosher RA, Savage MP, Chan S-C (1979). Cyclic AMP derivatives stimulate the chondrogenic differentiation of the mesoderm subjacent to the apical ectodermal ridge in the chick limb bud. J Exp Zool 209:221.

Le Douarin N (1973). A biological cell labeling technique and its use in experimental embryology. Develop Biol 30:217.

Lev R, Spicer SS (1964). Specific staining of sulfate groups with Alcian blue at low pH. J Histochem Cytochem 12:309.

Levitt D, Dorfman A (1972). The irreversible inhibition of differentiation of limb-bud mesenchyme by bromodeoxyuridine. Proc Natl Acad Sci USA 69:1253.

Lewis J (1977). Growth and determination in the developing limb. In Ede DA, Hinchliffe JR, Balls M (eds): "Vertebrate Limb and Somite Morphogenesis," Cambridge: Cambridge Univ Press, p 215.

Linsenmayer TF, Hendrix MJC (1980). Monoclonal antibodies to connective tissue macromolecules: type II collagen. Biochem Biophys Res Commun 92:440.

MacCabe JA, Saunders JW, Pickett M (1973). The control of the anterior-posterior and dorsal-ventral axes in embryonic chick limbs constructed of dissociated and regenerated limb-bud mesoderm. Develop Biol 31:323.

Meier S, Solursh M (1978). Ultrastructural analysis of the
effect of ascorbic acid on secretion and assembly of extracellular matrix by cultured chick embryo chondrocytes.
J Ultra Res 65:48.

Moscona A (1956). Development of heterotypic combinations
of dissociated embryonic chick cells. Proc Soc Exp Biol
Med 92:410.

Newman SA, Pautou M-P, Kieny M (1981). The distal boundary
of myogenic primordia in chimeric avian limb buds and its
relation to an accessible population of cartilage progenitor cells. Develop Biol 84:440.

Owens EM, Solursh M (1981). In vitro histogenic capacities
of limb mesenchyme from various stage mouse embryos.
Develop Biol 88:297.

Owens EM, Solursh M (1982). Cell-cell interactions by
mouse limb cells during in vitro chondrogenesis: analysis
of the brachypod mutation. Develop Biol 91 (In press).

Reiter RS, Solursh M (1982). Mitogenic property of the
apical ectodermal ridge. Develop Biol (In press).

Rubin L, Saunders JW (1972). Ectodermal-mesodermal interactions in the growth of limb buds in the chick embryo:
constancy and temporal limits of the ectodermal induction.
Develop Biol 28:94.

Saunders JW (1948). The proximo-distal sequence of origin
of the parts of the chick wing and the role of the ectoderm. J Exp Zool 108:363.

Singley CT, Solursh M (1981). The spatial distribution of
hyaluronic acid and mesenchymal condensation in the embryonic chick wing. Develop Biol 84:102.

Solursh M (1980). Histogenic mechanisms in in vitro limb
chondrogenesis. In Pratt RM, Christiansen RL (eds):
"Current Research Trends in Prenatal Craniofacial Development," New York: Elsevier North Holland, p 315.

Solursh M (1982a). Cell-cell interaction in chondrogenesis.
In Hall BK (ed): "Chondrogenesis," Vol 2, New York:
Academic Press (In press).

Solursh M (1982b). Epithelia as determinants of skeletal
pattern. Differentiation (Submitted).

Solursh M, Ahrens PB, Reiter RS (1978). A tissue culture
analysis of the steps in limb chondrogenesis. In Vitro
14:51.

Solursh M, Jensen KL, Singley CT, Linsenmayer TF, Reiter RS
(1982a). Two distinct regulatory steps in cartilage
differentiation. Develop Biol (Submitted).

Solursh M, Linsenmayer TF, Jensen KL (1982b). Chondrogenesis by single limb mesenchyme cells. Develop Biol (Submitted).

Solursh M, Reiter RS (1975). Determination of limb bud chondrocytes during a transient block of the cell cycle. Cell Differ 4:131.

Solursh M, Reiter RS (1980). Evidence for histogenic interactions during in vitro limb chondrogenesis. Develop Biol 78:141.

Solursh M, Reiter RS, Ahrens PB, Pratt RM (1979). Increase in levels of cyclic AMP during avian limb chondrogenesis in vitro. Differentiation 15:183.

Solursh M, Reiter RS, Ahrens PB, Vertel BM (1981a). Stage and position-related changes in chondrogenic response of chick embryonic wing mesenchyme to treatment with dibutyryl cyclic AMP. Develop Biol 83:9.

Solursh M, Singley CT, Reiter RS (1981b). The influence of epithelia on cartilage and loose connective tissue formation by limb mesenchyme cultures. Develop Biol 86:471.

Thorogood PV, Hinchliffe JR (1975). An analysis of the condensation process during chondrogenesis in the embryonic chick hind limb. J Embryol Exp Morphol 33:581.

Umansky R (1966). The effect of cell population density on the developmental fate of reaggregating mouse limb bud mesenchyme. Develop Biol 13:31.

Zwilling E (1955). Ectoderm-mesoderm relationships in the development of the chick embryo limb bud. J Exp Zool 128:423.

Zwilling E (1966). Cartilage formation from so-called myogenic tissue of chick embryo limb buds. Ann Med Exp Fenn 44:134.

Limb Development and Regeneration
Part B, pages 149–158
© 1983 Alan R. Liss, Inc., 150 Fifth Avenue, New York, NY 10011

THE ELABORATION OF EXTRACELLULAR MATRIX BY CHICKEN CHONDRO-
CYTES IN CULTURE

Barbara M. Vertel, Ph.D.

Department of Biology
Syracuse University
Syracuse, New York 13210

The developing limb is characterized by the differentiation of chondrocytes, myocytes and loose connective tissue and
by accompanying transitions in the type and distribution of
extracellular matrix molecules (reviewed by von der Mark, Conrad 1979). Presumably, these changes reflect changes in gene
expression, for, clearly in the case of the collagens (Bornstein, Sage 1980), and perhaps for the proteoglycans, genetic-
ally distinct molecular species are synthesized. Some experiments suggest that specific matrix molecules may in fact direct
or regulate developmental events which follow (Nogami, Urist
1974; Meier, Hay 1974; Reddi, Huggins 1975; Lash, Vasan 1978).
The correct patterning of the cartilage model is essential for
the replacement by bone and further limb outgrowth. Appropriate
construction of the cartilage matrix appears to be a necessary
step in this process, for there are several mutants which are
characterized by shortened limbs and these either lack or contain a defective major structural component of the cartilage
matrix (Pennypacker, Goetinck 1976; Orkin et al 1976; Rittenhouse et al 1978; Kimata et al 1981).

Fibrous type II collagen (Miller, 1971) and aggregates of
chondroitin sulfate proteoglycan (CSPG) subunit and hyaluronic
acid, stabilized by link protein (reviewed by Roden 1980), are
the main constituents of cartilage extracellular matrix. Recent work suggests that other components, such as additional
collagen species (Burgeson, Hollister 1979; Reese, Mayne 1981),
chondronectin (Hewitt et al 1979) and perhaps fibronectin (Weiss,
Reddi 1981; Melnick et al 1981),may be involved in the formation
of cartilage matrix, but the roles of these molecules are not
yet clearly defined.

In our studies, both biochemical and morphological analyses have been used to examine the synthesis and deposition of the two most prominent cartilage matrix molecules, type II collagen and CSPG. Molecular studies were designed to characterize early steps in the synthesis and processing of these products while immunofluorescence studies, described later, were designed to characterize these synthetic events at the cellular level. Systems for cell-free protein synthesis were established and antibodies prepared which enabled us to identify the protein precursors to CSPG subunit and type II collagen translated directly from cartilage mRNA (Upholt et al 1979, 1981). Cell-free protein synthesizing systems permit the study of unmodified nascent polypeptides independent of modifications such as proteolytic processing, glycosylation and hydroxylation (for collagens) which characterize these secreted macromolecules in vivo. For both molecular and morphological studies of biosynthetic intermediates it is necessary that the antibodies to be used react with matrix molecules at all stages of synthesis. Immunofluorescence studies established that our antibodies reacted with intracellular biosynthetic intermediates and the highly modified, completed products in extracellular matrix (Vertel, Dorfman 1979).

Recent studies have utilized cartilage cells from embryonic chicken sterna in culture. Cells were pulsed in suspension for 5 min with ^{35}S methionine and chased with excess unlabeled methionine for varying times up to 3 hr in order to identify the first detectible precursors and biosynthetic intermediates in intact cartilage cells (Fig. 1B,C). These products were compared with cartilage mRNA-directed products synthesized in the wheat germ (Fig. 1D) and reticulocyte lysate (Fig. 1A) cell-free protein synthesizing systems using SDS-polyacrylamide gels electrophoresed to optimize resolution of products larger than 100,000 molecular weight. Nascent polypeptides for the following are indicated in Fig 1: the core protein of CSPG (CP), the α1 and α2 chains of type I collagen and the α1 chain of type II collagen. Since the collagens are known to migrate anomalously on gels, the molecular weight scale serves as a basis for discussion in these cases.

The core protein of CSPG, identified among cell-free products (Fig. 1A,D) by immunoprecipitation (Upholt et al 1979), migrates as a protein of 340,000. A product of approximately 355,000 was synthesized by chondrocytes (Fig. 1B,C) and could be immunoprecipitated by anti-CSPG as well (Vertel, unpublished results). This apparent increase presumably reflects post-trans-

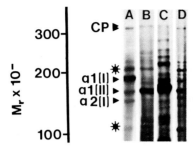

Fig. 1. Gel electrophoresis of ^{35}S methionine-labeled products of cell-free translation directed by cartilage mRNA using rabbit reticulocyte lysate (A) or wheat germ (D) systems and products of chondrocytes in suspension culture pulsed for 5 min (B) or pulsed for 5 min and chased with excess unlabeled methionine for 2 hr (C).

lational modifications which occur in vivo and not in the cell-free systems. Although further modifications are clearly involved in the synthesis of the completed 2-3 x 10^6 dalton CSPG subunit, no progressive increase in the size of the CSPG intermediate was observed as a function of chase time up to 3 hr. A similar intermediate, synthesized by rat chondrosarcoma cells, was described by Kimura et al (1981).

The type II collagen precursor, identified previously among translation products by its collagenase sensitivity and immunoprecipitability with anti-type II collagen (Upholt et al 1979), is a major immunoprecipitable (Vertel, unpublished results) product of intact chondrocytes and cartilage mRNA-directed cell-free protein synthesis in the wheat germ system (Fig. 1B,C,D). Although the type II collagen precursor is synthesized in the reticulocyte system, translation of α1(I) preprocollagen seems to be favored (Fig. 1A). The size of the earliest observable type II collagen precursor in chondrocytes appears equivalent to the size of the cell-free synthesized product while the size of intermediates observed at later chase times appears to increase progressively with time of chase (Fig. 1B,C), in contrast to the CSPG intermediates. Thus, the initial modifications of the type II collagen precursor do not result in apparent size changes, but later modifications, which should include hydroxylation (Cheah et al 1979), are accompanied by decreased electrophoretic mobility. Two other collagenase-sensitive products of cell-free translation reactions, migrating as proteins of 205,000 and 110,000 (Fig. 1, asterisks; Upholt et al 1979), are also synthesized by intact chondrocytes and exhibit chase-related increases in apparent molecular weight. The extension of these molecular studies should be useful to the understanding of progressive modifications of matrix products which occur during biosynthesis.

In a second approach, immunofluorescence reactions were used to examine both intracellular localization and extracellular accumulation of two matrix products simultaneously for individual cartilage cells in culture. Antibodies directed against the link protein of CSPG aggregate (Vertel, unpublished results) and fibronectin were used in addition to the type II collagen and CSPG subunit antibodies just described. Some of these results have been the subject of previous reports (Vertel, Dorfman 1979; Dessau et al 1981; Dorfman et al 1981).

Within 10 min after the trypsin release of cartilage cells from chicken sterna, fibronectin and secreted cartilage products can be detected at the cells' surfaces. Over the next 24 hr in suspension or monolayer culture, cartilage cells remain rounded and their products become organized in a characteristic matrix shown in Fig. 2. The proteoglycan aggregate components CSPG subunit and link protein are found in an amorphous extended matrix while type II collagen is deposited in fibers and fibronectin is tightly associated at the cell surface. Subsequently, chondrocytes shed these matrix shells and leave empty ones behind (Abbott, Holtzer 1966; Vertel 1981).

After several days, colonies of more flattened, polygonal chondrocytes are observed, surrounded by matrix of a different composition (Fig. 3A-F) and sometimes in association with empty matrix shells. Fibronectin now constitutes the extracellular filamentous substratum while type II collagen is observed intra-

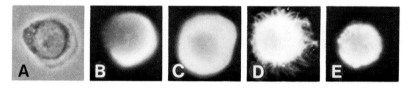

Fig. 2. Chondrocytes in monolayer culture for 24 hr stained with antibodies for link (B), CSPG (C), type II collagen (D), or fibronectin (E) and fluorescent anti-antibodies.

Fig. 3. Simultaneous double immunofluorescence staining of control (A-F) and ascorbate-treated (G-L) cultures with rabbit anti-CSPG (B,H) and guinea pig anti-link (C,I) or with rabbit anti-fibronectin (E,K) and guinea pig anti-type II collagen (F,L). Phase micrographs (A,D,G,J) are shown first for each set. Immunofluorescence methods have been described (Vertel, Dorfman 1979). ⟶

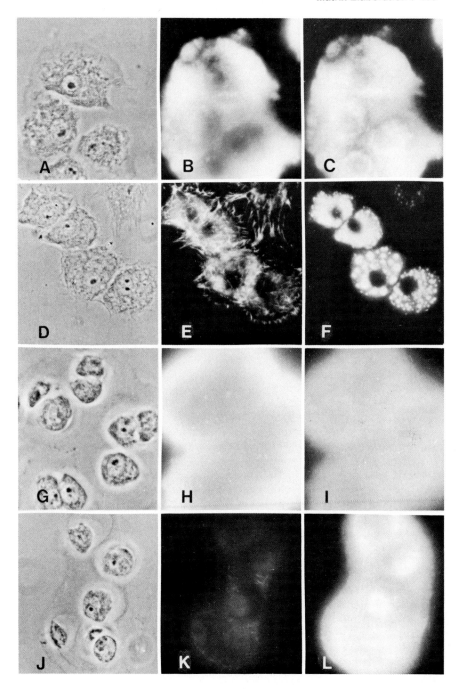

cellularly in discrete cytoplasmic vesicles within chondrocytes but not in the extended matrix. The proteoglycan aggregate components continue to be codistributed in an amorphous matrix.

If chondrocytes are grown for several days in medium supplemented with 50 μg/ml ascorbate, the cell morphology and accumulated matrix again appear different (Fig. 3G-L). Type II collagen is now deposited in a fine fibrous meshwork while fibronectin is no longer a major filamentous constituent of the extended matrix. CSPG subunit and link remain codistributed extracellularly, but in a more densely packed, though still amorphous, matrix. In this case, the shape of individual cartilage cells appears to be more constrained by the surrounding matrix, in a manner similar to matrix-embedded in vivo chondrocytes in lacunae.

Aggregated CSPG can be released by digestion with purified testicular hyaluronidase from the extracellular matrix surrounding chondrocytes grown without added ascorbate (Vertel, Dorfman 1979). CSPGs and related biosynthetic intermediates can then be distinguished within chondrocytes, which usually contain intracellular type II procollagen as well (Fig. 4A,C,E). Hyaluronidase digestion similarly releases extracellular link protein in proteoglycan aggregate to reveal intracellular link, also in discrete cytoplasmic vesicles (Vertel, unpublished results).

In contrast, in cultures grown with added ascorbate, immunoreactive portions of CSPG subunit and link protein remain in association with extracellular type II collagen, even after hyaluronidase digestion (Fig. 4B,D,F). Using electron microscopy, Meier and Solursh (1978) have shown that only indistinct remnants of proteoglycan granules remain in association with collagen fibers after hyaluronidase digestion. Perhaps protein interactions among the core protein of CSPG, link protein and fibrillar collagen serve to retain immunoreactive components in the extracellular matrix.

Under some conditions, not all extracellular CSPG is removed by hyaluronidase digestion of control cultures grown without added ascorbate. That portion of extracellular CSPG which remains appears to be codistributed with filamentous fibronectin (Fig. 4G,H). Perhaps cartilage CSPG can, like other proteoglycans, interact with fibronectin (Yamada et al 1980). Thus filamentous fibronectin could serve as a scaffolding for the extended matrix of proteoglycan aggregates in

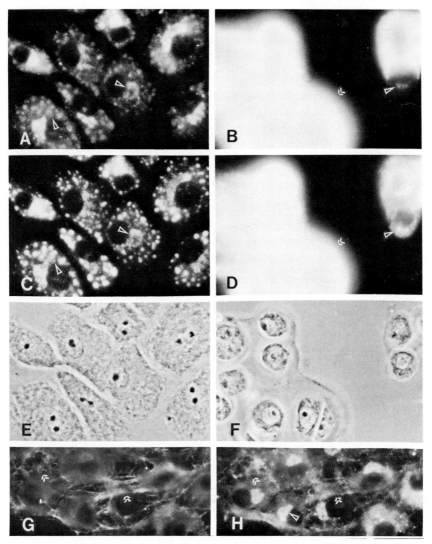

Fig. 4. Codistribution of matrix molecules remaining after hyaluronidase digestion of control (A,C,E; G,H) and ascorbate-treated cultures (B,D,F). Samples were reacted simultaneously with rabbit anti-CSPG (A,B) and guinea pig anti-type II collagen (C,D) and then fluorescent anti-antibodies. Phase micrographs (E,F) are beneath each set. Alternatively, digested controls were reacted with rabbit anti-fibronectin (G) and guinea pig anti-CSPG (H). Intracellular matrix molecules (▷) and areas of apparent codistribution (»») are indicated.

the absence of fibrous type II collagen. In light of the reported concentration of fibronectin in the precartilage region
of the limb mesenchyme (Dessau et al 1980; Melnick et al 1981),
one might speculate that such fibronectin-rich regions could
be utilized initially to retain and concentrate CSPG and there-
by stimulate chondrogenesis (Nevo, Dorfman 1972). However,
the demonstration that fibronectin suppresses expression of
the cartilage phenotype in culture (Pennypacker et al 1980;
West et al 1980) would argue against, but not exclude, this
possibility.

   Immunofluorescence studies demonstrate that chondrocytes
synthesize and secrete type II collagen, CSPG aggregate components and fibronectin in culture, yet organize these molecules
in an extracellular matrix which varies considerably as a
function of time and under different growth conditions. Steps
which occur subsequent to protein synthesis alone appear to be
involved in the creation of an extracellular matrix and the
control of its composition. The understanding of the entire
process of extracellular matrix formation is critical to the
understanding of the role of extracellular matrix during development.

ACKNOWLEDGEMENTS

   I wish to thank Dr. Albert Dorfman for first introducing
me to cartilage and extracellular matrix and for his encourage-
ment and support during my postdoctoral years in his lab. Ms.
Sandra D'Arcangelo is acknowledged for her valuable technical
assistance. The molecular studies involved collaboration with
Dr. William Upholt and some of the immunofluorescence studies
resulted from a collaboration with Drs. Waltraud Dessau and
Klaus von der Mark. Drs. Waltraud Dessau and Richard Hynes
kindly provided the fibronectin antibodies. The original research reported here was supported by NIH postdoctoral fellowship HD-05363 at Chicago and NIH grant AM-28433 at Syracuse.

Abbott J, Holtzer H (1966). The loss of phenotypic traits by
   differentiated cells. III. The reversible behavior of chondrocytes in primary cultures. J Cell Biol 28:473.
Bornstein P, Sage H (1980). Structurally distinct collagen
   types. Ann Rev Biochem 49:957.
Burgeson RE, Hollister DW (1979). Collagen heterogeneity in
   human cartilage: Identification of several new collagen
   chains. Biochem Biophys Res Commun 87:1124.

Cheah KSE, Grant ME, Jackson DS (1979). Translation of type II procollagen mRNA and hydroxylation of the cell free product. Biochem Biophys Res Commun 91:1025.

Dessau W, Vertel BM, von der Mark H, von der Mark K (1981). Extracellular matrix formation by chondrocytes in monolayer culture. J Cell Biol 90:78.

Dessau W, von der Mark H, von der Mark K, Fischer S (1980). Changes in the pattern of collagens and fibronectin during limb bud chondrogenesis. J Embyol Exp Morphol 57:51.

Dorfman A, Vertel BM, Schwartz NB (1981). Immunological studies of chondroitin sulfate proteoglycans. Curr Top Dev Biol 14 (II):169.

Hewitt AT, Kleinman HK, Pennypacker JP, Martin GR (1980). Identification of an adhesion factor for chondrocytes. Proc Natl Acad Sci USA 77:385.

Kimata K, Barrach H-J, Brown KS, Pennypacker JP (1981). Absence of proteoglycan core protein in cartilage from the cmd/cmd (cartilage matrix deficiency) mouse. J Biol Chem 256:6961.

Kimura JH, Thonar EJ-M, Hascall VC, Reiner A, Poole AR (1981). Identification of core protein, an intermediate in proteoglycan biosynthesis in cultured chondrocytes from the Swarm rat chondrosarcoma. J Biol Chem 256:7890.

Lash JW, Vasan NS (1978). Somite chondrogenesis in vitro. Stimulation by exogenous extracellular matrix components. Dev Biol 66:151.

Meier S, Hay ED (1974). Control of corneal differentiation by extracellular materials. Collagen as a promoter and stabilizer of epithelial stroma production. Dev Biol 38:249.

Meier S, Solursh M (1978). Ultrastructural analysis of the effect of ascorbic acid on secretion and assembly of extracellular matrix by cultured chick embryo chondrocytes. J Ultrastr Res 65:48.

Melnick M, Jaskoll T, Brownell AG, MacDougall M, Bessem C, Slavkin HC (1981). Spatiotemporal patterns of fibronectin distribution during embryonic development. I. Chicken limbs. J Embryol Exp Morphol 63:193.

Miller EJ (1971). Isolation and characterization of a collagen from chick cartilage containing three identical α chains. Biochemistry 10:1652.

Nevo Z, Dorfman A (1972). Stimulation of chondromucoprotein synthesis in chondrocytes by extracellular chondromucoprotein. Proc Natl Acad Sci USA 69:2069.

Nogami H, Urist MR (1974). Substrate prepared from bone matrix for chondrogenesis in tissue culture. J Cell Biol 62:510.

Orkin RW, Pratt RM, Martin GR (1976). Undersulfated chondroitin sulfate in the cartilage matrix of brachymorphic mice (bm/bm). Dev Biol 50:82.

Pennypacker JP, Goetinck PF (1976). Biochemical and ultra-structural studies of collagen and proteochondroitin sulfate in normal and nanomelic cartilage. Dev Biol 50:35.

Pennypacker JP, Hassell JR, Yamada KM, Pratt RM (1979). The influence of an adhesion cell surface protein on chondro-genic expression in vitro. Exp Cell Res 121:411.

Reddi AH, Huggins CB (1975). Formation of bone marrow in fibroblast transformation ossicles. Proc Natl Acad Sci USA 72:2212.

Reese CA, Mayne R (1981). Minor collagens of chicken hyaline cartilage. Biochemistry 20:5443.

Rittenhouse E, Dunn LC, Cookingham J, Calo C, Speigelman M, Dooker GB, Bennett D (1978). Cartilage matrix deficiency (cmd): A new autosomal recessive lethal mutation in the mouse. J Embryol Exp Morphol 43:71.

Roden L (1980). Structure and metabolism of connective tissue proteoglycans. In Lennarz WJ (ed): "Biochemistry of Glyco-proteins and Proteoglycans", New York: Plenum, p 267.

Upholt WB, Vertel BM, Dorfman A (1979). Translation and char-acterization of messenger RNAs in differentiating chicken cartilage. Proc Natl Acad Sci USA 76:4847.

Upholt WB, Vertel BM, Dorfman A (1981). Cell-free translation of cartilage RNAs. Alabama J Med Sci 18:35.

Vertel BM (1981). Chondrocytes "molt" in culture. J Cell Biol 91:151a.

Vertel BM, Dorfman A (1979). Simultaneous localization of type II collagen and core protein of chondroitin sulfate proteoglycan in individual chondrocytes. Proc Natl Acad Sci USA 76:1261.

von der Mark K, Conrad G (1979). Cartilage cell differentia-tion. Clin Orthopaed Relat Res 139:185.

Weiss RE, Reddi AH (1981). Appearance of fibronectin during differentiation of cartilage, bone and bone marrow. J Cell Biol 88:630.

West CM, Lanza R, Rosenbloom J, Lowe M, Holtzer H (1979). Fibronectin alters the phenotypic expression of cultured embryo chondroblasts. Cell 17:491.

Yamada KM, Olden K, Hahn L-HE (1980). Cell surface protein and cell interactions. In Subtelny S, Wessells NK (eds): "The Cell Surface, Mediator of Development", New York: Academic Press, p 43.

Limb Development and Regeneration
Part B, pages 159–166
© **1983 Alan R. Liss, Inc., 150 Fifth Avenue, New York, NY 10011**

AN IMMUNOLOGICAL STUDY OF CARTILAGE DIFFERENTIATION
IN CULTURES OF CHICK LIMB BUD CELLS: INFLUENCE OF
A TUMOR PROMOTER (TPA) ON CHONDROGENESIS AND ON
EXTRACELLULAR MATRIX FORMATION

Joachim Sasse, Klaus von der Mark,
Maurizio Pacifici and Howard Holtzer

Max-Planck-Institut für Biochemie
8033 Martinsried, Fed. Rep. of Germany
and Dept. of Anatomy, Medical School,
University of Pennsylvania
Philadelphia, PA 19104

Early chick limb mesodermal cells do not
constitute a homogeneous population but are precursors for at least two cell lineages, which yield
respectively either chondroblasts and fibroblasts
or myoblasts and fibroblasts (Dienstman et al.,
1974). Culture conditions can be varied to favour
the survival and terminal differentiation of
either presumptive chondroblasts or presumptive
myoblasts. This study focuses on cartilage differentiation and utilizes the high-density micromass culture system in which limb bud cells are
grown as "spots" in a culture dish (Ahrens et al.,
1977). The events leading to the appearance of
definitive chondroblasts in these cultures follow
a characteristic, predictable sequence: during
the first two days of culture, clusters of cells
arc formed which subsequently will differentiate
into cartilage nodules.
     This work continues our earlier studies on
extracellular matrix formation during chick limb
chondrogenesis in vitro (von der Mark and von der
Mark, 1977) and in vivo (Dessau et al., 1980).
By means of indirect immunofluorescence we describe 1)
the change in the distribution of two extracellular
matrix proteins, fibronectin and type I collagen,
and 2) the onset of synthesis and deposition of two
cartilage-specific macromolecules, type II collagen
and type IV proteoglycan, concomitant with the

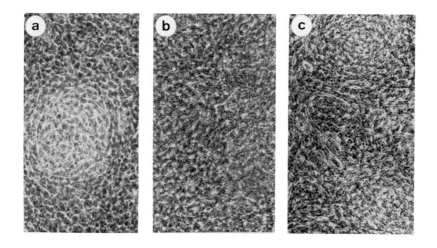

Fig. 1. Reversible effect of the tumor promoter
TPA on the differentiation of stage 23 chick limb
bud cells. A four-day-old control culture (a)
displays cartilage differentiation whereas a
culture reared in medium with 10^{-7}M TPA fails to
chondrify (b). A sister culture to that shown in b
after another 5 days in normal medium develops
cartilage nodules (c).

appearance of microscopically-identifiable chondroblast nodules. In addition we have examined the
influence of the tumor promoter 12-0-tetradecanoyl-
phorbol-13-acetate (TPA) which previously has been
shown to reversibly block the phenotypic synthetic
program of terminally differentiated chondroblasts
(Lowe et al., 1978).

Experimental Results

    In control cultures of stage 23 chick limb bud
cells, within 24 hours clusters with higher cell
density appear; these aggregates give rise to cartilage nodules (Fig. 1a). In contrast, cells reared in
medium containing 10^{-7}M TPA replicate, but fail to
form cell clusters or cartilage nodules (Fig. 1b)
even if maintained in culture as long as 24 days.
If cell cultures kept in TPA for up to four days

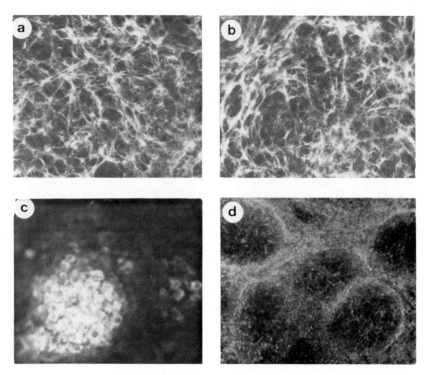

Fig. 2. Changes in the extracellular matrix during
cartilage differentiation. 2-day-old stage 23
chick limb bud cells display a dense network of
type I collagen (a) and fibronectin (b). After onset
of chondrogenesis cartilage matrix in four-day-old
cultures stains intensively for type II collagen,
concomitant with the disappearance of fibronectin
(d) from the center of the cartilage nodules.

are returned to normal medium, an almost normal
number of chondrogenic nodules is observed within
3-6 days (Fig. 1c). Cells exposed to TPA for 5-7
days require another 7-14 days in normal medium to
form cartilage nodules, which are highly reduced in
number. If applied for more than 8 days, TPA irreversibly suppresses chondrogenesis; likewise, the
TPA-treatment of chick limb bud cells seeded at
densities below those of micro-mass cultures is
not readily reversible (Holtzer, 1978).

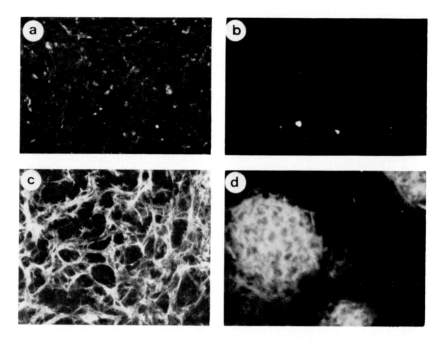

Fig. 3. Effect of TPA on matrix formation.
Chick limb bud cultures reared in medium containing
TPA for 3 days show weak staining for fibronectin (a)
and no reaction with anti-type II collagen antibodies
(b). Sister cultures, returned to normal medium,
after another three days display a dense network of
fibronectin (c). Two days later, cartilage nodules,
staining for type II collagen, form (d).

        Indirect immunofluorescence reveals that during
the initial 1-2 days of culture, chick limb bud
cells deposit a dense extracellular network consisting of type I collagen (Fig. 2a) and fibronectin
(Fig. 2b). These fibrils extend quite uniformly
throughout the culture - surprisingly, the density
of this meshwork is only slightly increased in areas
with cell aggregates, supposed to yield cartilage
within the next hours. While the first differentiated
chondroblasts in the center of cell aggregates
already can be recognized in 40-48-hours-old cultures
with antibodies against the cartilage-specific type II

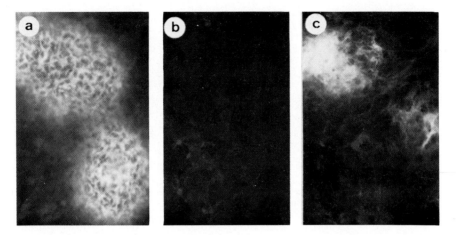

Fig. 4. Influence of TPA on the expression of the cartilage-specific type IV proteoglycan. 4-day-old control cultures show prominent staining of the cartilage matrix with antibodies against type IV proteoglycan (a); in contrast, its expression is completely suppressed in a sister culture reared in 10^{-7}M TPA (b). Parallel cultures to that shown in b, after another five days in normal medium develop cartilage staining for type IV proteoglycan (c).

collagen (not shown), another 1-2 days are required to observe a more complete deposition of the cartilage matrix (Fig. 2c). This process is accompanied by a gradual loss of fibronectin (Fig. 2d) and type I collagen (not shown) from the chondrogenic areas.

In contrast to control cultures, TPA-treated limb bud cells deposit a greatly reduced amount of fibronectin (Fig. 3a) and type I collagen (not shown). Lack of staining for type II collagen (Fig. 3b) demonstrates the suppression of cartilage differentiation. If TPA-treated cells are returned to normal medium, within 2-3 days they lay down a network of type I collagen (not shown) and fibronectin (Fig. 3c) similar to that of untreated cultures (compare Fig. 2b) subsequently followed by the expression of type II collagen (Fig. 3d).

The appearance of the cartilage-unique type IV proteoglycan parallels that of type II collagen.

Likewise, it is deposited in cartilage nodules of
control cultures (Fig. 4a), but its deposition is
reversibly blocked by TPA-treatment (Fig. 4b,c).

Discussion

We report here - in accordance with studies
performed in vivo (Linder et al., 1975; Dessau et
al., 1980; Silver et al., 1981; Manasek et al.,
1981) - that chick limb bud cells accumulate an
extracellular matrix consisting of a meshwork of
fibronectin and type I collagen prior to the onset
of chondrogenesis. This result is compatible with
a permissive role of this network, possibly acting
as a structural, organizing factor during morphogenesis (compare Chiquet et al., 1981). The fact
however, that these macromolecules are deposited
in both presumptive cartilage and presumptive non-
cartilage areas renders any "specific" function
during chondrogenesis unlikely. This holds true
also for the in vivo situation (Linder et al., 1975;
Dessau et al., 1980; Silver et al., 1981; Tomasek
et al., 1982).

After onset of chondrogenesis the developing
cartilage regions appear to lose fibronectin and
type I collagen gradually. Since the immunofluorescence stainings were obtained with culture dishes
preincubated with testicular hyaluronidase prior to
the antibody-treatment it seems unlikely that
macromolecules masked by cartilage proteoglycans
have escaped detection. Also in vivo studies
indicate that mature cartilage is devoid of fibronectin (Linder et al., 1975; Dessau et al., 1978;
Dessau et al., 1980).

The tumor promoter TPA inhibits extracellular
matrix formation in cultures of chick limb bud cells.
Similar effects have been observed in fibroblast
cultures where TPA reversibly reduces the synthesis
and the deposition of fibronectin (Blumberg et al.,
1976) and of type I collagen (Delclos and Blumberg,
1979).

TPA is known to reversibly block the differentiation program of definitive myoblasts (Cohen et al.,

1977) and of chondroblasts (Lowe et al., 1978).
Here we demonstrate that a short-term treatment
with TPA also influences the differentiation of
chondrogenic precursor cells in a reversible
fashion. The TPA-treated cells replicate and
covertly transmit their genetic program for many
cell generations. Therefore, the differentiation
program which a presumptive chondroblast inherits
from his mother is not dependent on his expression;
experiments with viral-transformed cells have led
to similar conclusions (for review, see Holtzer
et al., 1981). After removal of TPA these cells
then express their unique synthetic program,
characterized by the synthesis of the cartilage-
specific macromolecules, type II collagen and
type IV proteoglycan.

References

Ahrens P, Solursh M, and Reiter R (1977). Dev. Biol.
60:69.
Blumberg P, Driedger P, and Rossow P. (1976).
Nature 264:446.
Chiquet M, Eppenberger H, and Turner D (1981).
Dev. Biol. 88:220.
Cohen R, Pacifici M, Rubinstein N, Biehl J, and
Holtzer H (1977). Nature 266:538.
Delclos K, and Blumberg P (1979). Cancer Res.
39:1667.
Dessau W, Sasse J, Timpl R, Jilek F, and von der
Mark K (1978). J. Cell Biol. 79:342.
Dessau W, von der Mark H, von der Mark K, and
Fischer S (1980).   J. Embryol. Exp. Morphol.
57:51.
Dienstman S, Biehl J, Holtzer H, and Holtzer S
(1974). Dev. Biol. 39:83.
Holtzer H (1978). In: Stem Cells and Tissue
Homeostasis (eds. Lord, Potten and Cole) pp.1-28,
Cambridge University Press, Cambridge.
Holtzer H, Pacifici M, Croop J, Toyama Y, Dlugosz A
(1981). Fortschritte der Zoologie 26:207.
Linder E, Vaheri A, Ruoslahti E, and Wartiovara J
(1975). J. Exp. Med. 142:41.
Lowe M, Pacifici M, and Holtzer H (1978). Cancer
Res. 38:2350.

Silver M, Foidart J, and Pratt R (1981). Differen-
    tiation 18:141.
Tomasek J, Mazurkiewicz J, and Newman S (1982).
    Dev. Biol. 90:118.
Von der Mark K, and von der Mark H (1977). J.Cell
    Biol. 73:736.

Limb Development and Regeneration
Part B, pages 167–174
© 1983 Alan R. Liss, Inc., 150 Fifth Avenue, New York, NY 10011

THE GROWTH OF EMBRYONIC CHICK LIMB MESENCHYME CELLS IN
SERUM-FREE MEDIUM

John P. Pennypacker

Department of Zoology
University of Vermont
Burlington, Vermont   05405

Embryonic chick limb mesenchyme in cell culture has
provided an important model system for the study of chondrogenesis (Caplan, 1970; Goetinck et al., 1974). In this
system, mesenchyme cells from pre-chondrogenic limb buds
will multiply and differeniate into chondrocytes when plated
at high cell density (Umansky, 1968). This can be demonstrated by the analysis of proteoglycan synthesis over the
culture period. Mesenchyme cells synthesize low molecular
weight proteoglycans containing dermatan sulfate (see
Rosenberg, this volume). As chondrogenesis proceeds in
culture, the synthesis of a high molecular cartilage-
specific proteoglycan containing chondroitin sulfate and
keratan sulfate is augmented (Goetinck et al., 1974;
DeLuca et al., 1980).

A common feature of most studies utilizing the limb
mesenchyme culture system has been the use serum-containing
medium. In some cases, this has involved the use of
several different sera as well as embryo extract (Goetinck
et al., 1974). This approach, while very useful, pre-
cludes a precise analysis of the requirements for chondrogenesis in culture. However, an important advance in cell
culture technology has been the development of chemically
defined serum-free media (Barnes and Sato, 1980). This
provides a powerful tool for the determination of the
requirements for the maintenance and normal behavior of
cultured cells. This approach has been applied with some
success to the limb mesenchyme system (Karasawa et al.,
1979), but in that study, the defined medium was unable to
support the differentiation of high density monolayer

cultures. Instead, cell pellets were generated by centrifugation and cultured in that form. In this study, the
serum-free conditions for high density monolayer cultures
are described.

METHODS

Limb buds were dissected from stage 23-24 chick
embryos and pooled in calcium and magnesium-free saline
G (CMF-sal G). The limb buds were then incubated in CMF-
sal G containing 0.1% trypsin (Grand Island Biological,
GIBCO) and 0.1% EDTA for 30 minutes at 37o. Under these
conditions, the limb buds are still intact after 30 minutes
and are transferred to CMF sal G containing 0.2% soybean
trypsin inhibitor (GIBCO). A cell suspension was obtained
by flushing the buds several times through a narrow bore
Pasteur pipet. The resulting suspension was filtered through
two layers of 20μ mesh Nitex monofilament, and the cell
number was then determined using a Coulter counter. The
cells were resuspended in the appropriate medium at
2×10^7 cell/ml and inoculated into Falcon 3008 multi-well
plates according to the procedure of Ahrens et al. (1977).
For this purpose, a 20-μl drop containing 4×10^5 cells is
pipeted onto the center of the dish and incubated for 2
hours to permit cell attachment. The cultures were then
flooded with 0.5 ml medium and incubated for 5 days.

The basal medium was Eagles' minimal essential medium
with added non-essential amino acids (GIBCO) and contained
0.5 μg/ml transferrin (Sigma), 5 μg/ml ascorbic acid,
10 mM Hepes buffer (GIBCO) and 50 μg/ml gentamicin
(Microbiological Associates). Other components were added
as indicated in the text. The medium was changed every 2
days by complete replacement.

In order to assess chondrogenesis in culture, the cell
layers were stained with alcian blue according to the procedure of Pennypacker et al. (1978). The amount of alcian
blue staining material was estimated by extraction of the
stain from the cultures with 4.0M guanidine hydrochloride
and measuring the absorbance at 600nm.

Proteoglycan synthesis was assessed quantitatively by
measuring $[^{35}S]$- SO_4 incorporation into cetylpyrridinium
chloride (CPC) precipitable glycosaminoglycans. For this

purpose, day 5 cultures were incubated with $Na_2[^{35}S]O_4$
(5μCi/ml. New England Nuclear) for five hours. After labeling, 10 μl papain (7.25 units, Sigma) and 50 μl 0.05M
cysteine, 0.05M EDTA were added to the medium. The samples
were incubated at 50° for 5 hours, and then NaOH was added
to a final concentration of 0.3N. The samples were hydrolyzed
overnight at room temperature, and then centrifuged at
3000 RPM for 10 minutes. An aliquot (0.2ml) of the supernatant fraction was added to 1.0ml 0.035M NaCl, carrier
chondroitin sulfate (0.1ml at 2mg/ml, Sigma) was added to
each sample, and the glycosaminoglycans precipitated with
CPC according to the procedure of Scott (1960). Glycosaminoglycans were isolated (Scott, 1960), and radioactivity
measured by liquid scintillation spectrometry. DNA content
of equivalent cultures was measured by the fluorometric
procedure of Kissane and Robins (1958).

Qualitative analysis of proteoglycans was carried out
by sucrose density gradient centrifugation of the extracted
proteoglycans. For this purpose, day 5 cultures were
labeled for 24 hours with 50μCi/ml $Na_2[^{35}S]O_4$. The medium
was removed, the cell layer rinsed twice with saline G,
and then extracted with a 4.0M guanidine hydrochloride
solution containing 0.01M sodium EDTA, 0.005M benzamidine-
HCl, and 0.05M sodium acetate, pH 5.8 (Oegema et al., 1975).
Free radioactivity was removed from the extract by repeated
precipitation with 95% ethanol, 1.3% potassium acetate
until radioactivity in the supernatant fraction was at
background levels. The proteoglycan samples (0.5ml) were
layered on 11ml linear gradients of sucrose (5-20%) containing 4.0M guanidine-HCl and 0.05M sodium acetate, pH 5.8,
formed on a 0.5ml cushion of 40% sucrose in the same sol-
vent (Kimata et al., 1974). The gradients were centrifuged in a SW488 rotor (International) at 4° and 38,000
RPM for 20 hours. The gradients were separated into
0.3ml fractions, and aliquots of each fraction were taken
for measurement of radioactivity.

RESULTS AND DISCUSSION

The major problem in establishing monolayer cultures of
limb mesenchyme cells is the initial and continued attach-

ment to the plastic substrate. When cultured in basal
medium containing transferrin, ascorbate, hepes buffer and
gentamicin, plating efficiency was low and the monolayer
continued to detach over the 5 days of incubation. This is
in agreement with the observations of Karasawa et al. (1979),
who observed a decrease in DNA content of the mesenchyme
cultures when grown as a monolayer in the absence of serum.
This problem was initially overcome by adding fibronectin
(10 µg/ml) and Pedersen's fetuin (10 µg/ml) to the culture
medium. Both preparations have previously been shown to
promote the attachment and spreading of cells (Yamada et al.,
1976; Fisher et al., 1958). Figure 1 shows that with added
fibronectin and fetuin (-BSA condition), plating efficiency
is 70% of serum controls, and DNA content increased slightly
over the five days.

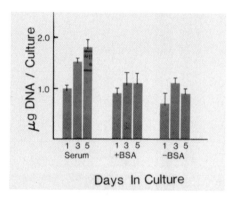

Fig. 1. DNA concentration per culture. The medium of the
-BSA and +BSA conditions contain fetuin (10 µg/ml) and
fibronectin (10 µg ml). Values represent the mean of 4
determinations ± S. D.

Under these conditions, $[^{35}S]$-SO_4 incorporation into
glycosaminoglycans increases over the culture period
(Figure 2), and alcian blue staining nodules are present on
day 5 (not shown). Initial attachment is further enhanced
by addition of bovine serum albumin (BSA, 0.5 mg/ml) to the
culture medium (Figure 1). While the addition of BSA does
not promote the subsequent growth of the monolayer to serum
control levels, the incorporation $[^{35}S]$-SO_4 is significantly
increased (Figure 2), and there is greater staining with
alcian blue.

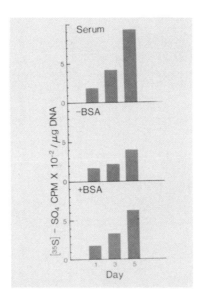

Fig. 2. $[^{35}S]$-SO$_4$ incorporation/µg DNA into total cell layer and medium glycosaminoglycans. Values represent the mean of 4 determinations.

Proteoglycans synthesized in day 5 cultures were characterized by sucrose density gradient centrifugation in order to further assess chondrogenesis in the absence of serum. In serum controls, $[^{35}S]$-SO$_4$ incorporation into the high molecular weight proteoglycan species represented 82% of the total, while in the absence of BSA, incorporation into that fraction was reduced to 27% of the total (Figure 3).

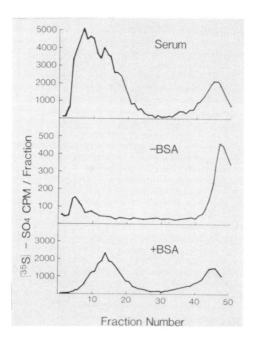

Fig. 3. Sucrose density gradient centrifugation of the
$[^{35}S]$-SO$_4$ -labeled proteoglycans extracted from the cell
layer.

The addition of BSA was found to increase the cartilage
proteoglycan fraction to 63% of the total (Figure 3),
indicating the BSA significantly enhances chondrogenesis
in these serum-free conditions.

The primary effects of BSA are most likely related to
initial survival after trypsinization. In the absence of
BSA, the cell density of the monolayer on day 1 is less,
and there is considerable cell debris. In the presence of
BSA, there is little debris, and as is indicated by the
DNA content, the cell density on day 1 is greater. Since
chondrogenesis in culture is density dependent (Umansky,
1968), the changes in proteoglycan synthesis may be related
to the greater initial cell density.

Various hormones and growth factors were also tested for their effects on chondrogenic expression. Initial screening by measuring alcian blue extractable stain and DNA content showed that fibroblast growth factor, insulin, thyroxine, multiplication stimulating activity and dexamethasone have no effect. The lack of effect by insulin or fibroblast growth factor is particularly interesting since chondrocytes isolated from 15 day chick embryo sterna are responsive to both (Pennypacker, in prep.). Therefore, given the proper conditions for substrate attachment, it appears that chondrogenesis proceeds normally under serum-free conditions and does not require factors necessary for the growth and normal behavior of diffenentiated chondrocytes in culture.

Ahrens PB, Solursh M, Reiter RS (1977). Stage-related capacity for limb chondrogenesis in cell culture. Develop Biol 60: 69.

Barnes D, Sato G (1980). Methods for the growth of cultured cells in serum free medium. Anal Biochem 102: 255.

Caplan AI (1970). Effects of the nicotinamide-sensitive tertogen 3-acetylpyridine on chick limb cells in culture. Exp Cell Res 62: 341.

Deluca S, Lohmander S, Nilsson B, Hascall VC (1980). Proteoglycans from chick limb bud chondrocyte cultures. J Biol Chem 255: 6077.

Fisher HW, Puck TT, Sato G (1958). Molecular growth requirements of single mammalian cells: The action of fetuin in promoting cell attachment to glass. Proc Natl Acad Sci 44: 4.

Goetinck PF, Pennypacker JP, Royal PP (1974). Proteo-chondroitin sulfate synthesis and chondrogenic ex-pression. Exp Cell Res 87: 241.

Karasawa K, Kimata K, Ito K, Kato Y, Suzuki S (1979). Morphological and biochemical differentiation of limb bud cells cultured in chemically defined medium. Develop Biol 70: 287.

Kimata K, Okayama M, Aohira A, Suzuki S (1974). Hetero-geneity of proteochondroitin sulfates produced by chon-drocytes at different stages of cytodifferentiation. J. Biol Chem 249: 1646.

Kissane JM, Robins E (1958). The Fluorometric measurement of DNA in animal tissues with special reference to the central nervous system. J Biol Chem 233: 184.

Oegema TR, Hascall VC, Sziewitkowski DD (1975). Isolation

and characterization of proteoglycans from the Swarm rat chondrosarcoma. J Biol Chem 250: 6151.

Pennypacker JP, Lewis CR, Hassell, JR (1978). Altered proteoglycan metabolism in mouse limb bud mesenchyme cell cultures treated with vitamin A. Arch Biochem Biophys 186: 351.

Pennypacker JP, Shonk LA (in prep). Growth and phenotypic expression of embryonic chick chondrocytes in serum-free medium.

Scott J (1960). Aliphatic ammonium salts in the assay of acidic polysaccarides from tissues. In Glick D (ed): "Methods of Biochemical Analysis," New York: Wiley, Vol 8: 145.

Umansky R (1966). The effect of cell population density on the developmental fate of reaggregating mouse limb bud mesenchyme. Develop Biol 13: 31.

Yamada KM, Yamada SS, Pasten I (1976). Cell surface protein partially restores morphology, adhesiveness, and contact inhibition of movement to transformed fibroblasts. Proc Natl Acac Sci 73: 1217.

Limb Development and Regeneration
Part B, pages 175–182
© **1983 Alan R. Liss, Inc., 150 Fifth Avenue, New York, NY 10011**

GENE EXPRESSION DURING CHICK LIMB CARTILAGE DIFFERENTIATION

William B. Upholt, Dean Kravis, Linda Sandell,
Val C. Sheffield, and Albert Dorfman
Dept. of Pediatrics and the Joseph P. Kennedy, Jr.
Mental Retardation Research Center, The University
of Chicago, Chicago, IL   60637

Stage 24 chicken limb mesenchyme, when dissociated and cultured either at high cell densities on tissue culture dishes or as aggregates over agar, differentiates into cartilage in 4-10 days (Caplan, Zwilling, Kaplan 1968; Levitt, Dorfman 1974). This phenomenon parallels normal embryonic development and therefore furnishes an ideal model system for studying the process of cartilage differentiation in cultured cells. Initially, stage 24 limb mesenchymal cells synthesize type I collagen (von der Mark, von der Mark 1977). Those cells which differentiate into chondrocytes cease synthesis of type I collagen and begin synthesis of polypeptides characteristic of cartilage extracellular matrix including type II collagen (von der Mark, von der Mark 1977), the core protein of cartilage chondroitin sulfate proteoglycan (Levitt, Dorfman 1974), and the link protein of proteoglycan aggregate (Vasan, Lash 1977).

We are studying the mechanisms by which genes for these proteins are regulated during cartilage differentiation. We have concentrated thus far on identifying the unprocessed nascent chains synthesized in cell-free protein synthesizing systems, characterizing the mRNAs for these proteins and obtaining recombinant DNAs for these genes.

RNA has been prepared from embryonic chicken calvaria, stage 24 limb buds, high-density cultures of stage 24 limb mesenchyme differentiating into cartilage and embryonic chicken sterna (Upholt, Vertel, Dorfman 1979). Figure 1 shows cell-free translation products directed by the latter two RNAs. We have identified bands corresponding to the

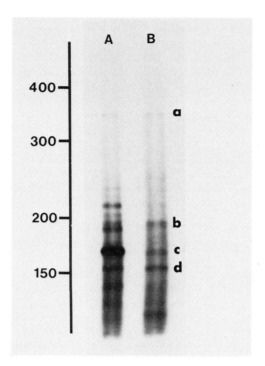

Figure 1. Gel electrophoresis of cell-free translation products directed by RNA isolated from sterna and differentiating limb bud cultures. RNA was translated in a wheat germ cell-free extract, and polypeptides were labeled with ^{35}S-methionine as described previously (Upholt, Vertel, Dorfman 1979). Samples were electrophoresed on a 0.1% sodium dodecyl sulfate/5% polyacrylamide gel. Lanes: A, translation products of total sternal RNA; B, translation products of total RNA from 8-day high-density limb bud cultures. Bands a,b,c, and d are cell-free translation products corresponding to proteoglycan core protein and the $\alpha 1$(I), $\alpha 1$(II), and $\alpha 2$(I) procollagen chains respectively. A scale of molecular weight x 10^{-3} is shown to the left of the figure. (Fig. reproduced from Upholt, Vertel, Dorfman 1981 Alabama J Med Sci 18:35).

core protein of cartilage chondroitin sulfate proteoglycan, the $\alpha 1$ and $\alpha 2$ chains of type I procollagen and the $\alpha 1$ chain of type II procollagen using a variety of techniques including immunoprecipitation, collagenase digestion and com-

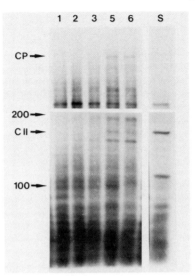

Figure 2. Cell-free translation of total RNA prepared from high-density cultures of stage 24 chick limb mesenchymal cells after 1,2,3,5 or 6 days of culture. Translation products using sternal RNA (S) are included for comparison. RNA was translated and electrophoresed as described in fig. 1. A longer exposure of the portion of the gel containing polypeptides larger than 200,000 daltons is shown so that both the proteoglycan core protein (CP) and type II procollagen (C II) nascent translation products are visible. Arrows at 100 and 200 indicate the migration positions for polypeptides of 1 x 10^5 and 2 x 10^5 daltons respectively as determined from molecular weight standards. (Fig. reproduced from Vuorio et al. 1982. Nucleic Acids Res 10:1176).

parison of translation products using RNAs from a variety of tissues which have characteristic polypeptide compositions (Upholt, Vertel, Dorfman 1979; 1981).

Figure 2 shows cell-free translation of RNAs prepared from stage 24 limb mesenchymal cells after culture for 1-6 days at high density. The cell-free translation products corresponding to the core protein and type II procollagen are not detected when RNA from day 1 or day 2 cultures is translated. Translatable mRNA for these products begins to appear on day 3 of culture. The time course of appearance of core protein and type II procollagen mRNAs closely paral-

lels the differentiation of the high-density cultures as
determined previously by a variety of techniques measuring
the synthesis or accumulation of extracellular matrix products characteristic of cartilage (Levitt, Dorfman 1974).
Figure 2 also shows that increased amounts of type I procollagen mRNAs are present at later days of culture. The
increase in both type I and type II procollagen mRNAs correlates with measurements made by von der Mark and von der Mark
(1977) of actual amounts of type I and type II collagen synthesized by high-density limb bud cultures on tissue culture
dishes. Since chondrocytes in high-density cultures have
been shown by immunological techniques to synthesize primarily type II collagen (von der Mark, von der Mark 1977), the
increase in type I procollagen mRNA most likely occurs in
other cell types present in the rather heterogeneous high-
density limb mesenchyme cultures.

We and others (Upholt, Vertel, Dorfman 1977; Herson,
Schmidt, Seal, Marcus, van Vloten-Doting 1979; Kabat,
Chappell 1977) have found that cell-free translation, although in general reliable for the measurement of relative
changes in the amount of a single mRNA species, cannot be
used to compare relative amounts of two different mRNAs. In
particular, changes in either the source of cell-free extract or cell-free translation conditions can significantly
alter the relative amounts of type I and type II procollagen
chains synthesized using the same mixture of RNAs. In order
to directly measure the amounts of specific mRNAs present
in differentiating chondrocytes, it is necessary to prepare
DNA probes which can be hybridized to the mRNAs of interest.
We have partially purified the mRNAs for the type I and
type II procollagen chains and used these as templates with
AMV reverse transcriptase and E.coli DNA polymerase I to
prepare double-stranded complementary DNAs (cDNAs) (Vuorio,
Sandell, Kravis, Sheffield, Vuorio, Dorfman, Upholt 1982).
The double-stranded cDNAs were inserted into plasmid vector
pBR322 and cloned in E.coli. Recombinant DNAs containing
sequences corresponding to portions of the mRNAs for the
$\alpha 1$ chain of type II procollagen and the 1 and 2 chains
of type I procollagen were identified by DNA sequence analysis and by hybridization to gel blots of RNAs from tissues
characteristically synthesizing either type I or type II
collagen (Vuorio, et al. 1982; Vuorio, Kravis, Upholt,
unpublished results). In addition, restriction enzyme
cleavage site maps of our type I procollagen recombinant

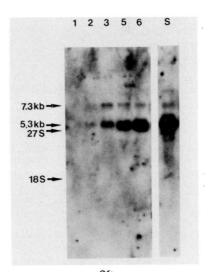

Figure 3. Hybridization of ^{32}P-labeled type II procollagen cDNA probe to a glyoxal gel transfer of total RNA prepared from differentiating limb bud cultures after 1,2,3,5
and 6 days of culture. Sternal RNA (S) is shown for comparison. The positions of the 18S and 27S ribosomal RNAs
are indicated. (Fig. reproduced from Vuorio et al. 1982.
Nucleic Acids Res 10:1175).

cDNAs were compared with previously published maps of type I
collagen cDNAs obtained by other investigators (Sobel,
Yamamoto, Adams, DiLauro, Avvedimento, de Crombrugghe,
Pastan 1978; Lehrach, Frischauf, Hanahan, Wozney, Fuller,
Boedtker 1979).

We have used these cloned DNA sequences to directly
examine the relative changes in mRNA levels during limb
cartilage differentiation in culture (Vuorio et al. 1982).
Equal aliquots of the RNA preparations used for cell-free
translation in figure 2 were denatured by glyoxal treatment,
electrophoresed on a 1% agarose gel, and transferred to
nitrocellulose paper. DNA probes were prepared by purifying the inserted cDNA sequences from the bacterial vector
and labeling them with ^{32}P by nick translation. The labeled
probes were then hybridized to the RNA transfer and subsequently detected by autoradiography. Results using pCAR2,
one of the two type II procollagen mRNA probes, are shown

in figure 3. Type II procollagen mRNA is not detected in
RNA prepared from day 1 cultures, appears at day 2, and
increases until about day 5 or 6. These results are consistent with those obtained using cell-free translation (Fig. 2)
and confirm by direct analysis that the amount of type II
procollagen mRNA increases in parallel with the differentiation of chick limb mesenchyme into cartilage.

Relative changes in the amount of a single mRNA species
can be directly assessed by these methods, but determination
of the number of copies per cell or relative comparisons
between different mRNA species require different probes and
techniques. Probes prepared from double-stranded recombinant DNAs are not ideal for quantitative DNA-RNA hybridization techniques since only 50% of the label (one strand) can
be hybridized to the mRNA and since self-association of the
2 complementary labeled DNA strands competes with DNA-
RNA hybridization. In order to avoid these difficulties,
we have recloned our cDNA inserts in M13 single-stranded
DNA vectors (Messing, Crea, Seeburg 1981). Probes can be
prepared from these recombinant DNAs which contain only
the DNA strand complementary to the mRNA. Results from
preliminary experiments with such probes (Kravis, unpublished
results) are consistent with those shown in figures 2 and
3 suggesting that the major control mechanisms in the regulation of type II procollagen synthesis must be those determining the rate of accumulation of type II procollagen
mRNA.

In order to assess rates of transcription and to investigate possible primary or secondary modifications of
the type II procollagen gene involved in regulation, it
is necessary to have cloned DNA sequences corresponding to
the actual gene, including intervening sequences and the
promoter region. We have used our cloned sequences complementary to type II procollagen mRNA to screen a Charon 4A
library of the chicken genome and have obtained a genomic
clone containing approximately 50% of the coding sequences
of the 3' end of the type II procollagen gene (Sandell,
Yamada, Dorfman and Upholt, unpublished results). DNA sequence analysis shows that this clone contains sequences
corresponding exactly to those present in our cDNA clones
and also contains sequences which, when translated into
amino acids, correspond to known amino acid sequences for
type II collagen. These cloned DNAs will be used to isolate
the complete type II procollagen gene and to extend our

studies of the regulation of this gene.

Our preliminary results using a variety of techniques suggest that expression of the type II procollagen gene is regulated by mechanisms controlling the rate of type II procollagen mRNA accumulation. Such mechanisms include rates of processing and degradation of mRNAs and their precursors as well as rates of transcription. Transcription rates may be influenced by secondary modifications of the DNA in the gene or by changes in the organization or composition of proteins associated with chromatin containing the gene. The availability of a variety of probes for the genes for type II procollagen and other cartilage proteins will allow these possibilities to be investigated.

Supported by NIH grants: AM 05996, HD 09402, and HD 04583. The expert technical assistance of Gloria Enriquez, James Mensch and Ida Schaefer is gratefully acknowledged.

REFERENCES

Caplan AI, Zwilling E, Kaplan NO (1968). 3-Acetylpyridine: Effect in vitro related to teratogenic activity in chicken embryos. Science 160:1009.
Herson D, Schmidt A, Seal S, Marcus A, van Vloten-Doting L (1979). Competitive mRNA translation in an in vitro system from wheat germ. J Biol Chem 254:8245.
Kabat D, Chappell MR (1977). Competition between globin messenger ribonucleic acids for a discriminating initiation factor. J Biol Chem 252:2684.
Lehrach H, Frischauf AM, Hanahan D, Wozney J, Fuller F, Boedtker H (1979). Construction and characterization of pro α1 collagen complementary deoxyribonucleic acid clones. Biochemistry 18:3146.
Levitt D, Dorfman A (1974). Concepts and mechanisms of cartilage differentiation. Curr Top Dev Biol 8:103.
Messing J, Crea R, Seeburg PH (1981). A system for shotgun DNA sequencing. Nucleic Acids Res 9:309.
Sobel ME, Yamamoto T, Adams SL, DiLauro R, Avvedimento VE, de Crombrugghe B, Pastan I (1978). Construction of a recombinant bacterial plasmid containing a chick pro-α2 collagen gene sequence. Proc Natl Acad Sci USA 75:5846.
Upholt WB, Vertel BM, Dorfman A (1979). Translation and characterization of messenger RNAs in differentiating chicken cartilage. Proc Natl Acad Sci USA 76:4847.

Upholt WB, Vertel BM, Dorfman A (1981). Cell-free translation of cartilage RNAs. Alabama J Med Sci 18:35.

Vasan NS, Lash JW (1977). Homogeneity of proteoglycans
in developing chick limb cartilage. Biochem J 164:179.

von der Mark K, von der Mark H (1977). Immunological and
biochemical studies of collagen type transition during
in vitro chondrogenesis of chick limb mesodermal cells.
J Cell Biol 73:736.

Vuorio E, Sandell L, Kravis D, Sheffield VC, Vuorio T,
Dorfman A, Upholt WB (1982). Construction and partial
characterization of two recombinant cDNA clones for
procollagen from chicken cartilage. Nucleic Acids Res
10:1175.

Limb Development and Regeneration
Part B, pages 183–192
© **1983 Alan R. Liss, Inc., 150 Fifth Avenue, New York, NY 10011**

SYNTHESIS AND CHARACTERIZATION OF CHICKEN CARTILAGE COLLAGEN cDNA

Yoshifumi Ninomiya, Allan M. Showalter,
and Bjorn R. Olsen
Department of Biochemistry
UMDNJ-Rutgers Medical School
Piscataway, New Jersey 08854

The collagens comprise a family of large proteins in connective tissues (for review, see Bornstein, Sage 1980). The members of this family are similar in that they are composed of polypeptide chains, α-chains, with a repeating structure containing glycine in every third position and proline or 4-hydroxy-proline frequently preceding the glycine residues. Most collagen molecules are composed of three α-chains arranged in a triple-helical conformation. The biosynthetic precursors of collagen molecules are procollagens (see Olsen 1981). The presence of large peptide extensions (propeptides) at the amino and carboxyl termini of the α-chains distinguishes procollagens from collagens. With the exception of a small region within the amino propeptide, the propeptides do not contain the characteristic repetitive structure of α-chains. During collagen biosynthesis, procollagen molecules are converted to collagen molecules by the action of specific endoproteases that remove the amino and carboxyl propeptides.

Several types of collagen, containing polypeptide products of different gene loci, have been identified (see Bornstein, Sage 1980). Different proportions of collagen types confer on various tissues characteristic mechanical properties. For example, while skin contains Types I, III and V collagens, hyaline cartilage contains a different set of collagens (see Reese, Mayne 1981; von der Mark et al. 1982; Burgeson et al. 1982). These variations among tissues reflect the modulated expression of the various collagen genes during tissue differentiation. For this reason, the isolation and characterization of collagen genes is of considerable interest. Most of the available data deal with the pro α 2(I)gene, the gene that codes for one of the two polypeptide chains of Type I collagen in tendon, skin and bone (Ohkubo et al. 1980; Wozney et al. 1981). In particular, studies on cloned genomic DNA fragments have allowed a structural and functional analysis of the promoter region of the α 2(I) collagen gene (Vogeli et al. 1981; Merlino et al. 1981). To identify regulatory regions that function in differential transcription of collagen genes during tissue differentiation, it is important to isolate, characterize, and compare different collagen genes.

We have recently synthesized and are currently characterizing cDNAs to cartilage mRNAs. Our goals are a) to obtain cDNAs that can serve as probes for the isolation of the various collagen genes that are expressed in hyaline cartilage, b) to obtain structural information about various cartilage collagen polypeptides through nucleotide sequencing of such cDNAs and genomic clones, and c) to obtain probes for assays of collagen mRNA levels during cartilage differentiation.

MATERIALS AND METHODS

RNA was extracted from 17-day old chick embryo sternal cartilage by a modified guanidine hydrochloride extraction method (Cox, 1968; Adams et al. 1977). Poly(A$^+$) RNA was obtained by oligo(dT)-cellulose chromatography, and the poly(A$^+$) RNA was fractionated further by sucrose gradient centrifugation. In several experiments, about 30-40 μg of high-molecular weight poly(A$^+$) RNA (>28S) was obtained from 36 dozen sterna.

The mRNA activity was assayed by cell-free translation using a commercial (Amersham) rabbit reticulocyte lysate. Translation products were labelled with ^{35}S-methionine and analyzed by polyacrylamide gel electrophoresis in the presence of sodium dodecyl sulfate.

cDNA to sternal mRNA was synthesized using a modification of the methods described by Goodman and MacDonald (1979). The mRNA was transcribed into single-stranded cDNA using AMV reverse transcriptase (Life Sciences, Inc.) and oligo(dT) (Collaborative Research) as a primer. After purification of the single-stranded cDNA on an alkaline sucrose gradient, the second strand was synthesized with AMV reverse transcriptase as described previously (Myers et al. 1981).

The double-stranded cDNA was trimmed with S1-nuclease, fractionated on a sucrose gradient, and blunt-end ligated to synthetic Hind III-linkers (Collaborative Research) according to Goodman and MacDonald, (1979). The cDNA was inserted between two Hind III-sites of the expression plasmid pGW 134 (generously provided by Dr. Geoffrey Wilson) using standard procedures. The recombinant DNA was used to transform E. coli K 802 (λ), and transformants were selected on Luria agar plates supplemented with ampicillin (50μ g/ml).

Transformants were grown in liquid broth culture, and their DNA was isolated by the rapid boiling method of Holmes and Quigley (1981). This DNA was restricted with Hind III and electrophoresed on 1% agarose gels to screen for the presence of inserts.

DNA from transformants harboring inserts was screened by the dot hybridization method of Kafatos et al. (1979) using alkali-fragmented ^{32}P-labelled poly(A$^+$) RNA from 17-day old chick sterna and calvaria.

For Northern-blot analysis, poly(A$^+$) RNAs were electrophoresed on 0.8% agarose gels, transferred onto nitrocellulose filters, and hybridized to nick-translated DNA according to the methods described by Thomas (1980) with some modifications. DNA used for RNA blot hybridizations was ^{32}P-labelled in vitro by nick-translation (Maniatis et al. 1975) using α-^{32}P-deoxynucleotide triphosphates (800 Ci/mmole) purchased from New England Nuclear.

Restriction endonucleases were obtained from Bethesda Research Laboratories and New England Biolabs, and digests were carried out according to the conditions described by Davis et al. (1980). DNA sequence analysis of the recombinant clones was performed using the method of Maxam and Gilbert (1977; 1980).

RESULTS AND DISCUSSION

High-molecular weight mRNA from chick embryo sterna directed the synthesis of several collagenous polypeptides when assayed by cell-free translation in a reticulocyte lysate (Fig. 1). One of the major translation products was identified as the pre-pro α1(II)-chain based on its sensitivity to bacterial collagenase and its reactivity with antibodies specific for the carboxyl propeptide of Type II procollagen (S. Curran and M. Sobel, unpublished). The translation products also included polypeptides with mobilities similar to those of pre-pro α1(I)- and pre-pro α2(I)-chains, as well as additional collagenase-sensitive and collagenase-resistant polypeptides within the same molecular weight range. Of interest was the presence of a collagenase-sensitive translation product below the pre-pro α2(I) band (see Fig. 1, asterisk). The identity of this polypeptide is presently unknown, but it may represent one of the minor cartilage collagens (Gibson et al. 1982).

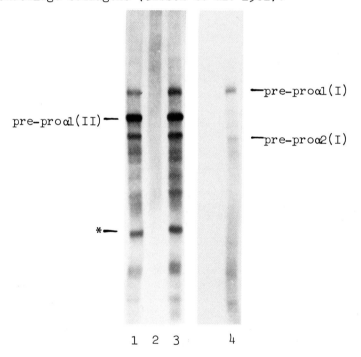

Fig. 1. Polyacrylamide gel electrophoresis of cell-free translation products. Lane 1: Products obtained with 0.07 μg sternal cartilage mRNA. Lane 2: Bacterial collagenase digestion of products obtained with 0.07 μg sternal cartilage mRNA. Lane 3: Products obtained with 0.035 μg sternal cartilage mRNA. Lane 4: Products obtained with 1 μg calvarial bone mRNA.

cDNA was synthesized and amplified by molecular cloning as described above. Plasmid DNA was isolated from 179 transformants and analyzed by agarose gel electrophoresis. Of those screened, 48 recombinants with inserts of 150-800 base-pairs were selected for further analysis. When their DNA was examined by dot hybridization using ^{32}P-labelled poly(A$^+$) RNA from sterna and calvaria, four recombinants hybridized strongly to sternal RNA but showed no hybridization with calvarial RNA (Fig. 2). Further analysis of these four recombinants, including detailed restriction enzyme mapping and nucleotide sequencing, demonstrated that the four recombinants were identical. Only one of them, pYN40, was therefore used in further experiments.

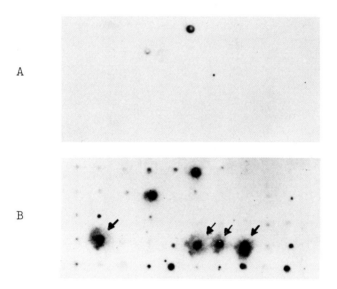

Fig. 2. Dot hybridization of DNA isolated from 48 transformants to ^{32}P-labelled poly(A$^+$) RNA from calvaria (A) and sterna (B). The arrows indicate four identical transformants that showed hybridization with sternal RNA only.

As seen in Fig. 2, two other recombinants showed strong hybridization with both sternal and calvarial RNA. Subsequent analysis, including DNA sequencing, showed that the two recombinants were identical. One of them, pYN143, was used in further experiments.

Northern-blot analysis showed that pYN40 hybridized to a sternal cartilage poly(A$^+$) RNA species migrating slightly slower than the RNA species of α1(I) in a 0.8% agarose gel (Fig. 3). pYN143 hybridized to a species of RNA that electrophoresed with a slightly higher mobility than chicken α1(I) RNA. The identity of pYN143 is presently unknown; however, Northern-blot analysis as well as the nucleotide sequence of the pYN143 insert appears to rule out the possibility that it is either an α1(I) or α2(I) cDNA. In fact, one of the 48 clones we have analyzed, pYN156, has been shown by Northern-blot analysis, Southern hybridization to a pro α1(I) cDNA probe, and by nucleotide sequencing to code for the carboxy-terminal portion of the chicken α1(I) carboxyl propeptide and extend through the untranslated 3'-region of the mRNA into the poly A- segment (Showalter et al. 1980; Fuller, Boedtker 1981). This clone, pYN156, is clearly different from pYN143.

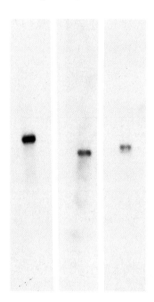

Fig. 3. Northern-blot analysis of pYN40 and 143. Chick sternal cartilage poly(A$^+$) RNA was run on a 0.8% agarose gel and hybridized to α -^{32}P-nick-translated inserts from pYN40 (lane 1), pYN143 (lane 2), and pCg54 (lane 3). pCg54 is a chicken pro α1(I) cDNA clone generously provided by Dr. H. Boedtker.

We believe the insert of pYN40 represents a partial copy of α1(II) collagen mRNA for several reasons:

1. pYN40 hybridizes to a species of RNA present in sternal cartilage but absent in calvarial bone.

2. The size of the mRNA hybridizing to pYN40 is consistent with its coding for a polypeptide of pro α1(II) size.

3. pYN40 does not hybridize to RNA from mesenchymal cells obtained from 4-day old chick embryo limb buds, but it does hybridize to RNA extracted from limb bud cells allowed to differentiate into chondroblasts in mass culture (S. Argraves and P. Goetinck, personal communication, data not shown).

4. Finally, a detailed restriction map of pYN40 (Fig. 4) shows remarkable similarity to a cartilage cDNA, pCAR2, reported by W. Upholt and his co-workers (Vuorio et al. 1982). In addition, comparison of a partial sequence determined for pYN40 (Fig. 5) and a partial sequence of pCAR2 showed complete identity over a stretch of 58 contiguous nucleotides (W. Upholt, personal communication).

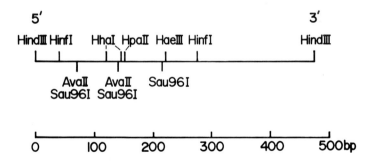

Fig. 4. Restriction map of the insert of pYN40.

```
AGCTTGGTTTTTTTTTTTAAAAAAAGAAAGGAATCCAGCCCAATCCCATAAA
        10        20        30        40        50

AGCAAACCAGTCCCACCCCTAGGACCCGCACGTTCCCAGCACAACTTCTG
        60        70        80        90        100

CACTGAACGGATGGCACGACCCCGCGCCCCTTCGGACCCTCCGGCGCCGT
       110       120       130       140       150

CACCGGGCAGACTGCGAAATACAACCACGGGCTTATATTTATTTATTGCC
       160       170       180       190       200

TTCCTGGAAGGCTTGG - 134bases - GTATCATACAACAGAAAAGG
       210                              360       370

AGCATTACAGTAAAACAAGTCTGTATTTTTAACAACAGTTGATATAAAAA
       380       390       400       410       420

CAACAAAAAAAAAATTCTTTTGGTGGAAAGTAAAAAAAAAAAAAAAAAAAA
       430       440       450       460       470

AAAAAAAAAAAACCA
       480
```

Fig. 5. Partial nucleotide sequence of the insert of pYN40.

Since pCAR2 has been demonstrated to represent a portion of the chicken α1(II) collagen gene through isolation and partial sequence analysis of a genomic α1(II) DNA fragment (W. Upholt, data presented at this meeting), this indicates that pYN40 also represents a Type II collagen cDNA. The insert of pYN40 appears to include most of the untranslated 3'-region of the α1(II)mRNA and part of the poly A-segment. The 5'-end of the insert probably starts several bases downstream from the translation stop codon of the mRNA. pYN40 should therefore be a useful probe for analysis of Type II collagen mRNA levels in cells as well as a probe for further isolation of chicken Type II collagen gene sequences.

ACKNOWLEDGEMENTS

We thank Ms. M. Gordon for generous assistance with the cell-free translation and Mr. G. Vasios for providing mRNA from chick embryo calvaria. This study was supported in part by research grant AM 21471 from the National Institutes of Health of the United States Public Health Service and by a Johnson and Johnson graduate student fellowship (to A.M.S.).

REFERENCES

Adams SL, Sobel ME, Howard BH, Olden K, Yamada KM, de Crombrugghe B, Pastan I (1977). Levels of translatable mRNAs for cell surface protein, collagen precursors and two membrane proteins are altered in Rous sarcoma virus-transformed chick embryo fibroblasts. Proc Natl Acad Sci USA 74:3399.

Bornstein P, Sage H (1980). Structurally distinct collagen types. Annu Rev Biochem 49:957.

Burgeson RE, Hebda PA, Morris NP, Hollister DW. (1982). Human cartilage collagens. Comparison of cartilage collagens with human type V collagen. J Biol Chem 257:7852.

Cox RA (1968). The use of guanidinium chloride in the isolation of nucleic acids. In Grossman L, Moldave K (eds): "Methods in Enzymology", Vol 12(B), Academic Press, p 120.

Davis RW, Botstein D, Roth JR (1980). Advanced bacterial genetics. Cold Spring Harbor Laboratory, p 227.

Fuller F, Boedtker H (1981). Sequence determination and analysis of the 3' region of chicken pro-$\alpha 1$(I) and pro-$\alpha 2$(I)collagen messenger ribonucleic acids including the carboxy-terminal propeptide sequences. Biochem 20:996.

Gibson GJ, Schor SL, Grant ME (1982). The effects of matrix macromolecules on chondrocyte gene expression. J Cell Biol, in press.

Goodman HM, MacDonald RJ (1979). Cloning of hormone genes from a mixture of cDNA molecules. In Wu R (ed): "Methods in Enzymology", Vol 68, Academic Press, p 75.

Holmes DS, Quigley M (1981). A rapid boiling method for the preparation of bacterial plasmids. Anal Biochem 114:193.

Kafatos FC, Jones CW, Efstratiadis A (1979). Determination of nucleic acid sequence homologies and relative concentrations by a dot hybridization procedure. Nucleic Acids Res 7:1541.

Maniatis T, Jeffrey A, Kleid DG (1975). Nucleotide sequence
of the rightward operator of phage λ . Proc Natl Acad Sci
USA 72:1184.

Maxam A, Gilbert W (1977). A new method for sequencing DNA.
Proc Natl Acad Sci USA 74:560.

Maxam A, Gilbert W (1980). Sequencing end-labelled DNA with
base-specific chemical cleavages. In Grossman L, Moldave K
(eds): "Methods in Enzymology", Vol 65, Academic Press,
p 499.

Merlino GT, Vogeli G, Yamamoto T, de Crombrugghe B, Pastan I
(1981). Accurate in vitro transcriptional initiation of
the chick α 2(type I) collagen gene. J Biol Chem 256:11251.

Myers JC, Chu ML, Faro SH, Clark WJ, Prockop DJ, Ramirez F
(1981). Cloning a cDNA for the pro-α2 chain of human type I
collagen. Proc Natl Acad Sci USA 78:3516.

Ohkubo H, Vogeli G, Mudryj M, Avvedimento VE, Sullivan M,
Pastan I, de Crombrugghe B (1980). Isolation and characterization of overlapping genomic clones covering the chick
α2(type I) collagen gene. Proc Natl Acad Sci USA 77:7059.

Olsen BR (1981). Collagen biosynthesis. In Hay ED (ed):
"Cell biology of extracellular matrix", Plenum Publ Corp,
p 139.

Reese CA, Mayne R (1981). Minor collagens of chicken hyaline
cartilage. Biochem 20:5443.

Showalter AM, Pesciotta DM, Eikenberry EF, Yamamoto T, Pastan I,
de Crombrugghe B, Fietzek PP, Olsen BR (1980). Nucleotide
sequence of a collagen cDNA-fragment coding for the carboxyl end of pro α1(I)-chains. FEBS Lett 111:61.

Thomas PS (1980). Hybridization of denatured RNA and small
DNA fragments transferred to nitrocellulose. Proc Natl Acad
Sci USA 77:5201.

Vogeli G, Ohkubo H, Sobel ME, Yamada Y, Pastan I,
de Crombrugghe B (1981). Structure of the promoter for
chicken α2 type I collagen gene. Proc Natl Acad Sci USA 78:
5334.

Von der Mark K, Van Menxel M, Wiedemann, H (1982). Isolation
and characterization of new collagens from chick cartilage.
Eur J Biochem 124:57.

Vuorio E, Sandell L, Kravis D, Sheffield VC, Vuorio T, Dorfman
A, Upholt WB (1982). Construction and partial characterization of two recombinant cDNA clones for procollagen
from chicken cartilage. Nucleic Acids Res 10:1175.

Wozney J, Hanahan D, Morimoto R, Boedtker H, Doty P (1981).
Fine structure analysis of the chicken pro α2 collagen gene.
Proc Natl Acad Sci USA 78:712.

Limb Development and Regeneration
Part B, pages 193–201
© **1983 Alan R. Liss, Inc., 150 Fifth Avenue, New York, NY 10011**

ULTRASTRUCTURAL AND BIOCHEMICAL PROPERTIES OF LIMB
CARTILAGE IN MICE WITH CHONDRODYSPLASIA

Robert E. Seegmiller

Department of Genetics and
Developmental Biology
Brigham Young University
Provo, UT 84602

Chondrodysplasia in mice is being studied as a model
for certain forms of human skeletal dysplasia, e.g.,
recessively inherited diastrophic dwarfism (Sillence et
al., 1979). This study shows promise not only of
elucidating the mechanism of certain skeletal dysplasias,
but also of increasing our understanding of gene
controlled growth and organization at the tissue, cell and
molecular levels of normal endochondral bone formation.
The present communication reviews our current
understanding of the molecular aspects of chondrodysplasia
in mice and presents observations made with the scanning
electron microscope which confirm the matrix defect.

The gene arose spontaneously in the mid-sixties in
the C57BL strain of mice. It is lethal, recessively
inherited and expressed in the extracellular cartilage
matrix of the appendicular and axial skeleton, trachea,
lower jaw and other hyaline cartilage (Seegmiller et al.,
1971; Seegmiller et al., 1972; Seegmiller and Fraser,
1977). The tubular bones of newborn cho/cho mice are
generally wider at the metaphyses and are approximately
half the length of normal bones (Table 1, Fig. 1). The
proliferative and hypertrophic cells of the cartilage
epiphyses of cho/cho mice are not aligned into columns but
are either disorganized or arrayed perpendicular to the
longitudinal axis of the bone.

The matrix shows decreased metachromasia when the
tissue is fixed in aqueous solutions (Seegmiller et al.,
1971). It is suggested that the staining difference is

TABLE 1   COMPARISON OF NORMAL AND CHO SKELETAL PHENOTYPES

	Length[1]			Width[1,2]		
	Normal	Mutant	Ratio[3]	Normal	Mutant	Ratio[3]
FORELIMB						
Humerus	2.6	1.3	0.5	1.0	1.1	1.1
Radius	2.4	1.2	0.5	0.3	0.6	2.0
Ulna	3.0	1.7	0.6	0.5	0.7	1.4
Metacarpals	0.4	0.2	0.5	0.2	0.3	1.5
HINDLIMB						
Femur	2.3	1.0	0.4	0.8	1.1	1.4
Tibia	2.6	1.4	0.5	0.6	1.0	1.7
Fibula	2.6	1.3	0.5	0.3	0.4	1.3
Metatarsals	0.7	0.3	0.4	0.3	0.3	1.0
RIBS						
a)	4.0	2.2	0.6	0.3	0.3	1.0
b)	3.9	2.2	0.6	0.3	0.3	1.0
c)	3.6	2.0	0.6	0.3	0.3	1.0
d)	3.2	1.7	0.5	0.3	0.3	1.0
e)	2.7	1.5	0.6	0.2	0.3	1.5
f)	2.2	1.1	0.5	0.2	0.2	1.0
g)	1.6	0.8	0.5	0.2	0.2	1.0

[1] Average ossified portion of tubular bones of two normal and two mutant day-18 fetuses in millimeters.
[2] Proximal end of bone.
[3] Ratio of mutant to normal.

due to leaching of proteoglycan during the fixation
procedure (Stephens and Seegmiller, 1976).

Under the electron microscope the mutant's cartilage
matrix shows atypically large, banded collagen fibrils and
a paucity of the typical smaller fibrils (Seegmiller et
al., 1971, 1972; Monson and Seegmiller, 1981). The
collagen fibrils are associated with the usual number of
ruthenium red stained granules (presumed to be
proteoglycan), but the granules are smaller than normal
and the pericellular space observed in the normal is
absent in cho/cho mice. Observations made with the
transmission electron microscope have been confirmed with
the scanning electron microscope (Monson and Seegmiller,
1981; Seegmiller and Monson, 1982). The matrix of the
mutant does not show well defined lacunae and lacks the
homogeneous network of delicate collagen fibrils observed
in normal cartilage (cf. Figs. 2a-b and 3a-b).

All of these findings support our original hypothesis
(Seegmiller et al., 1971) that the mutant's matrix lacks
the cohesiveness and rigidity necessary to promote column
formation of the mitotically active cells of the
proliferative zone. Thus, the absence of alignment of the
mutant's proliferative cells into longitudinal columns can
be attributed to a structural defect of the extracellular
matrix. All other aspects of the syndrome appear to be
secondary to the matrix defect.

Considering that for the most part the components of
the extracellular matrix are posttranslational products of
the chondrocytes, we have studied several of these
products biochemically to determine their normality.
Proteoglycan was first examined and found to be normal
both quantitatively and qualitatively (Stephens and
Seegmiller 1976). More recently we reported that the
individual constituents of proteoglycan, viz., chondroitin
sulfate, keratan sulfate, and core protein, are normal
(Seegmiller et al., 1981). Likewise, there is preliminary
evidence that the hyaluronic acid content of the mutant's
cartilage is normal and that proteoglycan when extracted
with guanidine chloride aggregates with commercial
hyaluronic acid (Myers and Simmons, unpublished).
Furthermore, collagen made by mutant chondrocytes is type
II and normally hydroxylated and glycosylated (Seegmiller
et al., 1981). Thus, biochemical studies provide evidence

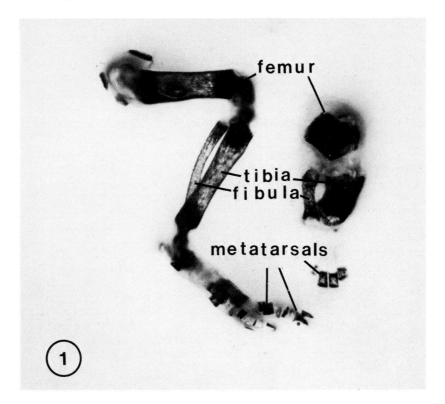

Fig. 1. Comparison of hindlimb skeletons of alizarin
red-stained day-18 normal (left) and mutant (right)
littermates illustrates the extreme shortening and
increased diameter of the mutant's tubular bones. X 13.

Fig. 2. Scanning electron micrographs of tibial cartilage
epiphyses from day-18 littermate fetuses. In the normal
(a) the lacunae are well defined and show considerable
depth whereas in the mutant (b) they are poorly defined
and quite shallow. The matrix of normal cartilage appears
more compact than that of the mutant. X 550.

Fig. 3. Scanning electron microscopy at higher
magnification again shows the poorly defined lacunae and
coarse fibrillar network of collagen of the mutant's
matrix (b) when compared with that of control (a).
X 2200.

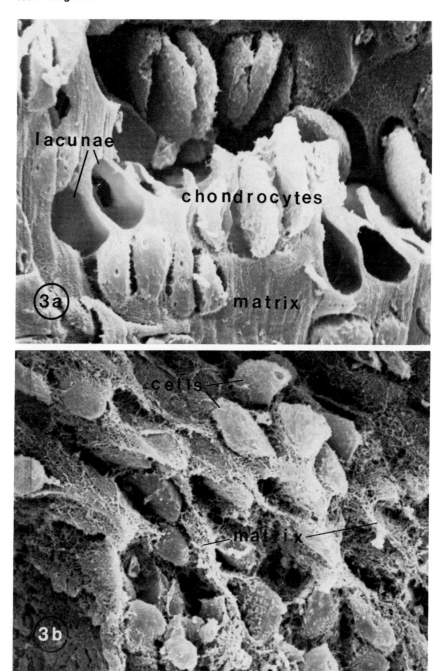

that the major macromolecules synthesized by cho/cho chondrocytes are quantitatively and qualitatively normal.

One concern has been whether the matrix defect is related to some humoral (hormonal) factor. Experiments designed to test this hypothesis involved culturing whole limb buds according to the method of Kochhar and Aydelotte (1974). The limbs were excised from day-11, prechondrogenic embryos which were offspring of heterozygous cho/+ parents. After six days in culture all limb buds showed toluidine blue-positive, cartilaginous skeletal rudiments. However, in approximately one in four limb buds (a ratio expected for homozygosity of a recessive gene) the cartilage matrix was histologically defective (cf. Fig. 4a-b). These results suggest that the mutant gene acts within the chondrocyte rather than through extrinsic factors.

The intriguing question remains: how does the cho gene express itself as a structurally defective matrix without the major matrix constituents being quantitatively or qualitatively altered. Certainly other components of the cartilage matrix need to be examined, viz. chondronectin, a protein believed to play a role in the association of chondrocytes with collagen fibrils (Hewitt et al., 1980), and link protein, a substance believed to stabilize the association of proteoglycan with hyaluronic acid (Hardingham, 1979). Another concern is whether the products of chondrocytes are being degraded prematurely by extracellular enzymes such as cathepsins. Certainly these and possibly other facets of chondrogenesis need to be explored before we will fully understand the cho/cho phenotype.

ACKNOWLEDGEMENTS

This study was supported in part by NIH Grant AM-21631. The assistance of Douglas R. Seegmiller in the morphometric analysis and John S. Gardner in the SEM analysis is greatfully acknowledged.

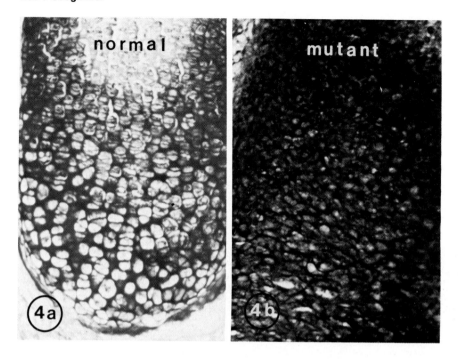

Fig. 4. Light micrographs of toluidine blue-stained limb
cartilage cultured <u>in vitro</u> from prechondrogenic, day-11,
whole limb buds. In the control (a) the lacunae are well
defined and the proliferative and hypertrophic
chondrocytes are aligned into columns. In the apparent
mutant (b) the lacunae are poorly defined, cells are not
aligned into columns, and the matrix is more hetero-
geneously stained compared with control. X 120.

REFERENCES

Hardingham TE (1979). The role of link-protein in the
    structure of cartilage proteoglycan aggregates.
    Biochem J 177:237.
Hewitt AT, Kleinman HK, Pennypacker JP, Martin GR (1980).
    Identification of an adhesion factor for
    chondrocytes. Proc Natl Acad Sci USA 77:385.
Monson CB, Seegmiller RE (1981). Ultrastructural studies
    of cartilage matrix in mice homozygous for chondro-
    dysplasia. J Bone Jt Surg 63A:637.
Seegmiller RE, Ferguson CC, Sheldon H (1972). Studies on
    Cartilage VI. A genetically determined defect in
    tracheal cartilage. J Ultr Res 38:288.
Seegmiller RE, Fraser FC (1977). Mandibular growth
    retardation as a cause of cleft palate in mice
    homozygous for the chondrodysplasia gene. J Embryol
    Exp Morph 38:277.
Seegmiller RE, Fraser FC, Sheldon H (1971). A new
    chondrodystrophic mutant in mice. Electron
    microscopy of normal and abnormal chondrogenesis. J
    Cell Biol 48:580.
Seegmiller, RE, Monson CB (1982). Scanning electron
    microscopy of cartilage in mice with hereditary
    chondrodysplasia. Scanning Electron Microscopy,
    in press.
Seegmiller RE, Myers RA, Dorfman A, Horwitz AL (1981).
    Structural and associative properties of cartilage
    matrix constituents in mice with hereditary
    chondrodysplasia (cho). Conn Tissue Res 9:69.
Sillence DO, Horton WA, Rimoin DL (1979). Morphologic
    studies in the skeletal dysplasias. Am J Pathol
    96:813.
Stephens TD, Seegmiller RE (1976). Normal production of
    cartilage glycosaminoglycan in mice homozygous for
    the chondrodysplasia gene. Teratology 13:317.

Limb Development and Regeneration
Part B, pages 203–213
© 1983 Alan R. Liss, Inc., 150 Fifth Avenue, New York, NY 10011

DIFFERENTIATION OF CARTILAGE AND BONE IN A MUTANT MOUSE
DEFICIENT IN CARTILAGE-SPECIFIC PROTEOGLYCANS

D.M. Kochhar, Ph.D. and John D. Penner, B.A.

Department of Anatomy
Jefferson Medical College
1020 Locust Street
Philadelphia, Pa. 19107

Neonatal mice homozygous for a recently described
mutation cartilage-matrix-deficiency (cmd/cmd) are
afflicted with a syndrome which includes short-limbed
dwarfism, cleft palate, short snout, and protruding abdomen
(Rittenhouse et al. 1978) (Fig. 1). The only molecular
abnormality found so far associated with this autosomal
recessive mutation is that cartilage-specific proteoglycans
in the fetal limbs are drastically reduced in amount and
that this is due to perhaps an absence or alteration in
the core-protein molecule (Kimata et al. 1981). Since
there are few opportunities to study the in situ developmental role of the extracellular matrix components, the
overall aim of our studies was to use this mutation as a
basis for drawing some conclusions regarding the role of
embryonic glycosaminoglycans (GAG) in limb morphogenesis.

Rittenhouse et al. (1978) reported that the newborn
appendicular skeleton was less than one-half of the normal
length. We found that the stunting of limbs as well as a
slight shortening of the trunk and cephalic dimensions
were present throughout the fetal stages, and in fact,
these external characteristics were used to identify the
mutant as early as on the 15th day of gestation. Further
differences between normal and mutant phenotypes were
brought out after staining of the fetal skeleton with
either alizarin red S or alcian blue dyes. Normal fetal
cartilaginous skeleton begins to stain positively with
alcian blue on the 14th day of gestation. In similarly
prepared mutant embryos the outline of cartilaginous
skeleton was also clearly visible but it failed to stain

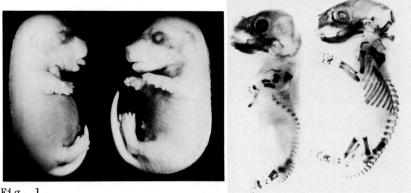

Fig. 1.
Near-term cmd/cmd and normal littermates, fixed in Bouin's
solution (left) or cleared in KOH and stained with
alizarin red S (right) showing a syndrome of defects
(see text).   2.5 X.

as densely as that of the normal littermates (Fig. 2).
The lack of staining was apparent in all parts of the
skeleton - chondrocranium, Meckel's cartilage, girdles
and limbs, vertebrae, ribs and tail.

The purpose of our investigation was to study the
mutant limb development during the period of initial
outgrowth and during precartilaginous and early chondrogenic stages so as to learn if there were other cellular or
biosynthetic derangements associated with this mutation.
The initial problem, of course, was to first identify the
homozygous mutant embryos at a stage before the onset of
overt changes in the developing cartilage. Neither
external features nor alcian blue staining were sufficient
to identify mutant embryos on the 12th and the 13th day of
gestation - at stages when no cartilage is as yet present.
These embryos were identified through the use of our organ
culture assay (Kochhar and Aydelotte 1974). A limb, a
segment of the tail, or a small clump of somites were
organ cultured on a nutrient medium for 5-6 days to a
point when well-developed cartilage with abundant alcian
blue-stainable matrix developed in the explant. As
expected, the explants derived from some of the 11th-13th
day embryos failed to show positive staining of their
cartilage matrix although grossly identifiable cartilage-
like tissue was present (Fig. 3).

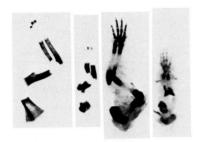

Fig. 2.
Fetal forelimbs (18th day)
from normal and cmd/cmd littermates stained with alizarin red
S (left) or alcian blue (right)
to compare extents of osteogenesis and chondrogenesis.
4 X.

Fig. 3
Embryonic forelimb buds of
13th day normal (left) and
cmd/cmd (right) littermates
cultured in vitro for 6
days and stained with al-
cian blue. Note a lack of
staining in cmd/cmd
cartilage. 15 X.

A total of 40 litters were collected on various
gestational days between the 11th to the 19th and the
phenotype of each embryo ascertained on the basis of
external features, alcian blue staining of the skeleton
or of the cultured tissue explants. Out of a total of 420
live embryos and fetuses recovered, 316 were normal, while
104 (25%) were cartilage-deficient; the latter were tenta-
tively identified as cmd/cmd. The homozygous condition
did not entail excessive prenatal death since the number
of fetuses per litter (10.5) or the ratio of normal to cmd
(3:1) did not vary greatly from gestational days 11th –
19th.

The organ culture assay was thereafter employed for
all biosynthetic studies described below.

Limb buds of proven cmd/cmd embryos and those of
normal littermates were labeled in organ culture with
^{3}H-proline for protein and collagen synthesis or with
^{35}S-sulfate for sulfated glycosaminoglycan (s-GAG)
synthesis (see Linsenmayer and Kochhar 1979 for methods).
The labeling period was either during the first three
days of culture or during the later stages of culture (3-6
days) when considerable amounts of cartilage forms in

Table 1

TOTAL PROTEIN, PROTEIN SYNTHESIS, AND COLLAGEN SYNTHESIS. LIMB BUDS OF NORMAL
AND MUTANT (CMD/CMD) LITTERMATES, EXPOSED TO ^{3}H-PROLINE IN ORGAN CULTURE.

		TOTAL PROTEIN µG	PROTEIN SYNTHESIS DPM/µG PROTEIN	COLLAGEN SYNTHESIS DPM/µG PROTEIN
(BLASTEMA STAGE, 0-3 DAY)	NORMAL	72 ± 10	26,680	5951
	MUTANT	81 ± 12	25,118	6352
(CHONDROGENIC STAGE, 3-6 DAY)	NORMAL	269 ± 22	27,625	6187
	MUTANT	164 ± 18 (39%↓)	21,426 (22%↓)	4552 (26%↓)

vitro. The growth of mutant limb buds, as determined by
total protein content, kept pace with normal limb buds at
first but lagged considerably behind the normal by six days
of culture at which time they had about 40% less protein.
Although overall protein synthesis was depressed in the
mutant limb buds, their collagen synthesis was not greatly
affected (Table 1). During the early blastema stage, the
total s-GAG synthesis in cmd/cmd was only slightly reduced
compared to normal limbs. The reduction, however, was
substantial – up to 73% – at advanced stages of chondrogenesis (Table 2). The depression was found only in the
synthesis of chondroitin sulfates, while dermatan sulfate
and other components (resistent to digestion by chondroitinase ABC) were, in fact, present at higher levels in
the mutant during the chondrogenic stages (Table 2).

Table 2

SULFATED GAG SYNTHESIS. LIMB BUDS OF NORMAL AND MUTANT (CMD/CMD)
LITTERMATES. EXPOSED TO ^{35}S-SULFATE IN ORGAN CULTURE.

		TOTAL ^{35}S-GAG CPM/4 LIMBS	CHONDROITIN SULFATES-4,6	DERMATAN SULFATE	RESIDUAL ^{35}S-GAG
(BLASTEMA, 0-3 DAY)	NORMAL	7755	53%	31%	16%
	MUTANT	6904 (21%↓)	62%	20%	17%
(CHONDRO- GENESIS 3-6 DAY)	NORMAL	64,006	85%	11%	4%
	MUTANT	14,575 (73%↓)	57%	28%	14%

Further studies on GAG synthesis revealed that the depression was not due to a sulfation defect in the mutant since similar depression was also found when experiments were repeated by using either ^{3}H-glucosamine or ^{3}H-acetate as labels (Tables 3 and 4).

One important outcome of studies on GAG synthesis was an indication that the level of synthesis in some of the phenotypically normal embryos was lower than that in the remaining normal embryos; in fact, the level was intermediate between the mutant and the normal limbs. Two groups of cmd/cmd and three groups of phenotypically normal embryos were used as donors for limb bud cultures which were labeled with ^{3}H-glucosamine during the chondrogenic stage. The samples were digested with protease, dialyzed and run on DEAE-52 (Whatman) columns to separate hyaluronate (HA) and S-GAG fractions (Silbert and DeLuca 1969). Twenty thousand counts were loaded on the columns in each instance. The fractions under each peak were pooled and counted, and ratios of counts between HA and S-GAG computed (Table 3). As anticipated, the HA: S-GAG ratios in cmd/cmd limbs were always lower than in each of the normal groups. In addition, the ratios in the three normal groups were different from each other indicating different levels of biosynthetic activity.

In a subsequent experiment we used ^{3}H-acetate to label limb buds from several litters of 13th day embryos. Since cmd/cmd animals always have pigmented eyes and coat-color (with the exception of two instances where mutant embryos with unpigmented eyes were encountered from among more than 700 heterozygous matings carried out so far in our laboratory), labeled limb buds in normal groups were pooled according to eye-color as shown (Table 4). According to the data, early chondrogenic stage was characterized by HA: S-GAG ratio of 1:2 in the normal and 1:1 in cmd/cmd. In the late chondrogenic stage, the ratio did not change in cmd/cmd but increased to 1:3 or 4 in favor of S-GAG in the normal limbs. It should be noted that in either instance phenotypically normal embryos with pigmented eye color had lower HA: S-GAG ratios than presumably +/+ embryos with unpigmented eyes. In the mixed sample which contained limbs from both types of normal embryos, the ratio was intermediate. We suggest that donors of a majority of limbs in our pigmented-eye samples were carriers of the recessive gene hence could be denoted as +/cmd.

Table 3

3H-GLUCOSAMINE-LABELED GAG FROM CULTURED
LIMB BUDS.   LABELING FOR 48 HR. (5-7 DAY OF
CULTURE).   ALL EMBRYOS WITH PIGMENTED EYES.

COUNTS IN PEAKS SEPERATED
BY DEAE-CELLULOSE COLUMNS
(20,000 COUNTS LOADED)

PHENOTYPE	GLYCO-PEPTIDES	HA	S-GAG	RATIO HA : S-GAG
CMD/CMD	2391	6901	9277	1 : 1.34
CMD/CMD	2287	6317	9713	1 : 1.54
NORMAL	1926	5086	11,540	1 : 2.27
NORMAL	1371	4465	13,024	1 : 2.92
NORMAL	1686	4194	12,813	1 : 3.06

GAG synthesis and chondrogenic expression were also
monitored in high density, micromass cultures of limb bud
mesenchymal cells obtained from cmd/cmd and phenotypically
normal embryos (see Ahrens et al. 1977 for method).   These
studies confirmed our contention that heterozygous embryos
have reduced ability to synthesize enhanced levels of S-GAG
typically associated with normal embryos (Fig. 4). In
addition, we found that cmd/cmd cells, although lacking the
ability to synthesize alcian blue-stainable cartilage
matrix, were still fully capable of undergoing cell-cell
aggregation forming precartilaginous nodules in the high-
density spots.

Table 4

3H-ACETATE-LABELED GAG FROM CULTURED LIMB BUDS

COUNTS IN PEAKS SEPERATED
BY DEAE-CELLULOSE COLUMNS
(20,000 COUNTS LOADED)

CHONDROGENIC STAGE	PHENOTYPE	EYE-COLOR	GLYCO-PEPTIDES	HA	S-GAG	RATIO HA : S-GAG
EARLY	CMD/CMD	PIGMENTED	3931	7722	8354	1 : 1.08
(DAY 1-2)	NORMAL	PIGMENTED	2250	7192	10,556	1 : 1.57
	NORMAL	UNPIGMENTED	2919	5513	11,548	1 : 2.09
LATE	CMD/CMD	PIGMENTED	5115	8410	6506	1 : 0.77
(DAY 5-6)	NORMAL	PIGMENTED	2702	5609	11,696	1 : 2.09
	NORMAL	MIXED	2375	4507	13,090	1 : 2.90
	NORMAL	UNPIGMENTED	1717	3869	14,416	1 : 3.73

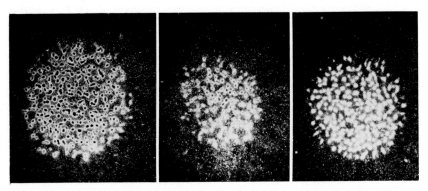

Fig. 4. High density spot cultures of limb bud mesenchymal cells from phenotypically normal (left and middle) and cmd/cmd (right panel) embryos. Reduced alcian blue staining of cartilage nodules (middle panel) permitted us to denote some donor embryos as heterozygous (+/cmd) carriers.

In vivo studies on morphological development showed that the cmd/cmd limb bud progressed normally through early stages of outgrowth including the appearance and spatial organization of mesenchymal cell condensations

Fig. 5.
A section of forelimb from a 13th day cmd/cmd embryo (left), stained with azure II, shows no deviation from normal forelimb (right) in formation and spatial pattern of regions of cell condensations. 40 X.

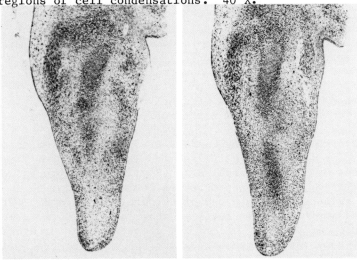

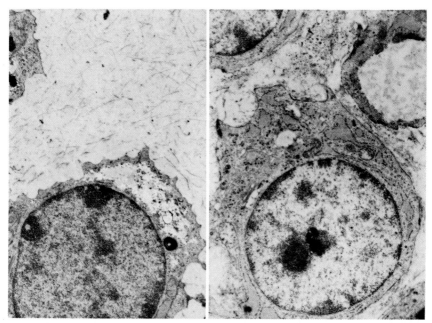

Fig. 6.
Limb cartilage of a normal (left) and a cmd/cmd (right)
embryo from cultures as shown in Fig. 3. Note drastic
reduction in the amount of extracellular matrix in
cmd/cmd cartilage.

(Fig. 5). The first deviation became apparent as soon
as normal chondrogenic cells began to accumulate extracellular matrix; cmd/cmd chondrogenic cells remained close
to each other and lacked the abundant extracellular matrix
which accumulated between normal cells (Fig. 6). It is
important to note that the mutation does not seem to
interfere with the process of determination of cartilage
even though it interrupts virtually completely one aspect
of the chondrogenic cell differentiation.

By the time primary centers of ossification appeared
in the limbs, cmd/cmd cartilage models were already stunted
to about 50% of the normal size (Fig. 7). The process of
ossification, however, progressed normally as far as it
could be ascertained in the histological specimens. Our
initial impression, in fact, is that there is a precocious
formation of bone in the cmd/cmd diaphyses; the presence of
spongy bone was noticed throughout the width of the shaft

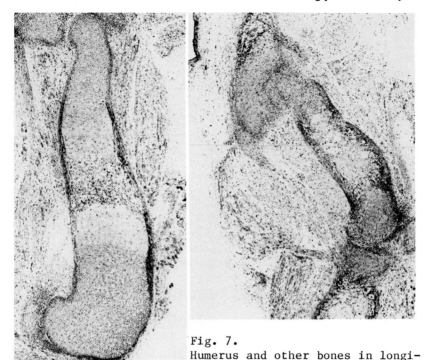

Fig. 7.
Humerus and other bones in longi-
tudinal sections of a normal (left) and a cmd/cmd (right)
limb from 16th day embryos. Note disorganized epiphyses
and smaller size of cmd/cmd cartilage model.

unlike that in the normal limbs at equivalent stages where
its presence was only limited to the periosteum (Fig. 7).

Microscopy revealed another important aspect of the
cmd/cmd cartilage. Embedded as they were in an abnormal
matrix, cmd/cmd chondrocytes began to undergo degenerative
changes typical of calcifying cartilage (Fig. 8). We
surmized that this process was also premature since cmd/cmd
chondrocytes did not first undergo hypertrophy typical of
normal cells.

Summary and Conclusions: Autosomal recessive mutation cmd
(cartilage matrix deficiency) is associated with short-
limbed dwarfism in late fetal and newborn mice. The only
molecular abnormality characterized well so far is in the
core protein of cartilage-specific proteoglycans. Additional-
al derangements were sought in this investigation on early
embryos during the period of limb outgrowth and the onset

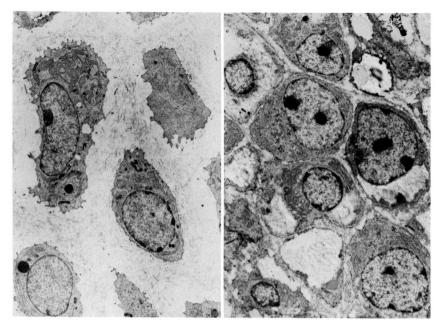

Fig. 8.
Limb cartilage of a normal (left) and a cmd/cmd (right)
16th day embryos. Resting chondrocytes and their surrounding matrix show vacuolation and other signs of degeneration
in the cartilage.

of chondrogenesis. We have found that (1) quantitatively
normal levels of chondroitin sulfate are synthesized by
the mutant limbs during precartilaginous stages, (2) the
ratio of hyaluronate:sulfated glycosaminoglycans
(HA:S-GAG), which is about 1 in both mutant and normal
limbs at the onset of chondrogenesis changes to 3 or more
in favor of S-GAG in the normal but remains essentially
unchanged in the mutant, (3) based on HA:S-GAG ratios,
phenotypically normal embryos may be denoted as either +/+
or +/cmd, (4) although early limb outgrowth and onset of
cell condensation occurs normally in the mutant, cell
proliferation patterns associated with these events need to
be investigated, (5) mutant chondrocytes, embedded as they
are in the proteoglycan-deficient matrix, begin to degener-
ate prematurely without first undergoing hypertrophy, and
(6) the process of ossification begins precociously in the
shortened cartilage models of the mutant, thus resulting in
overall shortening of limbs.

This investigation was supported by N.I.H. Grant #HD-10935. We are indebted to Sandra Parsons for expertly composing this manuscript, and to Barbara Kuczynski for skillful technical assistance.

References

Ahrens PB, Solursh M, Reiter RS (1977). Stage related capacity for limb chondrogenesis in cell culture. Develop Biol 60:69.

Kimata K, Barrach H-J, Brown KS, Pennypacker JP (1981). Absence of proteoglycan core protein in cartilage from the cmd/cmd (cartilage matrix deficiency) mouse. J Biol Chem 256:6961.

Kochhar DM, Aydelotte MB (1974). Susceptible stages and abnormal morphogenesis in the developing mouse limb, analyzed in organ culture after transplacental exposure to vitamin A (retinoic acid). J Embryol Exp Morph 31:721.

Linsenmayer TF, Kochhar DM (1979). In vitro cartilage formation: Effects of 6-Diazo-5-Oxo-L-Norleucine (DON) on glycosaminoglycan and collagen synthesis. Develop Biol 69:517.

Rittenhouse E, Dunn LC, Cookingham J, Calo C, Spiegelman M, Dooher GB, Bennett D (1978). Cartilage matrix deficiency (cmd): a new autosomal recessive lethal mutation in the mouse. J Embryol Exp Morph 43:71.

Silbert JE, DeLuca S (1969). Biosynthesis of chondroitin sulfate III. Formation of a sulfated glycosaminoglycan with microsomal preparation from chick embryo cartilage. J Biol Chem 244:876.

Limb Development and Regeneration
Part B, pages 215–227
© 1983 Alan R. Liss, Inc., 150 Fifth Avenue, New York, NY 10011

MORPHOLOGICAL AND BIOCHEMICAL TRANSFORMATION OF SKELETAL
MUSCLE INTO CARTILAGE

Mark A. Nathanson, Ph.D.

Department of Anatomy
New Jersey Medical School
Newark, NJ   07103

In vivo development is characterized by a high degree
of order and reproducibility. These characteristics are
exemplified during the development and regeneration of limb
tissues, insofar as skeletal muscle and cartilage develop
adjacent to one another, from a pool of similar-appearing
mesenchymal cells, and yet display divergent morphological
and biochemical characteristics. It would seem reasonable
to assume that such a clear-cut difference, as exists between
muscle and cartilage, is due to the action or occurrence of
specific factors. Indeed, a variety of theories have been
postulated to explain the development of limb tissues. These
include theories of myogenic and chondrogenic lineages
(Holtzer *et al* 1975), positional information (Wolpert 1969),
and the biochemical modulation of phenotypic expression.
The array of substances which modulate expression of the
cartilage phenotype is too extensive to list at the present
time. The existance of many theories and biochemical modulators suggests that possibly all are valid in the context
of their description, but that none can fully describe the
means by which embryonic limb mesenchyme forms both muscle
and cartilage.

Several studies suggest that limb differentiation does
not depend upon the presence of cells predetermined for
either muscle or cartilage phenotypes, but that their appearance is governed by local factors which are poorly understood
at the present time. It is possible that many of the biochemical modulators alluded to above constitute these factors
and that their presence conveys the appearance of cells committed to either phenotype. In much the same fashion, the
presence of a cell in a position which exposes it to "factors"

conducive to chondrogenesis would give the appearance of cartilage "fields" within the limb.

Cloneable myoblasts and chondroblasts do not appear until stage 19 in the chick (3-3½ days; Ahrens *et al* 1977). Blocks of cells from the chondrogenic region of stage 24 chick limbs (4½ days) remain as cartilage in only 10-13% of the cases when implanted into the prospective myogenic region prior to stage 25 (4½-5 days; Searls, Janners 1969). While the *in vitro* methodology of the former experiments could alter subsequent cell differentiation, the results of the latter experiments support the conclusion that phenotypic stability is an acquired rather than endowed characteristic. The demonstration that somitic mesoderm contributes to limb musculature and cartilage (Chevalier *et al* 1977), and that 4-day dissociated and reaggregated limb cells could form normal limb tissues (Kieny *et al* 1981), further suggests that limb mesoderm cells are not predetermined with respect to their ultimate fate. They most likely develop from mesoderm of equivalent potency, in response to a set of environmental cues, which appear and exert an effect between stages 19 and 25 in the chick. It is reasonable to assume that these same cues appear during development of mammalian limbs at appropriate stages.

Reports from the laboratories of Urist and Reddi, in which demineralized bone (bone matrix) elicits the formation of hyaline cartilage (*in vitro)* and bone *(in vivo)* from differentiated skeletal muscle, support the hypothesis that muscle tissue retains the innate ability to form cartilage. (Urist 1970; Reddi, Huggins 1975). In these experiments, bone matrix was either implanted into the abdominal musculature of adult rats *in vivo,* or skeletal muscle (triceps) was explanted onto bone matrix *in vitro*. In each case muscle tissue in the vicinity of bone matrix was replaced by bone or bone and cartilage. Such tissue would never give rise to cartilage during normal development.

This investigator has utilized the *in vitro* methodology to determine the nature of cells responding to bone matrix and to elucidate the patterns in which cartilage-specific extracellular matrix macromolecules are produced by a non-chondrogenic source tissue. The working hypothesis is that these patterns will accurately mimick those which occur during normal chondrogenesis and lead to an understanding of conditions which evoke the differentiation of cartilage versus skeletal muscle. The results of some of these investigations are discussed in this chapter.

Initial experiments were designed to answer two questions: 1) does hyaline cartilage arise from muscle tissue or from its associated fibrous connective tissue? and 2) does the ability to form cartilage reside solely within somatic mesoderm? Skeletal muscle cells and fibroblasts from skeletal muscle were grown in clonal culture and after a suitable growth interval the clones were transferred to bone matrix. Bone matrix was prepared and used as described by Urist (see: Nathanson *et al* 1978). For technical reasons the skeletal muscle was isolated from 11-day chick embryos (stage 37) and fibroblasts from 19-day rat embryo skeletal muscle. On bone matrix *in vitro*, both cell types formed cartilage. Insofar as the source tissue was derived from a stage well past that at which skeletal muscle is thought to be terminally differentiated (*ie.* stage 25), these results demonstrate that differentiated cells are capable of altering their phenotype when presented with an appropriate stimulus. The transformation is significant towards our understanding of development because it clearly demonstrates the ability of selected cell types to alter biosynthetic patterns which have previously been considered to be a fixed property of a particular cell type. The mesodermally-derived connective tissue capsules of embryonic chick thyroid and lung also formed hyaline cartilage, whereas their endodermally-derived parenchyma did not. These results additionally demonstrate that somatic and visceral mesenchyme are identical with respect to their ability to form cartilage and that "cartilageness" is not a fixed property of a particular tissue.

MORPHOLOGY OF THE CULTURES

The following experiments were performed with minced skeletal muscle derived from 19-day embryonic rats, and dissected free of tendons, nerves, and blood vessels. It is noteworthy that this embryonic tissue is largely composed of clusters of myoblasts surrounding a centrally-located, developing myotube (Nathanson, Hay 1980a). The clusters are surrounded by an open extracellular space, containing only few fibroblasts. Electron microscopic examination of this muscle, grown on bone matrix, has shown that only mononucleate cells survive the trauma of being placed into culture. Some mononucleate cells have slightly heterochromatic nuclei and a cytoplasmic compartment rich in free ribosomes (r; Fig. 1),

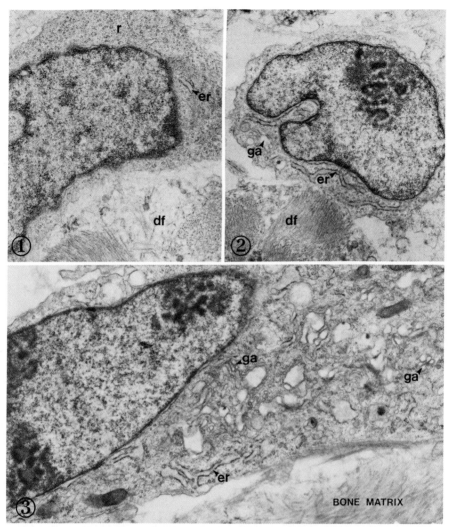

Figs. 1-3. Ultrastructure of skeletal muscle cultured on bone matrix 2-3 days. The source tissue contains mainly myoblasts and myotubes. True myoblasts contain an organelle-poor cytoplasm with abundant free ribosomes (r; Fig. 1). After 2 days on bone matrix, many mononucleate cells now contain endoplasmic reticulum and Golgi apparatus (er, ga; Fig. 2) similar to fibroblasts. After 3 days, mononucleate cells are entirely fibroblast-like (Fig. 3). Acquisition of fibroblast-like morphology plus proximity to degenerating myofilaments (df; Figs. 1,2) demonstrates transformation of myoblasts of the source tissue. (Fig. 1, x10,700; Fig. 2, x9,700; Fig. 3, x16,600. From: Nathanson, Hay 1980a).

but otherwise organelle-poor. These cells have the morphology of myoblasts. However, they are not as numerous as in the uncultured source tissue. As early as 24 hr *in vitro*, and continuing throughout the first two days, an increasing number of mononucleate cells display characteristics intermediate between those of myoblasts and fibroblasts. For example, the cytoplasmic compartment of these cells contains a variable amount granular endoplasmic reticulum (er) and Golgi apparatus (ga; Fig. 2). Their nuclei are largely euchromatic and have dispersed nucleoli. The cell shown in Fig. 2 resembles a fibroblast by virtue of its nuclear and cytoplasmic morphology, but note that its cytoplasmic compartment is small and the nuclear to cytoplasmic area is large; both of these latter characteristics are found in myoblasts. Finally, note that both cells are located near remnants of myotubes (df; Figs. 1,2) which populate the cultures at this stage and appear at the center of myogenic clusters. It is unlikely that the fibroblast-like cells detected at 24 hr *in vitro* could have arisen by proliferation. In the absence of detectable cell death, these observations show that many of the cells, previously appearing as myoblasts, have taken on the characteristics of fibroblasts. After 3 days *in vitro* the cultures consist almost entirely of fibroblast-like cells (Fig. 3). Each of these observations lead one to conclude that the myoblasts acquire traits reminiscent of the mesenchyme from which they arose.

Parallel cultures, serving as controls, consisted of aliquots of the same skeletal muscle mince grown on gels of type I collagen. This muscle was organ cultured in an identical fashion to those explants grown on bone matrix. Control cultures do not display transitions to a fibroblast-like morphology and regenerate skeletal muscle within 4-5 days *in vitro*.

If no further morphogenesis were to occur on bone matrix, then the tissue may have "dedifferentiated," however, beginning on the sixth day *in vitro* chondrocytes are detected (Fig. 4). This observation raises the question whether "dedifferentiation" as detected in cell culture is the same phenomenon as detected during limb regeneration and in the present experiments? It is likely that dedifferentiation actually represents two processes, one occurring in cell culture and representing unproductive growth conditions, and one which occurs in the limb blastema, and on bone matrix, and represents a necessary step in the differentiation of limb

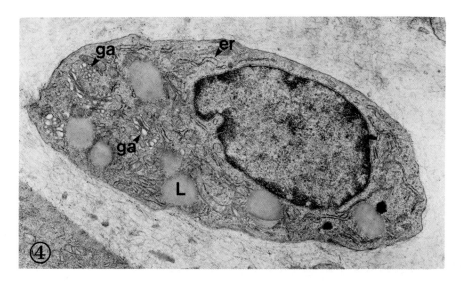

Fig. 4. A typical chondroblast, detected in a 12-day culture
of skeletal muscle on bone matrix. Euchromatic nuclei and a
cytoplasm rich in the secretory organelles endoplasmic reticulum (er) and Golgi apparatus (ga) are prominent features
of these cells. Lipid is often present (L). The extracellular
space is filled with 12-23 nm cartilage-type collagen
fibrils. (x15,100; from: Nathanson 1982).

tissues. The fibroblast-like cells observed in the present
experiments are thus believed to be the first step in the
differentiation of cells destined to become secretory in
function.

EXTRACELLULAR MATRIX SYNTHESIS

The source tissue, embryonic skeletal muscle, and also
embryonic cartilage incorporate labeled sulfate exclusively
into sulfated glycosaminolycans (GAG) (Nathanson, Hay 1980b).
Typically, the major sulfated GAG of cartilage, chondroitin-
4-sulfate (Ch-4-S) and chondroitin-6-sulfate (Ch-6-S) constitute 88% (Ch-4-S) and 6.7% (Ch-6-S) of the total sulfated
material synthesized by this tissue. Skeletal muscle synthesizes these isomeric forms as 17.7% (Ch-4-S) and 62.5%
(Ch-6-S), a clearly opposite pattern of synthesis. The percentages vary slightly between experiments, but the ratio of
Ch-4-S/Ch-6-S for skeletal muscle is quite stable at 0.28.

On bone matrix the ratio rapidly reverses and after only
24 hr *in vitro* the ratio becomes approximately 3.0, with
Ch-4-S accounting for 52% of the sulfated GAG (Fig. 5A).
On a per-cell basis the total sulfate incorporation does
not exceed that of control cultures grown on collagen gels.
These data demonstrate a direct increase in synthesis of
the 4'-sulfated isomer (Nathanson, Hay 1980b). Total
sulfate incorporation and synthesis of Ch-4-S does not rise
appreciably during the first 3 days, but by day 4 total sulfate incorporation increases 3-fold and is followed by a
rapid increase in Ch-4-S synthesis. By the termination of
the experiment at 10 days, Ch-4-S accounts for 75% of the
total sulfated material. Synthesis of Ch-6-S accounts for
17% of the total sulfated material at 24 hr and continues
to decline thereafter to reach 5% at 10 days.

An even more remarkable finding is that control cultures
similarly synthesize Ch-4-S in increased amounts, and Ch-6-S
in decreased amounts, after 24 hr *in vitro* (Fig. 5B). These
cultures cannot sustain this pattern of synthesis and Ch-4-S
levels begin to decline by day 2. Synthesis of Ch-6-S does
not rise, but is maintained between 14-20% of the total sulfated material.

The synthesis of Ch-4-S and Ch-6-S has been found to
approximate that of embryonic cartilage and their appearance
in vitro correlates well with the morphological data at
later time points. An early shift in the synthesis of sulfated GAG, towards Ch-4-S, does not correlate insofar as the
tissue contains only fibroblast-like cells and cells in transition to this phenotype. It is likely that the environmental conditions *per se* were sufficient to alter the ratio
of Ch-4-S/Ch-6-S synthesis towards that of cartilage. Further experiments performed *in vivo* have confirmed that operative trauma alone may elicit such a shift (Nathanson, Hay
1980b), but it is clear that while environmental conditions
may predispose a tissue towards cartilage-typical syntheses,
other factors must augment and stabilize this response during
a period of actual chondrogenesis.

In a separate experiment, a similar series of cultures
were exposed to media containing labeled sulfate plus [3]H-
glucosamine. Total glucosamine incorporation follows a pattern which differs from that of sulfate. Whereas sulfate
incorporation does not increase over control levels during
the initial 24 hr *in vitro,* glucosamine incorporation by

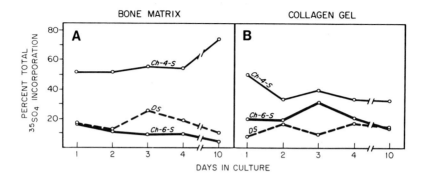

Fig. 5. Incorporation of $^{35}SO_4Na_2$ into chondroitin sulfate isomers by skeletal muscle cultured on bone matrix and collagen gels (Ch-4-S, chondroitin-4-sulfate; DS, dermatan sulfate; Ch-6-S, chondroitin-6-sulfate) (data from: Nathanson, Hay 1980b).

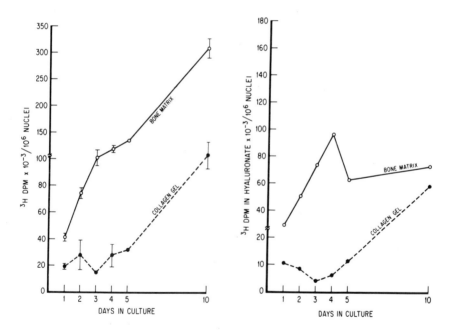

Figs. 6,7. Incorporation of ^{3}H-glucosamine by cultures on bone matrix and collagen gels. Fig.6 (left), total glucosamine incorporation, ± standard error. Fig. 7 (right), hyaluronic acid synthesis determined from chondroitinase digests of material depicted in Fig. 6. (data from: Nathanson 1982).

cultures on bone matrix is twice that of controls (Fig. 6).
From 24 hr onwards, glucosamine incorporation continues to
increase on bone matrix. Control cultures, in contrast,
incorporate increasing amounts of glucosamine only after
3 days. On bone matrix, the synthesis of hyaluronic acid (HA)
initially mimics total glucosamine incorporation insofar as
increasing amounts are synthesized during the first 3 days
in vitro (Fig. 7). However, prior to the period in which
chondrocytes appear, (*ie.* 4-5 days) a decrease in HA synthesis occurs and this decreased level of synthesis is maintained throughout the remainder of the experiment. The
occurrence of this decrease is similar to that detected with
the onset of chondrogenesis in chick vertebrae and axial skeleton (Toole 1972). Control cultures do not decrease their
synthesis of HA, but display a later appearing increase which
follows the pattern of total glucosamine incorporation.
Clearly, a major difference between cultures which will form
cartilage and those which will not occurs in the decrease of
HA synthesis which precedes chondrogenesis in these cultures.

It is well established that cartilage extracellular
matrix contains the sulfated GAG Ch-4-S and Ch-6-S in a bound
form, bound via a protein core to form a proteoglycan monomer.
Proteoglycan monomers are further bound to HA of high molecular weight to form a proteoglycan aggregate (Hascall,
Heinegard 1979). Whereas alterations in sulfated GAG synthesis must involve the monomer and aggregate forms of proteoglycans if they are to relate to the process of chondrogenesis, their appearance in cartilage typical patterns is not
specific to cultures undergoing chondrogenesis. The data do
implicate HA synthesis as a major controlling factor and it
is proposed that alterations in synthesis of this GAG is a
rate-limiting step in proteoglycan synthesis.

Proteoglycans were extracted by dissociative means in
4.0M guanidinium hydrochloride and fractionated by chromatography on Sepharose CL-2B. A control sample of proteoglycan from 19-day embryonic rat sternal cartilage fractionated
into three peaks, a high molecular weight, void-volume peak
which migrates as proteoglycan aggregate, an included peak
of proteoglycan monomer (Fig. 8, arrow at $K_{av}=0.36$), and a
peak of lower molecular weight material, which elutes just
prior to the column total volume and is characteristic of
non-chondrogenic tissue (Fig. 8, arrow at $K_{av}=0.87$) (see:
Palmoski, Goetinck 1973). Similar extraction of the source
tissue, embryonic rat skeletal muscle, results in the appearance of a single, low molecular weight peak (Fig. 8, arrow at
$K_{av}=0.87$).

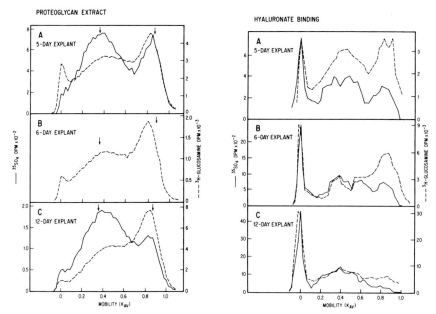

Figs. 8,9. Proteoglycan synthesis and binding of proteoglycans to added hyaluronic acid. Sepharose CL-2B chromatograms.
Fig. 8 (left), proteoglycans synthesized by 5,6 and 12-day
cultures of skeletal muscle on bone matrix. Arrows indicate
elution of sternal cartilage monomer (Kav=0.36) and low molecular weight proteoglycans (Kav=0.87). Fig. 9 (right),
elution profile of aliquots of proteoglycans shown in Fig. 8
after addition of Cock's Comb hyaluronic acid. (Nathanson,
manuscript in preparation).

Explants grown on bone matrix for 5 days contain material which elutes in three peaks and which have mobilities
similar to those detected for sternal cartilage proteoglycan
(Fig. 8A). It is clear that the source tissue not only alters
its synthesis of sulfated GAG, but begins to synthesize a
proteoglycan monomer which is not found in skeletal muscle.
While a shift from 4'-sulfation to 6'-sulfation could be
interpreted as a nonspecific inhibition of 6'-sulfation, the
data presented above, plus the synthesis of cartilage-typical
proteoglycans, evidences the genetic and biosynthetic changes
which accompany the transformation of the source tissue.
Most of the sulfated GAG (65% of the total sulfated material)
occurs in the monomer fraction and the formation of this

monomer cannot be rate-limiting for proteoglycan synthesis.
Data presented in Fig. 8 also demonstrate the new synthesis
of high molecular weight ^{3}H-glucosamine-containing material,
presumably HA. This difference is again indicative of chondrogenesis and a clear departure from the biosynthetic
specificity of skeletal muscle. Compared to sternal cartilage, the 5-day explants contain less material migrating at
the column void volume and greater amounts which migrate at
the total volume. This pattern appears intermediate between
those of skeletal muscle and cartilage and suggests that the
proteoglycan is somewhat immature.

After 12 days on bone matrix the synthesis of proteoglycans remains essentially unchanged, with monomer still
accounting for the greatest percentage of sulfated material
(69%; Fig. 8C). The monomer fractions do vary between 5-12
days in that 29% more of the ^{3}H-glucosamine label appears in
these fractions by 12 days. In the absence of an increase
in sulfated material, the additional glucosamine-labeled
material is most likely HA. This must be of relatively low
molecular weight since if it were as large as that found in
mature cartilage, aggregation with monomer would result in
species migrating at the void volume. Ultrastructural data
have shown that chondrocytes and a beaded extracellular material, similar to cartilage, form by 12 days, but this matrix
must also be relatively immature in a biochemical sense.

The present data suggests that HA synthesis declines
prior to cartilage differentiation on bone matrix, but that
at this same time, HA synthesis increases in control cultures.
Insofar as HA detected in explants onto bone matrix may be
of lower molecular weight, the functional differentiation
of cartilage must reside in the appearance of high molecular
weight HA. This may be tested by asking whether proteoglycan
monomer synthesized in this system is capable of binding to
high molecular weight HA if it were present.

Hyaluronate binding studies were carried out by adding
HA from Cock's Comb to aliquots of proteoglycan extracts
under dissociative conditions and allowing aggregates to form
by changing to associative conditions. Where only 4.76% of
the sulfate DPM migrate as proteoglycan aggregate in 5-day
extracts, exogenous HA causes an additional 7.0% to appear at
the column void volume (Fig. 9A). The monomer peak appears
reduced by 10.8% and the peak near to the total volume is
increased 3.84% after the addition of HA. A small portion

of the monomer may have been degraded during the experiment, but clearly the remainder becomes bound to HA. It is unclear why more of the monomer did not participate in this binding. 12-day proteoglycan extracts differ in that 28.13% additional sulfated material binds to exogenous HA (Fig. 9C). It is clear from a comparison of Figs. 8 and 9 that this additional sulfated material came from monomer as well as lower molecular weight material. The 12-day extract contains material which has an increased ability to bind to HA and reflects the synthesis of new or different components characteristic of cartilage. Clearly some of these components exist within the low molecular weight material which is thought not to participate in formation of cartilage extracellular matrix. The continued presence of this material in cartilage extracts suggests that it is not merely a remnant from an earlier undifferentiated state, nor a contaminant from cells which are not chondrocytes, but an aspect of cartilage differentiation which has been little investigated. These data additionally demonstrate that a rate-limiting factor in chondrogenesis is the appearance of high molecular weight HA.

In summary, these studies show that the potential to form cartilage is not restricted to a defined population of limb tissues, but resides within limb mesenchyme as one of its several potencies. A theoretically novel aspect of these studies is the relative ease with which the synthesis of sulfated GAG and proteoglycan monomer may be evoked. But, what appears to control the appearance of a fully differentiated cartilage matrix is a shift from the synthesis of low molecular weight HA to species of greater molecular weight.

Ahrens PB, Solursh M, Meier S (1977). The synthesis and localization of glycosaminoglycans in striated muscle differentiating in cell culture. J Exp Zool 202:375.
Chevalier A, Kieny M, Mauger A, Sengel P (1977). Developmental fate of the somitic mesoderm in the chick embryo. In Ede DA, Hinchliffe JR, Balls M (eds): "Vertebrate Limb and Somite Morphogenesis," Cambridge (UK): Cambridge University Press, p 421.
Hascall VC, Heinegård DK (1979). Structure of cartilage proteoglycans. In Gregory JD, Jeanloz RW (eds): "Glycoconjugate Research," New York: Academic Press, p 341.
Holtzer H, Rubenstein N, Fellini S, Yeoh G, Chi J, Birnbaum J, Okayama M (1975). Lineages, quantal cell cycles, and the generation of cell diversity. Quart Rev Biophys 8:523.

Kieny M, Pautou M-P, Chevalier A (1981). On the stability of the myogenic cell line in avian limb bud development. Arch Anat Micros Morph Exp 70:81.

Nathanson MA (1982). Analysis of cartilage differentiation from skeletal muscle grown on bone matrix. III. Regulation of glycosaminoglycan and proteoglycan synthesis by environmentally-induced, low molecular weight hyaluronic acid. Develop Biol Submitted.

Nathanson MA, Hay ED (1980a). Analysis of cartilage differentiation from skeletal muscle grown on bone matrix. I. Ultrastructural aspects. Develop Biol 78:301.

Nathanson MA, Hay ED (1980b). Analysis of cartilage differentiation from skeletal muscle grown on bone matrix. II. Chondroitin sulfate synthesis and reaction to exogenous glycosaminoglycans. Develop Biol 78:332.

Nathanson MA, Hilfer SR, Searls RL (1978). Formation of cartilage by non-chondrogenic cell types. Develop Biol 64:99.

Palmoski MJ, Goetinck PF (1973). Synthesis of proteochondroitin sulfate by normal, nanomelic, and 5-bromodeoxyuridine-treated chondrocytes in cell culture. Proc Nat Acad Sci USA 69:3385.

Reddi AH, Huggins CB (1975). The formation of bone marrow in fibroblast transformation ossicles. Proc Nat Acad Sci USA 72:2212.

Searls RL, Janners MY (1969). The stabilization of cartilage properties in the cartilage-forming mesenchyme of the embryonic chick limb. J Exp Zool 170:365.

Toole BP (1972). Hyaluronate turnover during chondrogenesis in the developing chick limb and axial skeleton. Develop Biol 29:321.

Urist MR (1970). The substratum for bone morphogenesis. In Runner MN (ed): "Changing Synthesis in Development," New York: Academic Press, p 125.

Wolpert L (1969). Positional information and the spatial pattern of cellular differentiation. J Theoret Biol 25:1.

Limb Development and Regeneration
Part B, pages 229-238
© **1983 Alan R. Liss, Inc., 150 Fifth Avenue, New York, NY 10011**

THE DEVELOPMENT OF LONG BONES OF THE LIMB: CELL AND MATRIX INTERACTIONS OF OSTEOCLASTS AND MONOCYTES

Philip Osdoby*, Mary Martini, and Arnold I. Caplan

*School of Dentistry, Washington University, St. Louis, MO 63110 and Biology Department, Case Western Reserve University, Cleveland, OH 44106

INTRODUCTION

Bone development is initiated at precise focal points throughout the embryo and progresses in a distinctive developmental pattern. In the embryonic chick limb, initial cytodifferentiation associated with bone formation begins by stage 28 (day 5). The appearance of cell associated alkaline phosphatase activity (Osdoby and Caplan, 1981) and Type 1 collagen associated with osteoid (Von der Mark et al., 1976) has been interpreted to indicate osteoblast activity. The initial bone forms at the mid-diaphyseal region and expands along the longitudinal periosteal area. Since osteoblast expression begins close to the onset of muscle and cartilage expression and since cultured stage 24 phenotypically unexpressive limb mesenchyme will differentiate into osteoblasts in culture (Osdoby and Caplan, 1979, 1980, 1981) the initial osteoblast population is most likely of mesenchymal origin.

The cells responsible for the precise skeletal architecture, the bone forming osteoblasts and the bone resorbing osteoclasts, appear to have distinctly different developmental histories. The large multinucleated osteoclast is not observed in the developing chick limb until stage 35 (day 8.5). The emergence of the osteoclast phenotype is coincident with vascular invasion of the hypertrophic mid-diaphyseal cartilage. Vascular infiltration transports monocyte-macrophage precursors into the limb core (Sorrel and Weiss, 1980; Szenburg, 1977). Walker (1973, 1975) and

Mundy et al. (1977) implicated the blood monocyte as a potential osteoclast precursor when parabiotic treatment or marrow transplantation cured a genetic osteopetrotic condition
in rats. Osteopetrosis in many forms is the result of osteoclast malfunction. The vascular origin for the osteoclast
is additionally supported by the studies of Kahn and Simmon's
(1975) and Joetereau and Le Douarins (1978) using chick-quail
chimerae which demonstrated host blood-borne origin for the
osteoclast and that osteoclasts appeared shortly after vascu-
larization of the cartilage core. Kahn et al. (1981) recently reported that osteoclast precursors may be present
in the chick circulation as early as day 3 of embryogenesis.

Importantly, studies on macrophages (osteoclast surro-
gates) illustrated that bone matrix components elicited a
chemotactic response (Malone et al., 1982). The developing
cartilage and bone matrices may therefore influence osteoclast recruitment, attachment to bone, and differentiation.
This paper discusses observations related to osteoclast and
monocyte interactions with developing bone cells and matrices. Primary cell isolates of osteoclasts and monocytes
have been utilized to determine if programmed developmental
matrix changes influence osteoclast attachment to bone.
Preliminary observations also will be presented suggesting
that osteoblast activity may be a prerequisite and influen-
tial in osteoclast differentiation.

CELL ISOLATION AND CULTIVATION

The isolation and cultivation of osteoclasts has proved
to be difficult. A number of osteoclast-like cell populations
have been described (Luben et al., 1976; Nelson and Bauer,
1977; Chambers, 1979). The heterogeneity and/or indistinct
morphology of these cells has made it difficult to assess
their osteoclastic characteristics. We have recently described an osteoclast cell culture system to study osteoclast
phenotypic parameters (Osdoby et al., 1982). Osteoclasts
are derived from day 19 embryonic chick tibias by trypsin
release. The resultant cell suspension is passed through
nitex screening to remove matrix debris. Subsequently,
cells are exposed to collagenase to remove adherent matrix
and then separated by Percoll discontinuous buoyant density
gradient centrifugation. Osteoclasts band in a density
range between 1.030 and 1.054 g/ml. The separation produces
cell populations containing 50-75% osteoclasts. Figure 1A

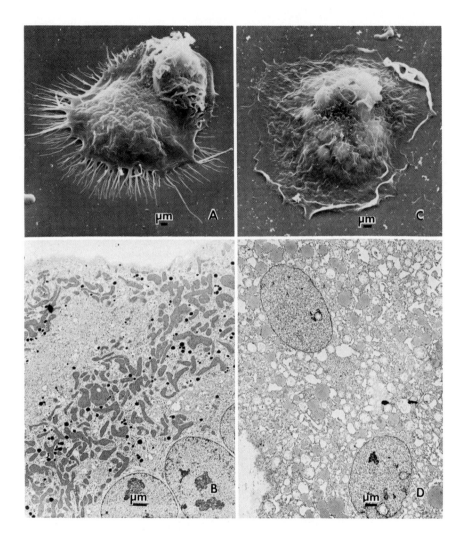

Fig. 1. (A) SEM of day 3 cultured osteoclast. (B) TEM of day 3 cultured osteoclast illustrating multinucleation and rich mitochondrial cytoplasm. (C) SEM of day 7 cultured giant cell derived from circulating monocytes. Notice smooth surface compared to cultured osteoclast. (D) TEM of day 7 giant cell illustrating vacuolated cytoplasm.

illustrates an isolated osteoclast that has been cultured
for three days. The long cytoplasmic processes may be indicative of the osteoclast ruffled border or alternatively represent a manifestation of culture conditions. Figure 1B
reveals the characteristic multinucleated cytoplasm and mitochondrial rich features of these cells.

Chick hatchling circulating monocytes were isolated by
Ficoll hypague gradient centrifugation with subsequent adhesion to culture dishes. In culture, monocytes fuse and form

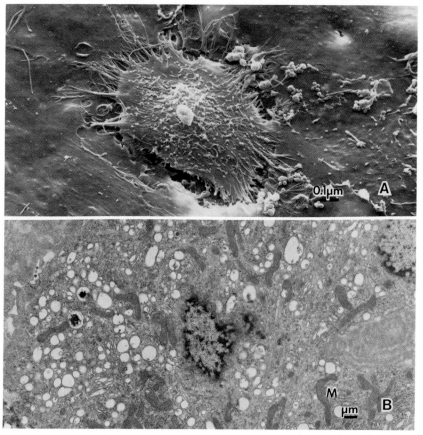

Fig. 2. (A) SEM of monocyte grown for 4 days and then over-
plated with stage 24 mesenchyme for an additional 8 days.
Notice plasma membrane processes. (B) TEM of cells grown
under condition identical to 2A. Notice the large numbers
of mitochondria (M).

large multinucleated giant cells (Figure 1C). Under specific culture conditions (Osdoby et al, 1982) these multinucleated giant cells display plasma membrane processes similar to, but not identical to, those observed on cultured osteoclasts. However, ultrastructural analysis exposes a less compartmentalized, less mitochondrial laden cytoplasm in comparison to osteoclasts (Figure 1D). Monocytes cultured over devitalized bone do not have morphological features different than those grown on plastic. However, when monocytes are co-cultured with stage 24 mesenchymal cells (osteoblasts) giant cells develop surface morphologies and ultrastructures indistinguishable from osteoclasts (Figure 2A and 2B) (Osdoby et al., 1982).

ADHESION EXPERIMENTS

The objective of the following experiments was to determine if developing cartilage and/or bone matrices were a contributory factor for osteoclast recruitment and differentiation. Isolated cells were labeled with (^{32}P)-KH$_2$P to quantitate cell number. Experiment I was designed to determine if osteoclasts and monocytes had an attachment preference for either the endosteal or periosteal bone surface. Experiment II was undertaken to determine if the developing matrix was a factor in osteoclast differentiation. In experiment I labeled osteoclast, monocytes, stage 24 mesenchymal cells, and embryonic skin fibroblasts, (the latter two serving as control cell populations), were inoculated into chambers containing, paired split day 19 tibias with either endosteal or periosteal surfaces exposed. The results are illustrated in figure 3. Twice as many osteoclasts and monocytes adhere to bone compared to fibroblasts and stage 24 mesenchyme. This would suggest that osteoclasts and monocytes have a high affinity towards bone. Furthermore, the ability to attach to bone may be a property of the precursor cell (monocyte) as well as the fully differentiated phenotype. Figure 3 also indicates that there does not appear to be a preference for the endosteal surface opposed to the periosteal bone surface. Since a majority of osteoclasts are located on the endosteal surface in vivo, we hypothesized that a potential matrix factor was involved. This does not appear to be the case. Rather, it is quite possible that since the endosteal surface is in apposition to the marrow cavity, presumptive osteoclast precursors attach there first and are therefore found in higher numbers.

In Experiment II, labeled cells were inoculated into chambers containing day 6 (cartilage cores), day 12 (initially calcifying bone and cartilage), or day 19 embryonic tibias (bone) (Martini et al., 1982). Figure 4 summarizes these results. The data indicate that only a few of each cell type attaches to day 6 cartilage cores. If any attach, it is the

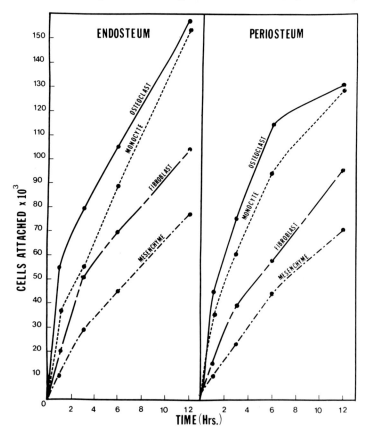

Fig. 3. Osteoclasts, monocytes, fibroblasts, and isolated stage 24 limb mesenchymal cells were labelled with 100 µCi of (^{32}P) monopotassium phosphate and inoculated over dissected day 19 embryonic tibias. 12 x 10^6 cells were inoculate on each dish with either endosteal or periosteal bone surfaces exposed. The specific activities of each cell type were: Osteoclast = 0.1 cpm/cell, monocytes = 0.9 cpm/cell, fibroblasts = 0.224 cpm/cell and mesenchyme = 0.9 cpm/cell. Each point is the average of duplicate samples.

fibroblast and mesenchymal cell populations. In contrast, large numbers of osteoclasts and monocytes adhere to day 12 and day 19 ossified matrices. Osteoclasts and monocytes adhere more to day 12 tibias than to 19 day tibias at both 2.5 hours and 5.0 hours. Osteoclasts had the highest attachment levels with monocytes attaching somewhat less, especially at 2.5 hours. It was quite possible that osteoclasts and monocytes were naturally a more adherent cell population without a specific substrate. This does not appear to be the

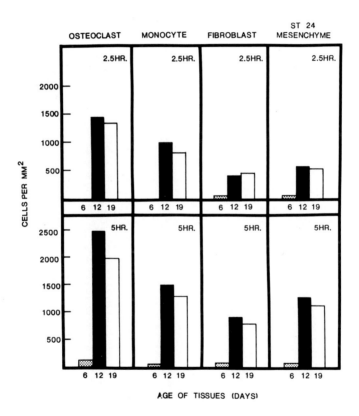

Fig. 4. Four cell types: Monocytes, osteoclasts, fibroblasts and stage 24 limb mesenchymal cells were labeled with $100\,\mu c$ (^{32}P) monopotassium phosphate and inoculated into vials containing 6, 12, or 19 day embryonic limb cores for 2.5 or 5.0 hrs. The cells/mm area were determined utilizing a geometric equation.

case if cell attachment to plastic is monitored. Three times as many mesenchymal cells and seven times as many fibroblasts attach to plastic in a similar 2.5 hour period compared to osteoclasts and monocytes (data not shown.)

The results of Experiment II indicate that osteoclasts and monocytes attach to bone (day 19 tibias) and bone containing hypertrophic chondrocytes (day 12 tibias). However, hyaline cartilage (day 6 limb cores) appears to be a poor attachment substrate. It is interesting that hyaline cartilage also contains anti-invasion factors, AIF, (Kuettner et al., 1976) that prevent vascular penetration. Slightly more cells attach to the 12 day cores compared to day 19 bones. This observation may be the result of complex matrix differences between the 12 and 19 day embryonic tibias. It is possible that the hypertrophic chondrocytic matrix provides the initial or stronger signal for attachment. We are presently pursuing these possibilities by looking at attachment and chemotaxis to various matrix components. For example, proteolytic products of Type I collagen serve as chemotactic elements for monocytes (Postlethwaite and Kang, 1976). Therefore, it is quite possible that a chemotactic factor may be involved in the cell attachment phenomenon. The hydrolysed matrix associated with cartilage hypertrophy may be an important factor in osteoclast recruitment in vivo and would explain the higher attachment levels for day 12 cases we observe in our in vitro assays.

SUMMARY

The adhesion experiments described indicate that osteoclasts, and their presumed progenitor the monocyte, selectively adhere to bone but show no adhesive preference for the endosteal versus periosteal surfaces. In the limb, osteoclasts attach to ossified 12 and 19 day embryonic cores but not to totally cartilaginous 6 day cores. These results support the general hypothesis that cellular interactions between bone or pre-bone matrix and osteoclasts are temporally regulated. This regulation appears to involve specific cellular-matrix recognition and attachment events. Monocytes as osteoclast precursors seem to have the machinery necessary to recognize and attach to bone before the differentiation event occurs or alternatively, the attraction and attachment to bone may initiate this differentiation process.

Bone matrix is composed of a complex molecular organizaation which includes Type 1 collagen and phosphoproteins (Cohen-Solal et al., 1979), bone proteoglycan (Reddi et al., 1978), and osteocalcin (Hauschka and Gallop, 1977). Any one or several components as well as their breakdown products may serve as chemotactic factors and/or attachment sites for the osteoclast and its precursor. The mechanism of chemotaxis and attachment are complex. They most likely involve separate processes which may be responsible for the initial stages of osteoclast recruitment and differentiation and appear to be dependent on the developmentally regulated matrix changes. However, it also appears that besides the matrix produced by osteoblasts, other osteoblast-mediating factors may contribute to osteoclast differentiation and activation. We are continuing to pursue this line of inquiry to unravel the mechanism of osteoblast-osteoclast interactions that ultimately result in a precisely patterned skeleton.

REFERENCES

Chambers TJ (1979). Phagocytosis and trypsin-resistant glass adhesion by osteoclasts in culture. J Path 127:55.

Cohen-Solal L, Lian J, Kossiva D, and Alimeher M (1979). Identification of organic phosphorous covalently bound to collagen and non-collagenous proteins of chick bone matrix. Biochem J 177:81.

Hauschka P and Gallop P (1977). Purification and calcium-binding properties of osteocalcin, the -carboxyglutamate containing protein of bone: "Calcium Binding Proteins and Calcium Function," Elsevier North Holland, New York, p 338.

Joetereau F and Le Douarin N (1978). The developmental relationship between osteocytes and osteoclasts. A study using quail-chick nuclear markers in endochondral ossification. Develop Biol 63:253.

Kahn AJ and Simmons DJ (1975). Investigation of cell lineage in bone using a chimaera of chick and quail embryonic tissue. Nature 258:325.

Kahn AJ, Simmons DJ, and Krukowski M (1981). Osteoclast precursor cells are present in the blood of preossification chick embryos. Develop Biol 84:230.

Kuettner KE, Hiti J, Essenstein R, and Harper E (1976). Collagenase inhibition by cationic proteins derived from cartilage and aorta. Biochem Biophys Res Comm 72:40.

Luben R, Wong G, and Cohn D (1976). Biochemical characterization with parathormone and calcitonin of osteoclasts and osteoblasts. Endocrinol 99:526.

Malone JD, Teitelbaum SL, Griffin GL, Senior RM, and Kahn AJ (1982). Recruitment of osteoclast precursors by purified bone matrix constituents. J Cell Biol 92:225.

Martini M, Osdoby P, and Caplan AI (1982). Adhesion of osteo clasts and monocytes to developing bone. Submitted to J Exp Zool.

Mundy G, Varni J, Orr W, Gouder M, and Ward P (1977). Resorbing bone is chemotactic for monocytes. Nature 275:132.

Nelson R and Bauer G (1977). Isolation of osteoclasts by velocity sedimentation at unity gravity. Calcif Tiss Res 22:303.

Osdoby P and Caplan AI (1979). Osteogenesis in cultures of limb mesenchymal cells. Develop Biol 73:84.

Osdoby P and Caplan AI (1980). A scanning electron microscopic investigation of in vitro osteogenesis. Calcif Tiss Intl 30:43.

Osdoby P and Caplan AI (1981). Characterization of a bone-specific alkaline phosphatase in chick limb mesenchymal cell cultures. Develop Biol 86:136.

Osdoby P and Caplan AI (1981). First bone formation in the developing chick limb. Develop Biol 86:147.

Osdoby P, Martini M, and Caplan AI (1982). Isolated osteoclasts and their presumptive precursor, the monocyte, in culture. Submitted to J Exp Zool.

Osdoby P, Weitzhandler M, and Caplan AI (1982). Interaction of osteogenic elements, in preparation.

Postlethwaite AE, and Kang AH (1976). Collagen and collagen peptide-induced chemotaxis of human blood monocytes. J Exp Med 143:1299.

Sorrel J and Weiss L (1980). A light and electron microscopic study of the region of cartilage resorption in the embryonic chick femur. Anat Rec 198:513.

Szenburg A (1977). Ontogeny of mylopoietic precursor cells "Advances in Exp Med and Biol" 88:3.

Von der Mark R, von der Mark H and Gay S (1978). Study of differentiation collagen synthesis during development of the chick embryo II. Localization of Type I and Type II collagen during long bone development. Develop Biol 53:153.

Walker DG (1973). Osteopetrosis cured by temporary para-biosis. Science 180:875.

Supported by grants from the N.I.H. and the Arthritis Foundation.

Limb Development and Regeneration
Part B, pages 239–248
© **1983 Alan R. Liss, Inc., 150 Fifth Avenue, New York, NY 10011**

THE RELATIONSHIP OF MONOCYTIC CELLS TO THE DIFFERENTIATION
AND RESORPTION OF BONE

Arnold J. Kahn, Ph.D., Steven L. Teitelbaum, M.D.,
J. David Malone, M.D., Marilyn Krukowski, Ph.D.

Washington University and the Jewish Hospital of
St. Louis, St. Louis, MO 63110

ABSTRACT

Osteogenesis in the developing limb is, as elsewhere,
the result of two functionally integrated processes, bone
matrix synthesis and matrix degradation. The latter process is a manifestation of the resorptive activity of osteoclasts (OCs), multinucleated giant cells which arise by
fusion of mononuclear, blood-borne precursors. For the past
several years, we have focused our efforts on several different aspects of OC development and differentiation. These
efforts have included observations on patients with osteopetrosis, an analysis of monocyte (OC precursor) chemotaxis
in response to bone matrix proteins, and the use of histochemical and bone grafting techniques to establish the
lineage and expression of the osteoclast phenotype in the
avian embryonic limb. Here, we (1) briefly review the evidence establishing the hematopoietic and probable monocytic
origin of OCs, (2) present new data on the role of L-γ-
carboxyglutamic acid in osteocalcin-evoked chemotaxis and
(3) describe the time course and likely developmental relationships between the appearance of circulating OC precursor cells, osteoclast differentiation, and the formation of
mineralized bone matrix in the chick embryo. The results
of the latter study indicate that OC differentiation is
initiated by contact between precursor cells and mineralized
bone, and suggest that the expression of the OC phenotype
is dependent upon proteins or glycoproteins closely associated with bone mineral. We note that osteocalcin is one
such protein. (Supported by grants DE-04629 and DE-05413).

INTRODUCTION

Overt osteogenesis in the embryonic limb begins with the formation of mesenchymal cell aggregates in positions corresponding to sites of future long bone development. The cells within these aggregates differentiate into chondrocytes (chondroblasts) and secrete a cartilage matrix shaped, overall, into "models" of specific long bones. These models then undergo a sequential series of changes beginning with chondrocyte hypertrophy in the diaphysis and culminating in the replacement of cartilage tissue by bone.

The removal of effete cartilage in the models and, subsequently, of bone during skeletal growth and development is a function of large, multinucleated giant cells nominally identified by use of the suffix "-clast"; chondroclasts for the giant cells responsible for degrading cartilage, osteoclasts (OCs) for the cells which resorb bone. The ontogeny of osteoclasts has been debated for many years, but recent evidence strongly suggests that they are derived from blood-borne precursors, belong to the mononuclear phagocyte family, and are recruited to sites of incipient cartilage or bone degradation by chemotaxis. In the present report, we (1) briefly review the evidence establishing the hematopoietic and probably monocytic origin of OCs, (2) present data on the role of L-γ-carboxyglutamic acid (L-gla) in osteocalcin-evoked chemotaxis and (3) describe the time course and likely developmental relationships between the appearance of circulating OC precursor cells, osteoclast differentiation, and the formation of mineralized bone matrix.

BACKGROUND

Hematopoietic Origin of Osteoclasts: After years of uncertainty, the origin of the osteoclast from blood-borne precursors now seems firmly established for both birds and mammals. In birds, Kahn and Simmons (1975) and, subsequently, Jotereau and Le Douarin (1978) showed that bone rudiments grafted, interspecifically, from the embryonic quail to the chorioallantoic membrane (CAM) of chick embryos, developed marrow cells and osteoclasts of host origin. Osteoblasts, on the other hand, originated from donor tissue. Since the principal mode of tissue interaction between CAM and graft was via a chick-derived blood vascular system, it was concluded that the osteoclasts (like the marrow cells) were generated from circulating hematopoietic precursor cells. More recently, Kahn et al. (1981), using a variant of the interspecific grafting technique, demonstrated that these precursor cells first appear in the peripheral circulation between days 3 and 6 of embryonic life and increase in number until days 9-12.

The evidence for a hematopoietic precursor of the mammalian osteoclast derives from several different sources (e.g. Buring 1975; Gothlin, Ericsson 1976) but, perhaps, nowhere more dramatically and persuasively than in the clinical study of Coccia et al. (1980). These investigators reported that a female patient with malignant osteopetrosis (marble bone disease), after receiving a marrow transplant from her HLA-MLC identical brother, developed osteoclasts containing nuclei of male (i.e., donor origin). Her osteoblasts, on the other hand, remained entirely her own, indicating (1) that 'clasts and 'blasts originate from different precursors (cf, the quail-chick studies) and (2) 'clasts are of bone marrow (hematopoietic) origin. What makes this case so dramatic is that malignant osteopetrosis is a fatal disease which responds poorly to conventional therapy; marrow transplantation, on the other hand, seems to significantly reverse, if not cure, the disorder.

Monocytes as Osteoclast Precursors: The notion that the monocyte is the precursor of the osteoclast follows from a number of observations, but probably most directly from the work of Gothlin and Ericsson (1976), Tinkler et al. (1981), and, most recently, Burger et al. (1982). Gothlin and Ericsson showed that following the injection of thorotrast-labeled peritoneal macrophages into suitably prepared host animals, the marker substance became evident in osteoclasts. Tinkler et al. injected [3]HTdr-labeled peripheral blood monocytes, i.p., into mice stimulated with 1 α-OH vitamin D and found subsequently that some osteoclasts contained thymidine-labeled nuclei. Finally, Burger et al. (1982) demonstrated that macrophages derived from bone marrow cultures and incubated, in vitro, with periosteum-free embryonic bone rudiments, gave rise to cells morphologically indistinguishable from osteoclasts.

MATERIALS AND METHODS

Chemotaxis: Chemotaxis of human peripheral blood monocytes is assessed in modified Boyden chambers using the double membrane technique described by Campbell (1977). The upper membrane (filter) is polycarbonate, pore size, 5 μm (Nucleopore Corp.); the lower membrane is cellulose acetate, pore size, 0.45 μm (Millipore Corp.). Mononuclear cells recovered from a Ficoll-Hypaque gradient are introduced into the upper compartment of the chambers and the assays run at 37° C for 2 hours. In addition to the test substances, each experiment includes the formylated tripeptide, f Met Leu Phe, or the complement degradation product, C5f, as a positive control to confirm the chemotactic responsivity of the monocyte populations. The membrane pairs are then removed, stained with hematoxylin, and mounted on microscope slides. The slides are coded and scored for cell movement by counting the number of cells which migrate into the interface between the membranes during the assay period. These values are corrected for the random movement of cells (counts derived from chambers lacking chemoattractants) and are expressed as net cells per high power microscope field. Each variable is run in triplicate and five random fields are counted per slide. Results are expressed as the mean ± SEM. Comparisons are made using Student's t test.

Microscopy: Tibias and forelimb rudiments are collected from chick embryos of various developmental stages and are fixed overnight in either cold 2.5% glutaraldehyde or 3.5% paraformaldehyde + 0.2% glutaraldehyde in cacodylate buffer, pH 7.4. The tissues are rinsed in buffer, dehydrated in alcohols and embedded in methyl methacrylate (JB-4 or DuPont). Two micron sections are cut, mounted on slides and stained with methylene blue-azure II, with alizarin red, by the von Kossa method, or, as outlined in Sigma Technical Bulletins 85 and 286, for alkaline or acid phosphatase.

RESULTS

Chemotaxis: Osteocalcin (bone gla protein) is one of several bone matrix-associated proteins which elicits a chemotactic response for peripheral blood monocytes, i.e., putative osteoclast precursors (Malone et al. 1982; Table IA). This molecule is synthesized by osteoblasts in the developing chick limb concurrently with the onset of mineralization (Hauschka, Reid 1978) and is one of several calcium-binding proteins characterized by the presence of γ-carboxyglutamic acid (gla) residues as part of their primary structure.

TABLE I. CHEMOTACTIC RESPONSE OF HUMAN PERIPHERAL MONOCYTES (PUTATIVE
OSTEOCLAST PRECURSORS) TO VARYING MOLAR CONCENTRATIONS OF
OSTEOCALCIN AND RELATED MOLECULES. VALUES REPRESENT MEAN NET
CELL MOVEMENT/HIGH POWER FIELD AND ARE PRESENTED $\pm$ S.E.M.
THREE REPLICATE CHAMBERS WERE PREPARED AND COUNTED FOR EACH
VARIABLE.

AGENT	1×10^{-6}M	2×10^{-6}M	4×10^{-6}M	8×10^{-6}M
A. PRIMATE OSTEOCALCIN	21 ± 6	56 ± 7	93 ± 7	82 ± 9
C5f	103 ± 15			

AGENT	10^{-10}M	10^{-8}M	10^{-6}M
B. BOVINE PROTHROMBIN	4.8 ± 2	12.3 ± 2	9.1 ± 3
PROTHROMBIN FRAG.1	7.7 ± 3	35.2 ± 2	17.9 ± 4
C5f (25 µl)*	79.8 ± 6		

AGENT	10^{-10}M	10^{-8}M	10^{-7}M
C. L-GLA	34.5 ± 4	44.5 ± 5	5.7 ± 3
L-GLU	IDFC	IDFC	IDFC**
C5f (25 µl)*	89.6 ± 4		

AGENT	10^{-14}M	10^{-10}M	10^{-8}M	10^{-6}M
D. L-GLA	44.4 ± 10	71.8 ± 6	---	3.3 ± 1
D-GLA	17.4 ± 3	19.0 ± 3	---	17.9 ± 4
f MET LEU PHE	---	---	72.6 ± 9	---

*25 µl IS EQUAL TO 2x THE AMOUNT NECESSARY TO ELICIT 50% MAXIMUM CHEMO-
TAXIS IN HUMAN MONOCYTES AS ASSAYED UNDER STANDARD CONDITIONS.

**IDFC - INDISTINGUISHABLE FROM CONTROLS RUN WITHOUT CHEMOATTRACTANT.

Several lines of evidence suggest that the gla residues
in osteocalcin play a pivotal role in initiating the chemotactic response in monocytes. First, other gla containing
proteins such as prothrombin and prothrombin Fragment 1 are
also chemotactic for monocytes (Table IB). The response is
most notable in Fragment 1, which contains all 10 of the gla
residues in the thrombin molecule and which presumably has a
less tightly coiled (more accessible?) tertiary configuration.
Second, L-gla itself is chemotactic, and optimally so at concentrations much lower than those needed to evoke a comparable
response to osteocalcin (Table IC). By contrast, the parent
molecule, L-glutamate, which lacks the carboxyl group, elicits
no detectable migratory response (Table IC). Finally, L-gla
is significantly (P < .001) more stimulatory than D-gla,
(Table ID) suggesting that not only is the carboxyl group important in evoking a migratory response but also the isomeric
configuration of the amino acid. (Note: The chemotactic

action of each of the above molecules was confirmed by demonstrating that enhanced cell movement occured only in the presence of a concentration gradient of the various substances.)

Osteoclastogenesis in the Developing Chick Limb: Alkaline phosphatase positive osteoblasts first appear in the tibial midshaft of the developing embryonic chick at stages 28-29. The cells secrete a microscopically visible matrix by stages 30-31 which, on the basis of von Kossa and alizarin red staining, begins to mineralize by stages 32-33 (Osdoby, Caplan 1981; Table II).

TABLE II. CHRONOLOGY OF EVENTS ASSOCIATED WITH OSTEOCLASTOGENESIS IN THE DEVELOPING CHICK TIBIA.

EVENT	TIME AND STAGE[1] OF ONSET
OSTEOCLAST PRECURSORS IN PERIPHERAL CIRCULATION[2]	5½ DAYS (STAGE 28)
ALKALINE PHOSPHATASE POSITIVE OSTEO-BLASTS	5½-6 DAYS (STAGE 28-29)
BONE MATRIX VISIBLE	5½-6 DAYS (STAGE 28-29)
MINERALIZATION OF BONE MATRIX	~8 DAYS (STAGE 33)
ACID PHOSPHATASE MONONUCLEAR CELLS ON BONE MATRIX	~8 DAYS (STAGE 33)
OSTEOCLASTS PRESENT; RELEASE OF ACID PHOSPHATASE	~9 DAYS (STAGE 35)

[1]AFTER STAGING SERIES OF HAMBURGER, HAMILTON 1951.
[2]FROM KAHN ET AL. 1981.

At stage 29, coincident with the onset of osteoblastic activity, large numbers of acid phosphatase positive mononuclear cells are found in the soft tissue adjacent to developing long bones. Such cells are noted to move progressively closer to the rudiment, through stage 32, as the bone matrix becomes increasingly more evident. It is not, however, until stage 33, when the matrix is undergoing mineralization, that occasional, strongly acid phosphatase-positive cells are found directly applied to bone surfaces (Fig. 1, Table II). Subsequently, at stages 35-36, multinucleated, acid phosphatase-positive cells (osteoclasts) are observed in the most mature, mid-diaphyseal region of the rudiment (Table II), and with their appearance the release of enzyme onto the matrix surface is noted (Fig. 2). We take the latter observations to

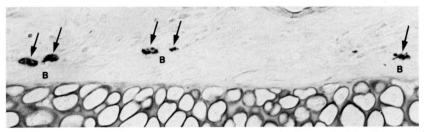

Fig. 1. Acid phosphatase-positive mononuclear cells (arrows) adjacent to periosteal bone (B) in a stage 35 (8-9 day) embryonic chick tibia. Such cells first appear apposed to bone surfaces at stage 33 coincident with bone matrix mineralization. 350X.

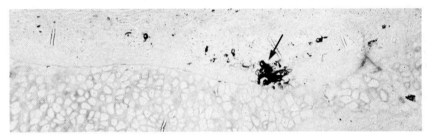

Fig. 2. Acid phosphatase (arrow) released onto the bone surface is the earliest sign of resorptive activity in a stage 35 rudiment. At this stage, enzyme reaction product is noted only in the most mature mid-diaphyseal region of the developing tibia. 150X.

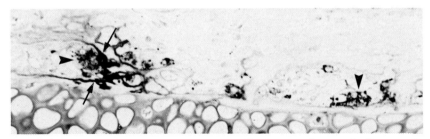

Fig. 3. Osteoclasts (arrow heads) in periosteal bone of a stage 36+ (~10 day) limb rudiment. Resorption is evidenced by bone seams coated with acid phosphatase (arrows). 350X.

indicate the onset of resorptive activity; a view confirmed by the presence, at stages 36-37, of enlarged canals rimmed by bone seams lined with acid phosphatase (Fig. 3).

DISCUSSION

The observations reviewed above confirm the hematopoietic origin of osteoclasts and identify cells within the monocyte family as the probable precursors. Monocytes are known to respond chemotactically to a wide variety of different substances and chemotaxis has been invoked as an explanation for the active movement of monocytes to sites of inflammation and tissue turnover (Wilkinson 1974). We (Malone et al. 1982) and others (Mundy et al. 1978, Minkin et al. 1981) have previously shown that monocytes exhibit a migratory response to bone-associated molecules and based upon these findings, it has been proposed that chemotaxis is also responsible for recruiting monocytes (as 'clast precursors) to regions of osteoclastogenesis.

Our most recent work on chemotaxis has been focused on osteocalcin (bone gla protein), a bone matrix constituent that may play some role in the mineralization process. Our data suggest that the migratory response to this molecule depends, in part, on osteocalcin's content of gla (γ -carboxyglutamic acid) and specifically show that L-gla is chemotactic for monocytes whereas D-gla and glu (glutamic acid) are not. Thus, the monocyte can apparently detect the presence of the additional carboxyl group on glutamic acid residues and distinguish between the isomeric forms of gla. This degree of sensitivity is reminiscent of the subtle structure-function relationships shown in the response of neutrophils to various synthetic tripeptides (Showell et al. 1976) and of the report by Mundy et al. (1980) that monocytes but not neutrophils are sensitive to di- and tri-peptides containing glycine and proline. To our knowledge, however, ours is the first report of chemotaxis in response to a single specific amino acid.

The accumulation of acid phosphatase-positive mononuclear cells adjacent to developing bone rudiments is presumably a manifestation of chemotactic activity in vivo. These putative osteoclast precursors first make contact with the rudiment surface at approximately the same time as the bone matrix begins to mineralize (This study; Osdoby, Caplan 1981) and osteocalcin synthesis is initiated (Haushka, Reid

1978). Osteoclast formation is noted shortly thereafter, but evidently only from acid phosphatase-positive mononuclear cells in contact with calcified bone. This apparent requirement for a mineralized matrix in osteoclast differentiation is in line with earlier reports (Irving and Handelman 1963, Glowacki et al. 1981) and our more recent observations of histiotypic osteoclasts developing in response to pulverized bone implanted onto chorioallantoic membranes of chick embryos (Krukowski and Kahn, 1982).

We have, in this report, presented a scenario for osteoclast differentiation which includes monocytes, chemotaxis and mineralized bone. This scenario is consistent with most of the available literature on osteoclastogenesis but is clearly not without some important unresolved issues. For example, osteoclast precursors are present in the circulation of the embryonic chick as early as days 3-6 (Kahn et al. 1981) but acid phosphatase-positive monocytes are not identifiable in blood until about day 19. Thus, the precursor cell which leaves the circulation in proximity to the bone rudiment is probably no more mature than a promonocyte and it is this cell which first differentiates into a more mature mononuclear phagocyte before fusing to form an osteoclast.

REFERENCES

Burger EH, van der Meer JWM, van de Gevel JS, Gribnau JC, Thesingh CW, van Furth R (1982). Origin of osteoclasts from immature mononuclear phagocytes. 5th Int Workshop Calcif Tiss p. 15.

Buring K (1975). On the origin of cells in heterotopic bone formation. Clin Orthop Rel Res 110:293.

Campbell PB (1977). An improved method for the in vitro examination of monocyte leukotaxis. J Lab Clin Med 90:381.

Coccia PF, Krivit W, Cervenka J, Clawson CC, Kersey JH, Kim TH, Nesbit ME, Ramsay NKC, Warkentin PI, Teitelbaum SL, Kahn AJ, Brown DM (1980). Successful bone marrow transplantation for infantile malignant osteopetrosis. N Eng J Med 302:701.

Glowacki J, Altobelli D, Mulliken JB (1981). Fate of mineralized and demineralized osseous implants in cranial defects. Calcif Tiss Int 33:71.

Gothlin G, Ericsson JLE (1976). The osteoclast. Review of ultrastructure. Clin Orthop Rel Res 120:201.

Hamburger V, Hamilton HL (1951). A series of normal stages in the development of the chick embryo. J Morphol 88:49.

Hauschka PV, Reid ML (1978). Timed appearance of a calcium-binding protein containing γ-carboxyglutamic acid in developing chicken bone. Devel Biol 65:426.

Irving JT, Handelman CS (1963). Bone destruction by multi-nucleated giant cells. In Sognnaes RF (ed): "Mechanisms of Hard Tissue Destruction," Washington DC: American Association for the Advancement of Science, p 515.

Jotereau FV, Le Douarin NM (1978). The developmental relationship between osteocytes and osteoclasts. A study using quail-chick nuclear marker in endochondral ossification. Devel Biol 63:253.

Kahn AJ, Simmons DJ (1975). Investigation of cell lineage in bone using a chimera of chick and quail embryonic tissue. Nature (Lond) 258:325.

Kahn AJ, Simmons DJ, Krukowski M (1981). Osteoclast precursor cells are present in the blood of preossification chick embryos. Devel Biol 84:230.

Krukowski M, Kahn AJ (1982). The inductive specificity of mineralized bone matrix in ectopic osteoclast differentiation. Calcif Tiss Int (in press).

Malone JD, Teitelbaum SL, Griffin GL, Senior RM, Kahn AJ (1982). Recruitment of osteoclast precursors by purified bone matrix constituents. J Cell Biol 92:227.

Minkin C, Posek R, Newbrey J (1981). Mononuclear phagocytes and bone resorption: identification and preliminary characterization of a bone-derived macrophage chemotactic factors. Metab Bone Dis Rel Res 2:363.

Mundy GR, DeMartino S, Rowe DW (1980). Monocyte chemotaxis in response to collagen. Calcif Tiss Int 31:57.

Mundy GR, Varani J, Orr W, Gondek MD, Ward PA (1978). Resorbing bone is chemotactic for monocytes. Nature (Lond) 275:132.

Osdoby P, Caplan AI (1981). First bone formation in the developing chick limb. Devel Biol 86:147.

Showell HJ, Freer RJ, Zigmond SH, Schiffmann E, Aswanikumar S, Corcoran B, Becker EL (1976). The structure-activity relations of synthetic peptides as chemotactic factors and inducers of lysosomal enzyme secretion for neutrophils. J Exp Med 143:1154.

Tinkler SMB, Linder JE, Williams DM, Johnson NW (1981). Formation of osteoclasts from blood monocytes during 1 α-OH Vit D-stimulated bone resorption in mice. J Anat 133:389.

Wilkinson PC (1974), "Chemotaxis and Inflammation," Edinburgh: Churchill Livingstone.

Limb Development and Regeneration
Part B, pages 249–259
© 1983 Alan R. Liss, Inc., 150 Fifth Avenue, New York, NY 10011

PHENOTYPIC MATURATION OF OSTEOBLASTIC OSTEOSARCOMA CELLS
IN CULTURE

Gideon A. Rodan and Robert J. Majeska
Department of Oral Biology
University of Connecticut Health Center
School of Dental Medicine
Farmington, CT 06032

Osteosarcoma is a heterogeneous tumor which contains
elements of bone, cartilage and fibrous tissue. If the
tumor originated from a single cell, these tissues represent
differentiated states of a common stem cell line perpetuated
in the tumor. The progeny of such stem cells either continue the stem cell line or undergo phenotypic maturation.
Environmental factors, local and systemic, presumably influence that choice. These assumptions can be tested by
separating cells from the tumor, growing them in culture,
characterizing their phenotypic properties, and exposing
them to various stimuli which may affect cytodifferentiation
in vivo. If successful, such models could be used to study
the effect of extracellular factors on differentiation or
phenotypic maturation. For the last several years, our
laboratory has developed such a system from rat osteosarcoma (Majeska et al. 1978; Majeska et al. 1980) for in-
vestigating the osteoblastic phenotype. Following is a
brief summary of recent findings.

PHENOTYPIC HETEROGENEITY OF CLONAL CELL LINES.

To establish cell lines in culture, cells dispersed
from tumors by collagenase digestion were cultured in MCDB-
103 or F-12 media supplemented with 10% fetal bovine serum
(Majeska et al. 1978; Majeska et al. 1980). Cell cultures
were screened for properties associated with the osteoblastic phenotype. Cytochemical studies of bone cells
in situ and biochemical studies of cells isolated from calvaria (Robinson et al. 1973; Luben 1976) have shown that
these include elevated levels of alkaline phosphatase (AP)

and receptors for parathyroid hormone (PTH). Alkaline
phosphatase was measured at confluence by hydrolysis of
para nitrophenyl phosphate (Majeska et al. 1980; Majeska,
Rodan 1982a) and functional receptors to PTH were assessed
by measuring the stimulation of adenylate cyclase in whole
cells or cell membranes (Rodan et al. 1980; Rodan, Rodan
1981). These osteoblastic properties were present to a
limited degree in primary cultures but were lost after a
number of passages (Majeska et al. 1980). We assumed that
the osteoblastic phenotype was expressed in vitro, but was
not favored by culture conditions and was reduced in subsequent cell generations. Green and Kehinde have shown
that specific clones of 3T3 fibroblasts (3T3-Ll and 3T3-L2)
have a higher probability of differentiating into fat cells
than other clones (Green, Kehinde 1976). In analogy to the
3T3 system we cloned the osteosarcoma cells and screened the
clonal cell lines for PTH responsiveness (Majeska et al.
1980). We found that certain cell lines (ROS 2/3) had a
much higher hormone sensitivity than the original mixed
culture, while other cell lines had none, and still others
had intermediate responses (Majeska et al. 1980). When
ROS 2/3 was reimplanted into animals, tumors developed and
when those were grown in culture and cloned, cell lines
with even higher responsiveness to PTH were obtained
(ROS 17/2). We noticed that the various cell lines also
differed in morphology. The highest PTH responders formed
a sheet of contiguous cuboidal cells, resembling calvaria
osteoblasts in situ (Jones, Boyde 1976); cells with intermediate responses were stellate, triangular or rhomboidal
in shape; whereas the non-responders looked fusiform
(fibroblastic), and grew in patterns of bands and swirls.
The alkaline phosphatase activity correlated closely with
the PTH response and the osteoblastic morphological appearance. Cell lines were examined for other putative osteoblastic properties, such as collagen type, synthesis of
γ-carboxyglutamic acid containing protein (osteocalcin),
presence of $1,25-(OH)_2D_3$ receptors, and those too correlated
with the other osteoblastic expressions (Majeska et al.
1980; Nishimoto, Price 1980; Manolagas et al. 1980; Aubin
et al. 1982). These findings, collected in several laboratories, are summarized in Table 1. The clonal cell
lines maintained these properties in culture for long periods of time (years), and many passages, when cells were
grown and subcultured under the same conditions.

TABLE 1

PHENOTYPIC HETEROGENEITY OF CLONAL RAT OSTEOSARCOMA
CELL LINES: CO-EXPRESSION OF "OSTEOBLASTIC" TRAITS.

	17/2·8	2/3	24/1	25/1
Morphology *	cuboid	mixed rhomboid	stellate	fusi-form
Alkaline Phosphatase * (μmol/min/mg protein)	1.62	0.32	0.27	0.02
PTH-stimulated adeny-late cyclase * (Treated/control)	11.3	4.2	1.2	1.0
$1,25(OH)_2D_3$ receptors † (fmol/mg protein)	24	+	–	N.D.
Bone Gla-protein synthesis Ψ (ug/cell/24 hr)	15.5	5.4	0	0
Collagen type §	I 98%	N.D.	N.D.	I 70% III 30%
Cyclooxygenase activity (ng $PGE/10^6$ cells/min at 100 μM arachidon-ate)	166	113	4.5	0.6
IN VIVO Osteo-genesis *	+	+	–	N.D.

* Majeska et al., 1980
† Manolagas et al., 1980
Ψ Nishimoto, Price, 1980
§ Aubin et al., 1982

PHENOTYPIC MATURATION IN CULTURE.

In analogy to 3T3-L1 differentiation into adipocytes, we assumed that the stronger expression of osteoblastic properties in certain osteosarcoma cell lines, evaluated at confluence, may represent a higher probability for osteoblastic differentiation following cell division. If this were the case, the expression of osteoblastic features should depend on seeding density and time in culture. We examined the effect of these variables on alkaline phosphatase specific activity in ROS 17/2·8, an "osteoblastic" cell line, which exhibits density dependent inhibition of growth (Majeska, Rodan 1982b). As seen in Fig. 1, AP specific activity showed a positive correlation with both, seeding density and time in culture, approaching a plateau around 2 μmol/min/mg protein.

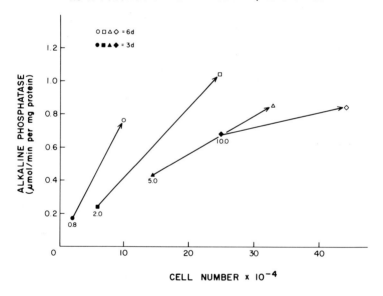

CHANGE IN ALKALINE PHOSPHATASE ACTIVITY IN ROS 17/2.8 CULTURES AS A FUNCTION OF PLATING DENSITY, CELL NUMBER AND TIME

Figure 1: ROS 17/2·8 cells were plated at the indicated cell densities ($0.8 - 10.0 \times 10^4$ cells/2cm^2) in F-12 plus 5% serum. Cultures were fed three days after plating and cells were assayed on days 3 and 6 for alkaline phosphatase and protein. Parallel cultures were trypsinized and counted by Coulter counter.

In cells seeded at relatively high density, AP levels were high at 3 days and did not increase much thereafter. Cells seeded at low density had low AP levels at 3 days, while actively dividing, and showed substantial increases in AP thereafter. In both cases, roughly equivalent AP levels were reached at confluence, with a tendency towards higher levels in cultures seeded at lower density. A more detailed examination of AP changes, as a function of time, showed that AP increases occurred in steps. Media were changed in these cultures every three days. Each change of medium and addition of fresh serum caused a proliferative burst, around 24 hours, accompanied by a small reduction in the specific activity of alkaline phosphatase (per protein or per cell). This drop was followed by a surge in AP above previous levels, 24 hours later. Above findings are consistent with the generally observed reciprocal relationship between proliferation and differentiation. This relationship was also noted in the effects of bone specific hormones, which modulated the two functions.

PARATHYROID HORMONE EFFECTS ON GROWTH AND ALKALINE PHOSPHATASE.

Hormones convey the needs of the organism to specialized tissues and control their activity accordingly. Similarly, hormones regulate the relative amount of specialized tissues, and thus influence growth and differentiation. This hormonal effect is most pronounced during development, but should also be present during tissue repair, remodeling or regeneration, when these processes occur in the adult. Osteogenesis can occur throughout life, during fracture repair, skeletal turnover and extraskeletal ossifications. It is of basic and practical interest to find out if hormones can affect these processes. In osteoblast-like cells, obtained from embryonic or newborn calvaria, PTH was shown to inhibit anabolic functions such as collagen synthesis and alkaline phosphatase activity (Luben et al. 1976; Wong et al. 1977). PTH was also shown to promote growth _in vivo_ as well as in organ cultures (Kalu et al. 1970; Tam et al. 1980; Howard et al. 1981). It is currently not known if these effects are due to direct hormone action on osteoblastic cells or secondary effects produced by local factors released during bone resorption (Howard et al. 1981).

In the osteoblast-like osteosarcoma cell lines,

(ROS 17/2 and ROS 2/3), PTH at physiological concentrations $(10^{-13} - 10^{-11}$ M) was mildly growth-stimulatory (approximately 20%) in the presence of 2% serum (Majeska, Rodan 1981a). Its effects could be larger in the absence of serum or endogenous growth factors released by the cells themselves. Cell cycle analysis indicated that PTH acted on the G_0-G_1 transition. At the same time, PTH prevented the rise in alkaline phosphatase associated with phenotypic maturation (Majeska, Rodan 1982a). Interestingly, the latter effect was much larger (50%) than growth stimulation (20%), suggesting that the two are not obligatorily coupled in a one to one relationship. Indeed, inhibition of proliferation by hydroxyurea did not result in alkaline phosphatase elevation. The PTH inhibition of AP elevation was concentration-dependent, with a K_m of around 10^{-11} M (Majeska, Rodan 1982a). This effect was first detectable 24 hours after hormone addition, and increased with time, peaking at about 72 hours. It was acting on de novo synthesis of alkaline phosphatase and was probably mediated by cyclic AMP, since isoproterenol, db cyclic AMP and 8Br-cyclic AMP produced similar effects. As in other systems, the K_m for PTH-adenylate cyclase activation was significantly higher (1-10 nM) than that for producing the (presumed) cyclic AMP-mediated effects (Dufau et al. 1977; Saez et al. 1978).

Glucocorticosteroids were shown to promote differentiation in several systems (Ono, Oka 1980; Kedinger et al. 1980). Treatment of ROS 17/2·8 cells with dexamethasone increased alkaline phosphatase about four-fold. The fractional inhibition by PTH and the dose-response curve were not changed. In the presence of steroids the alkaline phosphatase rose faster and growth was inhibited. Regulation of AP by the two hormones, most probably reflects action on different loci(Majeska, Rodan 1981b).

1,25(OH)$_2$D$_3$ EFFECTS ON GROWTH AND ALKALINE PHOSPHATASE.

The discovery of vitamin D was related to its ability to prevent and heal the rachitic changes in the skeleton, but its mode of action on skeletal tissue is still unknown. Osteoblastic cells, including those from osteosarcoma, possess high affinity receptors for the active metabolite of vitamin D 1,25(OH)$_2$D$_3$ (Manolagas et al. 1980; Chen et al. 1979). In calvaria cells this hormone was shown to inhibit collagen synthesis and alkaline phosphatase (Wong et al. 1977). In osteosarcoma derived osteoblasts, 1,25(OH)$_2$D$_3$

was reported both to stimulate (Manolagas et al. 1981) and inhibit (Majeska, Rodan 1982b) alkaline phosphatase, possibly a reflection of different effects of the hormones at different stages of differentiation. We examined this phenomenon in some detail in ROS 17/2·8 cells and found that: (i) $1,25(OH)_2D_3$ affected both proliferation and AP changes in opposite direction (ii) the hormone also had opposite effects in early, sparse and later dense cultures, and (iii) the concentration dependence curve was bell-shaped with maximum effects at less than maximum concentrations used. Following is a more detailed description of these findings.

Cells were seeded at 10,000 cells/cm^2 in 5% serum and were exposed to the hormone for three days immediately after seeding. Under this protocol $1,25(OH)_2D_3$ caused about two-fold increase in AP specific activity and 50% inhibition of growth. The K_m for this effect was approximately 10 nM. During that period AP rose to 0.1 μmol/min/mg protein in control cultures and to 0.2 in hormone-exposed cultures. When cells were seeded at higher density, 50,000 cells/cm^2, and hormone was added four days later with new medium containing 2% serum, $1,25(OH)_2D_3$ inhibited AP by about 50% and stimulated growth by about 20%, both measured three days after hormone addition. The hormone K_m for this effect was 0.01 nM, a concentration consistent with the receptor affinity measured by radioreceptor assays (Manolagas et al. 1980; Chen et al. 1979). In the latter experiments the AP specific activity rose to 0.4 μmol/min/mg protein in hormone treated cultures at the end of the experiment. Opposite effects of $1,25(OH)_2D_3$ as a function of maturation could also be demonstrated by exposing the same cell population, (parallel dishes), to 1 nM hormone between 0-3 days or 3-6 days, respectively. During the first three days, the hormone caused a 25% inhibition of growth and 40% enhancement of alkaline phosphatase, and during the last three days a 20% stimulation of growth and 45% inhibition of AP. The major interpretation of these findings is that the hormonal effects are pleiotropic and depend on the state of the responding cells. Since the cultures are not synchronized with respect to maturation state, the response of the dominant subpopulation will stand out.

Differences in the hormone sensitivity (K_m) of the AP stimulatory and inhibitory effect might reflect (i) differences in receptor status as a function of maturation

(Chen, Feldman 1981), (ii) low affinity interaction with receptors for another hormone or (iii) non-receptor mediated effects, for example, on membrane phospholipid turnover (Matsumoto et al. 1981).

1,25(OH)$_2$D$_3$ EFFECTS ON CELLS IN SERUM FREE MEDIUM.

In the experiments described above the presence of serum was a complicating factor since serum binds and contains vitamin D as well as many growth and maturation factors. Therefore, we started examining the effect of vitamin D and other hormones in serum-free medium (Johnson, Rodan 1982). ROS 17/2·8 cells attach and grow in F-12 medium containing 1.1 mM calcium. Growth is linear (rather than logarithmic) at a rate of approximately 10,000 cells/cm^2/day, following seeding at that density. Saturation density, reached at about 12 to 14 days, was approximately 100,000 cells/cm^2. 1,25(OH)$_2$D$_3$ increased the rate of proliferation and the saturation density three-fold. This effect was dose-dependent with a K$_m$ of 8 X 10^{-12} M and was accompanied by a concomitant reduction in alkaline phosphatase to about 30% of control levels. The concentration-dependent-curve was again bell-shaped, smaller effects being observed at 10^{-9} and 10^{-8} M as compared to 10^{-11} M. These findings reinforce the previous observations showing opposite effects on growth and alkaline phosphatase and various degrees of coupling between the two processes.

Studies in serum free medium provide the opportunity to examine the influence of individual hormones and factors, in isolation or in combination, on the growth and osteoblastic maturation of these cells in culture.

SUMMARY

Clonal cell lines established from rat osteosarcoma were shown to differ in the degree of expression of osteoblastic features and to maintain a stable phenotype in culture. A cell line with strong osteoblastic expression undergoes maturation in culture and this process is influenced by tissue specific hormones such as PTH, 1,25(OH)$_2$D$_3$ and glucocorticosteroids at physiological concentrations. The hormonal effects are pleiotropic and probably interdependent and could be further studied in these cells in defined culture media.

ACKNOWLEDGEMENTS

This work was supported by NIH Grants DE04327 and AM17848.

The authors thank David E. Johnson for skillful technical assistance and Geraldine Nebor for preparing the manuscript.

REFERENCES

Aubin JE, Heersche JNM, Merilees MJ, Sodek J (1982). Isolation of bone cell clones with differences in growth, hormone responses, and extracellular matrix production. J Cell Biol 92:452.
Chen TL, Feldman D (1981). Regulation of 1,25-Dihydroxyvitamin D_3 receptors in cultured mouse bone cells. J Biol Chem 256:5561.
Chen TL, Hirst MA, Feldman D (1979). A receptor-like binding macromolecule for 1α,25-Dihydroxycholecalciferol in cultured mouse bone cells. J Biol Chem 254:7491.
Dufau ML, Tsuruhara T, Horner EA, Podesta E, Katt KJ (1977). Intermediate role of adenosine 3',5'-cyclic monophosphate and protein kinase during gonadotropin-induced steroidogenesis in testicular interstitial cells. Proc Nat Acad Sci 74:3419.
Green H, Kehinde O (1976). Spontaneous heritable changes leading to increased adipose conversion in 3T3 cells. Cell 7:105.
Howard GA, Bottemiller BL, Turner RT, Rader JI, Baylink DJ (1981). Parathyroid hormone stimulates bone formation and resorption in organ culture: evidence for a coupling mechanism. Proc Nat Acad Sci 78:3204.
Johnson DE, Rodan GA (1982). The effect of 1,25(OH)$_2$D$_3$ on osteosarcoma ROS 17/2·8 cell growth and alkaline phosphatase in serum free medium. Calc Tiss Int in press.
Jones SJ, Boyde A (1976). Experimental study of changes in osteoblastic shape induced by calcitonin and parathyroid extract in an organ culture system. Cell Tiss Res 169:449.
Kalu DM, Pennock J, Doyle FH, Foster GV (1970). Parathyroid hormone and experimental osteosclerosis. Lancet i:1363.
Kedinger M, Simon PM, Raul F, Grenier JF, Haffen K (1980). The effect of dexamethasone on the development of rat intestinal brush border enzymes in organ culture. Dev Biol 74:9.

Luben RA, Wong GL, Cohn DV (1976). Biochemical character
ization with parathormone and calcitonin of isolated
bone cells: provisional identification of osteoclasts
and osteoblasts. Endocrinology 99:526.

Majeska RJ, Rodan GA (1981a). Low concentrations of parathyroid hormone enhance growth of clonal osteoblast-like
cells in vitro. Calcif Tiss Int 33:323.

Majeska RJ, Rodan GA (1981b). Hormonal regulation of alkaline phosphatase in an osteoblastic osteosarcoma cell line.
Calcif Tiss Int 33:297.

Majeska RJ, Rodan GA (1982a). Alkaline phosphatase inhibition by parathyroid hormone and isoproterenol in a clonal
rat osteosarcoma cell line. Possible mediation by cyclic
AMP. Calcif Tiss Int 34:59.

Majeska RJ, Rodan GA (1982b). The effect of 1,25(OH)$_2$D$_3$ on
alkaline phosphatase in osteoblastic osteosarcoma cells.
J. Biol Chem 257:3362.

Majeska RJ, Rodan SB, Rodan GA (1978). Maintenance of parathyroid hormone response in clonal rat osteosarcoma lines.
Exp Cell Res 111:465.

Majeska RJ, Rodan SB, Rodan GA (1980). Parathyroid hormone-
responsive clonal cell lines from rat osteosarcoma.
Endocrinology 107:1494.

Manolagas SC, Burton DW, Deftos LJ (1981). 1,25-Dihydroxy-
vitamin D$_3$ stimulates the alkaline phosphatase activity
of osteoblast-like cells. J. Biol Chem 256:7115.

Manolagas SC, Haussler MR, Deftos LJ (1980). 1,25-Dihy-
droxyvitamin D$_3$ receptor-like macromolecule in rat osteogenic sarcoma cell lines. J. Biol Chem 255:4414.

Matsumoto T, Fontaine O, Rasmussen H (1981). Effect of
1,25-Dihydroxyvitamin D$_3$ on phospholipid metabolism in
chick duodenal mucosal cell. J Biol Chem 256:3354.

Nishimoto SK, Price PA (1980). Secretion of the vitamin K-
dependent protein of bone by rat osteosarcoma cells.
J Biol Chem 255:6579.

Ono M, Oka T (1980). The differential actions of cortisol
on the accumulation of α-lactalbumin and casein in mid-
pregnant mouse mammary gland in culture. Cell 19:473.

Robinson RA, Doty SB, Cooper RR (1973). Electron microscopy of mammalian bone. In Zipkin I (ed): "Biological
Mineralization," New York: Academic Press, p 257.

Rodan SB, Egan JJ, Golub EE, Rodan GA (1980). Comparison
of bone and osteosarcoma adenylate cyclase. Biochem J
185:617.

Rodan SB, Rodan GA (1981). Parathyroid hormone and isoproterenol stimulation of adenylate cyclase in rat osteosarcoma clonal cells. Hormone competition and site heterogeneity. Biochim Biophys Acta 673:46.

Rodan SB, Rodan GA (1981). The role of Mg^{2+} in hormone
stimulation of rat osteosarcoma adenylate cyclase.
Biochim Biophys Acta 673:55

Rodan SB, Rodan GA, Simmons HA, Walenga RW, Feinstein MB,
Raisz LG (1981). Bone resorptive factor produced by
osteosarcoma cells with osteoblastic features is PGE_2.
Biochem Biophys Res Commun 102:1358.

Saez JM, Evain D, Gallet D (1978). Role of cyclic-AMP and
protein kinase on the steroidogenic action of ACTH pros-
taglandin E_1 and dibuturyl cyclic AMP in normal adrenal
cells and adrenal tumor cells from humans. J Cyc Nucl
Res 4:311.

Tam CS, Wilson DR, Harrison JE (1980). Effect of parathyroid extract on bone apposition and the interaction
between parathyroid hormone and vitamin D. Mineral
Electrolyte Metab 3:74.

Wong GL, Luben RA, Cohn DV (1977). 1,25-Dihydroxychole-
calciferol and parathormone: effects on isolated osteoclast-like and osteoblast-like cells. Science 197:663.

Limb Development and Regeneration
Part B, pages 261–268
Published 1983 by Alan R. Liss, Inc., 150 Fifth Avenue, New York, NY 10011

REGULATION OF LOCAL DIFFERENTIATION OF CARTILAGE AND BONE
BY EXTRACELLULAR MATRIX: A CASCADE TYPE MECHANISM

A.H. Reddi

Bone Cell Biology Section, Laboratory of
Biological Structure, NIDR, NIH
Bethesda, Maryland   20205

ABSTRACT
    The regenerative potential of the vertebrate skeletal
system is well known. Normal cartilage and bone development
can be mimicked by implantation of demineralized bone
matrix, which induces bone formation locally. The
sequential stages of the matrix-induced bone development and
recent advances in the underlying biochemical mechanisms are
described. The extracellular matrix components have
chemotactic, mitogenic and differentiative activities. It
would appear that a cascade type mechanism is operative
during local differentiation of endochondral bone.

INTRODUCTION
    The remarkable potential of the vertebrate skeletal
system for regeneration and repair is well known. The
sequence of events during limb regeneration in amphibians
and fracture healing in mammals recapitulates the stages of
limb development. A biochemical approach to explain
embryonic limb development has inherent difficulties such as
paucity of material, and problems of developmental
heterogeneity. However, these obstacles may be surmounted
in part by the investigation of bone differentiation using
experimental models such as extracellular matrix-induced
bone development. Subcutaneous implantation of
demineralized extracellular bone matrix in rats results in
local differentiation of cartilage, bone and bone marrow.
The aim of this article is to describe the recent progress
in the identification of chemotactic, mitogenic and
differentiative factors in the extracellular bone matrix.
The tight association of these crucial factors with

extracellular matrix helps to explain the localized response.

DEVELOPMENTAL CASCADE

On implantation of the extracellular bone matrix in subcutaneous sites, a sequential development of endochondral bone is initiated (Urist, 1965; Reddi and Huggins, 1972; Reddi and Anderson, 1976; Reddi, 1981). This developmental cascade is summarized in Fig. 1. There was an instantaneous formation of a blood clot at the site of implantation. On day 1, the implant was a button-like, plano-convex plaque consisting of a conglomerate of matrix, fibrin and neutrophils. Then there was chemotaxis for fibroblast-like cells in vivo (Reddi and Huggins, 1972). On day 3 the mesenchymal cells proliferated as observed by ^{3}H-thymidine incorporation by biochemical and radioautographic techniques (Rath and Reddi, 1979). Differentiating chondroblasts were seen on day 5. The implant consisted of hyaline cartilage on days 7-8 as observed by histology and $^{35}SO_4$ incorporation into proteoglycans (Reddi et al., 1978). On day 9 vascular invasion into the implant was noted and there was concomitant cartilage calcification. Osteogenic cells and osteoblasts were seen in the vicinity of sprouting capillaries on days 10-11. Bone development was marked by the appearance of Type I collagen (Reddi et al., 1977) and increased ^{45}Ca incorporation (Reddi, 1981). The newly formed bone was remodeled on days 12-18 as indicated by the activity of lysosomal enzymes (Rath et al., 1981). Finally, the remodeled ossicle was the site of hematopoietic bone marrow differentiation as monitored by ^{59}Fe incorporation (Reddi and Huggins, 1975). This experimental model is a prototype for the study of the role of extracellular matrix in vivo. In view of their biological importance we have explored the mechanisms underlying the matrix-cell interactions.

FIGURE 1

ENDOCHONDRAL BONE DIFFERENTIATION
A BIOLOGICAL CASCADE

- Chemotaxis of Progenitor Cells
- Proliferation of Mesenchymal Cells
- Differentiation of Chondrocytes
- Calcification of Cartilage Matrix
- Vascular Invasion
- Bone Differentiation and Mineralization
- Bone Remodeling and Marrow Differentiation

MECHANISM OF ACTION

The molecular mechanisms underlying the action of extracellular matrix on cells to initiate the developmental cascade are not known. The various properties of the matrix-induced bone development are presented in Fig. 2. The response to the extracellular matrix is specific and is elicited only by bone and tooth matrix but not by tendon and skin (Reddi, 1970). The surface charge on the matrix is crucial (Reddi, 1978). Perturbation of the charge characteristics of the matrix by chemical modifications such

FIGURE 2

PROPERTIES OF ECM INDUCED BONE DIFFERENTIATION

- Specificity
- Surface Charge
- Geometry of Matrix
- Acid Stable
- Alkali Labile
- Heat Stable
- Protease Sensitive
- Periodate Oxidation Sensitive
- Disulfide Reduction Sensitive
- Extractable Glycoprotein

FIGURE 3
A MODEL FOR MATRIX-CELL INTERACTION

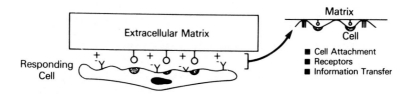

- Surface Charge Critical
- Geometry of Matrix
- Collagen-Fibronectin Interaction
- Specific Receptor Interaction
- Alteration of Gene Expression

as N-acetylation, carboxymethylation and modification of the
guanidino groups of arginine abolishes the inductive
response. Pretreatment of the matrix with heparin, dextran
sulfate and the anionic dye Evans blue inhibits bone cell
differentiation (Reddi, 1976). The physical dimensions of
the matrix particles have an important influence on the
quantitative response of new bone. The optimal size for
bone induction is in the range of 74-850 μm. These
observations are consonant with known requirements for the
dimensions of substratum in anchorage-dependent cell
proliferation (Stoker et al., 1968; Folkman and Moscona,
1978; Gospodarowicz et al., 1980; Reddi, 1976). The
inductive property is acid stable but alkali labile. The
biological activity is heat stable (65° for 8 h),
trypsin-sensitive, and appears to be a glycoprotein on the
basis of periodate sensitivity. The conformation of
disulfides is necessary for biological activity, as
reduction with mercaptoethanol abolishes the activity.

In view of the fact the implanted matrix is
predominantly collagenous we have examined the possible role
of fibronectin during early matrix-cell interactions (Weiss
and Reddi, 1980;1981). On implantation the matrix binds
fibronectin, which may constitute an important initial event
for cell attachment to the matrix. The present working
model is depicted in Fig. 3. The initial orientation of
cell surface to matrix is by electrostatic forces. However,
this is not sufficient for subsequent changes in the cells.
It is likely the collagen-fibronectin interaction helps to
bring the putative inductor in contact with focal cell
surface receptors. This results in transduction of cell
surface information to the genome. Current approaches in
our laboratory are directed towards the identification of
various factors that might regulate the developmental
cascade of matrix-induced bone development.

One of the first events in this cascade is chemotaxis
of cells towards the implanted matrix, followed by mitosis
and differentiation into cartilage and bone. In view of
this we have explored the potential of various dissociative
extractants to obtain factors that promote chemotaxis,
mitosis and differentiation. Treatment with 4 M guanidine
hydrochloride (Anastassiades et al., 1978) or 8 M urea
containing 1 M NaCl or 1% (w/v) sodium dodecylsulfate ⸱
extracted the bone inductive property from the matrix
(Sampath and Reddi, 1981). As summarized in Fig. 4 when the
lyophilized extract and the residue were tested alone they

FIGURE 4

DISSOCIATIVE EXTRACTION AND RECONSTITUTION

| Bone Matrix | →Dissociative Extraction→ | 4.0 M Guanidine HCl
8.0 M Urea 1M NaCl
1.0% SDS

pH 7.4, Protease Inhibitors |

Sample	Bone Differentiation
Bone Matrix	+
Dissociative Extract	—
Residue	—
Residue + Extract 'Reconstitution'	+

FIGURE 5

LOCAL REGULATION OF BONE DIFFERENTIATION BY EXTRACELLULAR MATRIX COMPONENTS

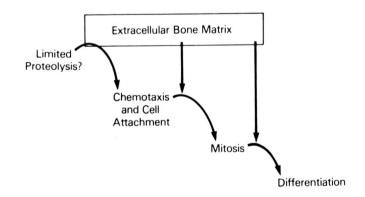

were devoid of biological activity. However, on
reconstitution of the respective residues with extracts by
dialysis against water, the bone inductive property was
restored. Gel filtration of the 4 M guanidine hydrochloride
extract on Sepharose CL-4B resulted in a broad heterogeneous
peak. Various pooled fractions were tested for bone
inductive property by reconstitution and it was found that
the fraction consisting of proteins less than 50,000 daltons
was active. We are currently involved in further
purification of these extracellular matrix components. In
view of the fact that chemotaxis and attachment of cells to
the matrix is an early step the chemotactic potency of 4 M
guanidine hydrochloride extracts was examined. Recent
experiments (Somerman et al., 1982) demonstrated potent
chemotactic activity in proteins eluted from a gel
filtration column of Sepharase CL-6B, in the region of
60,000 daltons. Examination of the mitogenic activity in
these extracts revealed that osteoinductively active
fractions (< 50,000 daltons) were growth promoting to
fibroblasts from rat-embryo and human skin (Sampath et al.,
1982).

CONCLUSION

Taken together these results imply that there might be
a cascade-type mechanism, as depicted in Fig. 5., that
governs extracellular matrix-induced bone differentiation.
The initial step in this scheme might be the release of
chemotactic factors by limited proteolysis. Then the
arrival of mesenchymal cells by chemotaxis is followed by
attachment to the matrix via fibronectin and related cell
adhesive proteins. The release of mitogenic factors from
the matrix promotes growth of cells. Finally the cells
differentiate into cartilage and bone in response to
specific matrix-derived factors. These observations may
have important implications for the local control of
skeletal differentiation and remodeling in postfetal life.
It is likely that extracellular matrix components might
specify positional information during embryonic development.
Advances in our understanding of the basic phenomena
regulating bone differentiation and growth should be
paralled by the exploration and development of the clinical
potential of these osteogenic factors in periodontal
diseases, orthopedic problems and congenital skeletal
anomalies.

Anastassiades T, Puzic O, Puzic R (1978). Effect of solubilized bone matrix components on cultured fibroblasts derived from neonatal rat tissues. Calcif Tiss Res 26:173.

Folkman J, Moscana A (1978). Role of cell shape in growth control. Nature 273:345.

Gospodarowicz D, Vlodavsky I, Savion N (1980). The extracellular matrix and the control of proliferation of vascular endothelial and smooth muscle cells. J. Supramol. Structure 13:339.

Rath NC, Reddi AH (1979). Collagenous bone matrix is a local mitogen. Nature 278:855.

Rath NC, Hand AR, Reddi AH (1981). Activity and distribution of lysosomal enzymes during collagenous matrix-induced cartilage, bone and bone marrow development. Develop Biol 85:89.

Reddi AH (1976). Collagen and cell differentiation. In Biochemistry of Collagen, Ramachandran GN, and Reddi AH pp 449-478. New York: Plenum Press.

Reddi AH (1981). Cell biology and biochemistry of endochondral bone development. Cell Res 1:209.

Reddi AH, Anderson WA (1976). Collagenous bone matrix-induced endochondral ossification and hemopoiesis. J Cell Biol 69:557.

Reddi AH, Huggins CB (1972). Biochemical sequences in the transformation of fibroblasts in adolescent rats. Proc Nat Acad Sci USA 69:1601.

Reddi AH, Huggins CB (1975). Formation of bone marrow in fibroblast transformation ossicles. Proc Nat Acad Sci USA 72:2212.

Reddi AH, Gay R, Gay S, Miller EJ (1977). Transitions in collagen types during matrix-induced cartilage, bone and bone marrow formation. Proc Nat Acad Sci USA 74:5589.

Reddi AH, Hascall VC, Hascall GK (1978). Changes in

proteoglycan types during matrix-induced cartilage and bone development. J Biol Chem 253:2429.

Sampath TK, Reddi AH (1981). Dissociative extraction and reconstitution of extracellular matrix components involved in local bone differentiation. Proc Nat Acad Sci USA 78:7599.

Sampth TK, DeSimone DP, Reddi AH (1982). Role of extracellular matrix in bone induction. Proc 5 Int Conference on Calcified Tissues. (in press).

Somerman M, Hewitt AT, Reddi AH, Seppa A, Varner H, Termine JD, Schiffman E (1982). The role of chemotaxis in bone induction. Proc 5th Int Conference on Calcified Tissues. (in press).

Stoker M, O'Neill C, Berryman S, Waxman V (1968). Anchorage and growth regulation in normal and virus transformed cells. Int J Cancer 3:683.

Urist, MR (1965). Bone: Formation by autoinduction. Science 150:893.

Weiss RE, Reddi AH (1980). Synthesis and localization of fibronectin during collagenous matrix-mesenchymal cell interaction and differentiation of cartilage and bone in vivo. Proc Nat Acad Sci USA. 77:2074.

Weiss RE, Reddi AH (1981). Role of fibronectin in collagenous matrix-induced mesenchymal cell proliferaton and differentiation in vivo. Exp Cell Res 133:247.

SECTION SEVEN
MUSCLE

Limb Development and Regeneration
Part B, pages 271–280
© **1983 Alan R. Liss, Inc., 150 Fifth Avenue, New York, NY 10011**

CELL DIVERSIFICATION: DIFFERING ROLES OF CELL LINEAGES AND
CELL-CELL INTERACTIONS

H. Holtzer, J. Biehl, R. Payette, J. Sasse, M.
Pacifici, S. Holtzer
Dept. of Anatomy, Medical School, University of
Pennsylvania
Philadelphia, PA 19104

This review focuses on the basic mechanisms generating
cell diversification. It stresses the theoretical
implications of our finding that relatively autonomous
founder cells for at least 5 different lineages are already
present in 15-17 hour dispersed chick blastodisc cells
(Payette et al., 1979; Holtzer et al., 1981a). Cell
diversification occurs in a simple culture medium and in the
absence of tissue-tissue interactions or migrations,
provided that the cells are permitted a minimal number of
quantal cell cycles. These observations render untenable
the existence at any stage of development of "undetermined",
"uncommitted", "undifferentiated" or even "multipotential"
cells. These findings also suggest that, in spite of a
massive literature to the contrary, tissue-tissue interactions
or the effects of exogenous molecules on embryonic cells are
critical only in that they permit already committed,
differentiated cells to express one or the other of their
binary options (Holtzer, 1978a; Holtzer et al., 1980).
Exogenous molecules do not act on "blank" cells to mediate
the inheritable rearrangements in chromosomal structure that
determine the limited differentiation-program of any given
cell. The generation of cell diversity requires "transit"
through a lineage, which in turn requires DNA synthesis and
passage through a sequence of quantal cell cycles. The
consequence of a quantal cell cycle is daughter cells with
relatively stable transcription complexes that differ
qualitatively from those in their mother cell. In contrast,
augmentation of numbers of cells within a given lineage-
compartment involves proliferative cell cycles - daughter
cells with the same transcription complexes as those of their

mother cell (Holtzer et al., 1972, 1975a, 1980).

This perspective of lineages has guided our experiments with limb buds and somites (Holtzer, 1968; Abbott et al., 1972, 1974; Dienstman et al., 1974; Holtzer et al., 1978a, 1980b). By cloning single cells or growing dispersed cells from different stages, we concluded that: (1) There are no "multipotential mesenchymal cells" in the sense that a single cell itself, without replicating, can become a definitive myoblast or chondroblast, or can yield, after a single cell cycle, both a definitive myoblast and a chondroblast: (2) Cells in early somites or limb buds have already bifurcated into either the myoblast-fibroblast or the chondroblast-fibroblast lineages: (3) Alterations in culture conditions can select for the replication, expression, or survival of cells in either lineage, but such manipulations do not establish founder cells or convert cells from one lineage into the other: (4) Cell-cell interactions control the number of proliferative cell cycles within a given lineage compartment. In this way the masses of cells required for the processes of morphogenesis are accumulated. Failure to distinguish between the extracellular influences that a) permit cells to express their unique metabolic options and b) regulate morphogenesis, from the intracellular mechanisms that generate cell diversity has lead to much confusion regarding differentiation (Holtzer et al., 1970, 1972, 1975a).

Founder cells in 15-17 hour chick blastodiscs: Trypsin-dissociated blastodisc cells have been cultured. Many of these yolk-rich cells die, but others adhere to the substrate and replicate. By days 3-4 the substrate is covered by a sheet of yolk-rich cells surrounding islands of small epithelioid cells lacking yolk. The yolk-free cells replicate 4 times more frequently than the yolk-rich cells.

Of the surviving cells, approximately 70% are vimentin- and 30% are cytokeratin-positive. Of the vimentin-positive cells, 20-40% are also desmin-positive. Desmin is synthesized only by cells in the terminal compartment of the smooth, cardiac, or skeletal myogenic lineages (Holtzer et al., 1981b, 1982a); neither vimentin nor desmin is synthesized in mouse morula cells (Jackson et al., 1981). Vimentin is synthesized in early compartments of the neurogenic, erythrogenic, myogenic, and chondrogenic lineages (Tapscott et al., 1981a,b; Woodcock, 1981).

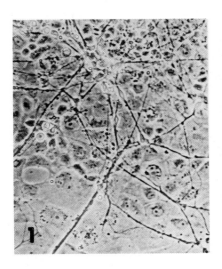 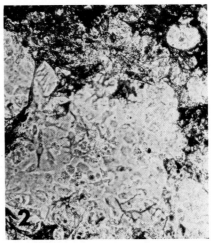

Figs. 1 and 2 are phase micrographs of day 10 cultures
prepared from dispersed 15-17 hour blastodiscs. Fig. 1
illustrates a cluster of postmitotic neurons with long
branching fascicles. Fig. 2 illustrates part of a focus of
pigmented, dendritic melanoblasts.

An invariant sequence of definitive phenotypes emerges
in these cultures. Within 24-36 hours, small clusters of
embryonic erythroblasts appear. These benzidine-positive
cells replicate 2-3 generations, become postmitotic, and die
(Weintraub et al., 1972; Holtzer et al., 1975a). The
addition of Ara-C at the time of plating blocks the emergence
of benzidine-positive cells. However, Ara-C added to 30
hour cultures does not block the emergence of erythroblasts.
Apparently, some sub-set of blastodisc cells will not
transcribe and/or translate globin mRNA unless the cells
undergo DNA synthesis during the first 30 hours of culture
(Campbell et al., 1974; Holtzer et al., 1975a; Holtzer, 1978a;
Groudine and Weintraub, 1981).

Day 3-4 cultures frequently display branching tubes
composed of endothelial-like cells. Often such tubes contain
erythroblasts. It is worth stressing that blood vessels are
one of the earliest structures to form during normal
embryogenesis.

Many postmitotic neurons are present in day 5-6 cultures

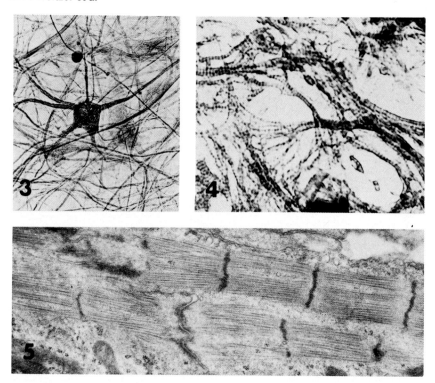

Fig. 3 is a micrograph of a neuron treated with anti-NIF$_{180}$
and then incubated with peroxidase-anti-peroxidase. Note
the numerous stained neurites from other nerve cells. Fig.
4 is a cluster of approximately 20 branching, striated cardiac
myoblasts treated with anti-myosin and then incubated with
peroxidase-anti-peroxidase. No skeletal myoblasts are present
in these cultures. Fig. 5 is an EM micrograph illustrating
the striations, intercalated disc and gap junctions
characteristic of immature cardiac myoblasts.

(Figs. 1 and 3). These neurons bind antibody to each of the
3 proteins that comprise the 10nm neurofilaments (Holtzer
et al., 1981b; Bennett et al., 1981; Tapscott et al., 1981a,b).
Adding Ara-C for 48 hours to day 3 cultures blocks the
emergence of nerve cells.

By day 7-8 dendritic and epithelioid pigmented
melanoblasts appear. They increase greatly in number. Foci
of melanoblasts of over 100 cells/focus are invariably

present in day 10 cultures. Ara-C added for 48 hours to day 3 cultures likewise blocks the emergence of pigmented cells (Fig. 2).

To determine whether myogenic cells were present, 14 standard day-10 cultures were stained with anti-myosin. This anti-myosin IgG binds to the A-bands of both skeletal and cardiac myoblasts; it does not bind to myosin in smooth muscles or presumptive myoblasts (Holtzer et al., 1957, 1972, 1980b, 1982b). Nine out of 14 cultures displayed patches of branching cardiac myoblasts (Fig. 4). That these were cardiac rather than skeletal myoblasts was confirmed by EM (Fig. 5). Ara-C added between days 2-4 blocked the emergence of cardiac myoblasts. It is noteworthy that skeletal myoblasts have never been identified in these cultures. These findings not only suggest an obligatory requirement for DNA synthesis for some sub-set of blastodisc cells to yield progeny that transit through the cardiac lineage, but that as early as 15-17 hours of incubation the skeletal and cardiac myogenic lineages have irreversibly diverged.

In brief, dispersed, replicating blastodisc cells can yield terminally-differentiated erythroblasts, neuroblasts, trunk and retinal melanoblasts, and cardiac myoblasts in the absence of inductive mesenchymal-epithelial tissue-tissue interactions.

Emergence of the skeletal myogenic and chondrogenic lineages: We have roughly determined the stage of embryogenesis at which cells in the myogenic and chondrogenic lineages become autonomous and are able, when cultured, to pass through all the compartments of their respective lineages and differentiate terminally. By definition, definitive myoblasts are mononucleate, postmitotic, striated cells (Holtzer et al., 1957, 1972, 1981a; Antin et al., 1981); definitive chondroblasts are cells which synthesize cartilage-specific Type IV sulfated proteoglycans (Okayama et al., 1976; Pacifici et al., 1977; Sparks et al., 1980). Dissociated whole embryos of 25-30 hours do not give rise either to definitive myoblasts or to chondroblasts. Similarly, dispersed cells from 30-35 hour embryos do not yield myoblasts or chondroblasts. Cells from 50-hour embryos do form small clusters of myoblasts and small myotubes: they do not form chondroblasts. Cells from 65-hour embryos will even form small numbers of definitive myoblasts in Ara-C. At low density cells from 65-hour embryos do not yield

chondroblasts, but do so at high density. High density favors
chondrogenesis and reasons for suspecting that these are
cranial rather than vertebral or limb chondroblasts are
discussed elsewhere (Holtzer, 1968; Holtzer and Abbott, 1968;
Holtzer et al., 1970, 1975). Of interest is the fact that
cells reared in F-10, a medium known to promote chondrogenesis,
blocks the emergence of definitive myoblasts. It is
probable that other culture conditions (e.g., addition of
ascorbate, cAMP, "growth factors", etc.) will allow the
emergence of myoblasts and chondroblasts from younger stages
than those reported here.

Cells in the myogenic and chondrogenic lineages in
early limb buds: When cells in later compartments of the
myogenic or chondrogenic lineages are cloned, they yield
populations containing definitive myoblasts and chondroblasts.
Cloned cells from stage 21 limb buds give rise to either
myoblast-fibroblast clones or chondroblast-fibroblast clones.
Mixed clones consisting of myoblasts and chondroblasts are
never observed. For a discussion of the constraints that
these findings impose on the number of generations which
separate a common "mesenchymal" precursor from its descendent
myoblasts and chondroblasts, see Dienstman et al. (1974) and
Holtzer (1978a, 1981a).

Recently, we cultured stage 20 limb buds at high and
low density. The major findings with respect to myogenesis
are: (1) At high and low density striated, postmitotic
mononucleated myoblasts appeared after 2-3 days (Holtzer et
al., 1980b). Ara-C did not preclude their emergence. That
these myoblasts share properties with their counterpart
myoblasts from 11-day breast muscle is shown by their
response to Cytochalsin-B (CB). CB arborizes replicating,
presumptive breast myoblasts, whereas it induces postmitotic
myoblasts to become spherical and prevents their fusing into
myotubes (Croop and Holtzer, 1975; Holtzer et al., 1975b).
Presumptive myoblasts and postmitotic myoblasts from stage
21 limb buds respond similarly to CB.

The major findings with respect to chondrogenesis are:
At low density chondroblasts did not emerge, although at
high density nodules of cartilage appeared on days 4-6.
With respect to cell-cell interactions it is worth stressing
that while single cells from stage 21 can yield chondrogenic
clones, cells from entire limb buds plated at low density do
not form chondroblasts. This apparent discrepancy is

interpreted by the "interference" experiments first performed
by Chacko et al. (1969) and Holtzer et al. (1970), and
recently confirmed by Solursh   et al. (1982). When
definitive chondroblasts are mixed with dispersed myogenic,
fibrogenic, liver, or even BudR-blocked chondroblasts at a
1:5 ratio and cultured, chondrogenesis is blocked. If such
blocked cultures are trypsinized and the cells cloned, many
cells yield chondrogenic clones days later. This suggests
that the microenvironment created by non-chondrogenic cells
reversibly depresses the expression of the chondrogenic
differentiation program of even fully-differentiated single
chondroblasts. Experiments with ts-RSV transformed
chondroblasts (Pacifici et al., 1977; Holtzer et al., 1981a)
yield complementary results. Reared at permissive
temperature for over 15 generations, infected chondroblasts
fail to express their chondrogenic phenotype; 30 hours after
being shifted to non-permissive temperature the infected
cells differentiate terminally. Clearly the differentiation
program of chondroblasts can be transmitted with fidelity to
their progeny for many generations without being expressed.
Presumably, the pp60 src kinase blocks expression of the
chondrogenic phenotype (as do inadequate culture media)
without changing the lineage status of the transformed cells.
Similar experiments have been performed with myogenic cells
(Holtzer et al., 1975c; Fizman and Fuchs, 1975; Moss and
Martin, 1979).

Conclusions: The major thrust of these experiments is
that: (1) The founder cell for all lineages is the zygote:
(2) Within hours after fertilization and consequent to a
small sequence of quantal cell cycles several remarkably
autonomous and already-restricted lineages are established:
(3) The formation of such lineages is independent of
embryogenesis, gastrulation, putative inductive interactions
or complex tissue migrations: (4) The neurogenic lineage
emerges earlier than the melanogenic lineage and neither is
dependent on neurulation: (5) The cardiac and skeletal
myogenic lineages emerge at different times, and the latter
arises earlier than the chondrogenic lineage: (6) Cells
enter and transit asynchronously through their respective
lineages: (7) Modest changes in the culture media, or in
the in vivo microenvironment, may select for cells in one
rather than another lineage.

Earlier data on cloned cells from stage 21-22 limb buds
(Dienstman et al., 1974), and the findings reported here,

are in excellent accord with the recent chick-quail grafting
experiments of Christ et al. (1977, 1979). They are
incompatible with the contention of many investigators that
the early limb bud or somites consist of "uncommitted,
multipotential, mesenchymal cells" that are induced by
inductive interactions or exogenous molecules to become
myoblasts or chondroblasts. If transit through a succession
of compartments in their respective lineage is obligatory,
then these simple observations imposes severe constraints
on the possible role(s) of exogenous molecules in the
generation of cell diversification. The major role of
tissue-tissue interactions is more probably associated with
events ordering morphogenesis.

References

Abbott J, Mayne R, and Holtzer H (1972). Devel. Biol. 28:
430.
Abbott J, Schiltz J, Dienstman S, and Holtzer H (1974). Proc.
Nat. Acad. Sci. 71:1506.
Antin P, Forry-Schaudies S, Tapscott S, and Holtzer H (1981).
J. Cell Biol. 90:300.
Bennett G, Tapscott S, Antin P, and Holtzer H (1981).
Science 212:567.
Campbell G, Weintraub H, and Holtzer H (1974). J. Cell
Physiol. 83:11.
Chacko S, Holtzer S, and Holtzer, H (1969). Biochem. Biophys.
Res. Comm. 34:183.
Christ H, Jacob H, and Jacob M.(1977). Anat. Embryol. 150:
171.
Christ H, Jacob H, and Jacob M (1979). Experientia 35:1376.
Croop J, and Holtzer H (1975). J. Cell Biol. 65:271.
Dienstman S, Biehl J, Holtzer S, and Holtzer H (1974).
Devel. Biol. 39:83.
Groudine M, and Weintraub H (1981). Cell 24:393.
Fizman M, and Fuchs P (1975). Nature 254:429.
Holtzer H, Marshall J, and Finck H (1957). J. Cell Biol. 3:
705.
Holtzer H (1968). In: Epithelial-Mesenchymal Interactions
(ed. Fleischmajer) pp. 152-164, Williams and Wilkins,
Baltimore.
Holtzer H, and Abbott J (1968). In: Stability of the
Differentiated State (ed. Ursprung) pp. 1-18, Springer-
Verlag, Berlin.
Holtzer H, and Matheson D (1970). In: Chemistry and
Molecular Biology of the Intercellular Matrix (ed. Balazs)

pp. 1753-1769, Academic Press.

Holtzer H, Chacko S, Abbott J, and Holtzer S (1970). In: Chemistry and Molecular Biology of the Intercellular Matrix (ed. Balazs) pp. 1471-1481.

Holtzer H, Weintraub H, Mayne R and Mochan B (1972). In: Curr. Topics Devel. Biology (eds. Moscona and Monroy) pp. 229-265, Academic Press.

Holtzer H, Sanger J, Ishikawa H, and Strahs K (1972). Cold Spring Harbor Symp. Quant. Biol. 37:549.

Holtzer H, Rubinstein N, Fellini S, Yeoh G, and Okayama M (1975a). Quart. Rev. Biophysics 8:523.

Holtzer H, Croop J, Dienstman S, and Somlyo A (1975b). Proc. Nat. Acad. Sci. 72:513.

Holtzer H, Biehl J, Yeoh G, and Kaji A (1975c). Proc. Nat. Acad. Sci. 72:4051.

Holtzer H (1978a). In: Stem Cells and Tissue Homeostasis (eds. Lord, Potten and Cole) pp. 1-28, Cambridge University Press, Cambridge.

Holtzer H, Okayama M, Biehl J, and Holtzer S (1978b). Experientia 34:281.

Holtzer H, Biehl J, West C, Boettiger D, and Payette R (1980a). In: Differentiation and Neoplasia (ed. Gehring) pp. 63-78, Springer-Verlag, Berlin.

Holtzer H, Croop J, Toyama Y, Bennett G, and Fellini S (1980b). In: Plasticity of Muscle (ed. Pette) pp. 133-146, Walter de Gruyter & Co., Berlin.

Holtzer H, Pacifici M, Croop J. Toyama Y, Dlugosz A (1981a). Fortschritte der Zoologie 26:207.

Holtzer H, Bennett G, Tapscott J, Dlugosz A, and Toyama Y (1981b). In: International Cell Biology 1980-1981 (ed. Schweiger) pp. 293-305, Springer-Verlag, Berlin.

Holtzer H, Bennett G, Tapscott S, Croop J, and Toyama Y (1982a). Cold Spring Harbor Symp. Quant. Biology 46:458.

Holtzer H, Forry-Schaudies S, Antin P, and Toyama Y (1982a). In: Molecular and Cellular Control of Muscle Development. Cold Spring Harbor (in press).

Jackson B, Grund C, Schmid E, Burki K, Franke W, and Illmensee K (1980). Differentiation 17:161.

Moss P, Honeycutt N, and Martin G (1979). Exp. Cell Res. 23:95.

Okayama M, Pacifici M, and Holtzer H (1975). Proc. Nat. Acad. Sci. 73:3224.

Pacifici M, Boettiger, D, Roby K, and Holtzer H (1977). Cell 11:891.

Payette R, Biehl J, and Holtzer H (1979). Anat. Rec. 193:647.

Solursh M, and Reiter R (1980). Devel. Biol. 78:141.
Sparks K, Lever P, and Goetinck P (1980). Arch Bio. and
    Biophys. 199:579.
Tapscott S, Bennett G, and Holtzer H (1981a). Devel. Biology
    86:40.
Tapscott S, Bennett G, Toyama Y, Kleinbart F, and Holtzer H
    (1981b). Nature 292:836.
Weintraub H, Campbell G, and Holtzer H (1972). J. Cell Biol.
    50:652.

This research was supported in part by the National
Institutes of Health grants HL-18708, HL-15835 (to the
Pennsylvania Muscle Institute) CA-18194 and 5-T32-HD07152
and by the Muscular Dystrophy Association.

Limb Development and Regeneration
Part B, pages 281–291
© **1983 Alan R. Liss, Inc., 150 Fifth Avenue, New York, NY 10011**

ON THE ORIGIN, DISTRIBUTION AND DETERMINATION OF AVIAN LIMB
MESENCHYMAL CELLS

Bodo Christ, H.J. Jacob, M. Jacob, F. Wachtler

Institute of Anatomy, Ruhr-University D-4630 Bochum, Federal Republic of Germany; Inst. of Histology and Embryology, University of Vienna, Austria

The avian limb bud is enclosed by an ectodermal envelope and consists of a mass of still undifferentiated mesenchymal cells representing the matrix of various tissues like muscles, cartilage, bone, tendons and soft connective tissue.

For a long time the limb bud mesenchyme has been thought to consist of a homogenous population of pluripotential somatopleural cells. Their developmental fate was considered to be a function of cellular interactions and positional information (ZWILLING, 1968; CAPLAN and KOUTROUPAS, 1973; SUMMERBELL et al., 1973).

Another way of explaining how diversification of limb bud mesenchymal cells occurs is to suppose that the mesenchyme is made up by distinct cell lines. Studies involving cell cultures (DIENSTMAN et al., 1974; NEWMAN, 1977) as well as the observation that somite cells do contribute to wing development (GRIM, 1970; GUMPEL-PINOT, 1974) have established the view that at least two subpopulations of mesenchymal cells already exist within the early limb bud.

Recent studies using the quail-chick marker system have shown clearly that the striated muscles of the limb originate from the somites, whereas cartilage, tendons and the whole of the connective tissue derive from the somatopleure (CHRIST et al., 1974a, 1977; CHEVALLIER et al., 1976, 1977).

In this paper we will give an outline of our current view on problems of the origin, distribution and determination of avian limb bud mesenchymal cells.

I. ORIGIN OF LIMB BUD MESENCHYMAL CELLS

After quail-to-chick replacements of the somites at the prospective limb levels which were carried out on embryos at HH-stages 12 - 15, the striated muscle cells of the developed limbs are constantly characterized by quail nuclei (Fig. 1), whereas the cartilage, the tendons and the soft connective tissue consist of cells showing chick nuclei. This distribution of quail (somitic) and chick (somatopleural) cells is seen to be generally established in all parts of the limbs (CHRIST et al., 1974b, 1979b).

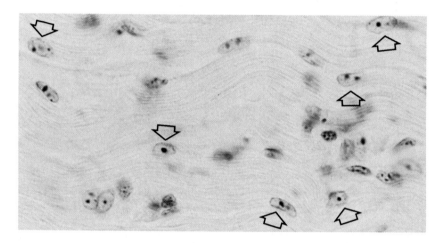

Fig. 1. M. flexor carpi ulnaris from a 2-week baby chicken after replacement of brachial somites. Arrows: quail nuclei within muscle fibres.   x 1100.

In a series of complementary experiments the somatopleural epithelium of the prospective limb regions in chick hosts was replaced by corresponding parts of quail somatopleure. The obtained limbs show cartilage, tendons and soft connective tissue of quail (somatopleural) origin. The striated muscle cells are always found to contain chick nuclei (Fig. 2), wherefore they should have originated from the host somites. It is interesting to note that the smooth muscle cells are found to be of somatopleural origin (CHRIST et al., 1979a; KIENY et al., 1979).

Another possible source of limb bud cells might be the neural crest. It is well known from studies on amphibian embryos that neural crest cells participate in limb mesenchyme formation (ROLLHÄUSER-TER-HORST, 1975).

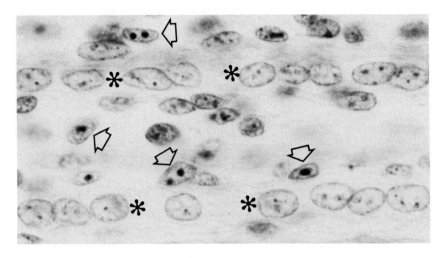

Fig. 2. Muscle blastema at the upper leg level seven days after replacement of somatopleure. Asterisks: myotubes containing chick nuclei; arrows: quail fibroblasts.   x 2000.

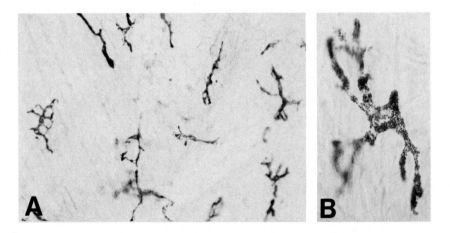

Fig. 3. Fascia of the M. flexor carpi ulnaris of a chick (White Leghorn) 18 days after replacement of the neural tube. Take notice of the distribution (A) and the shape (B) of the pigment cells.   x 140 (A),   x 1100 (B).

   To test this possibility for the avian limb bud the brachial part of the neural tube of 2-day chick embryos (White
Leghorn) was replaced by a piece of quail neural tube. The
developed wings show in addition to the Schwann cells and the
melanocytes situated within the epidermis, numerous branched
pigment cells of quail origin which are seen in the dermis
and the muscular connective tissue (Fig. 3). The cartilage,
however, is devoid of neural crest cells. Further studies
will have to ask for possible developmental functions of these
cells.

   We hold that the avian limb bud mesenchyme represents
cells from three specific lineages:
1. The somatopleural cells form the skeletal elements, tendons, all kinds of soft connective tissue and the smooth
   muscle cells.
2. The somite cells give rise to the striated muscle cells.
3. Neural crest cells constitute an integral part of the soft
   connective tissue.

II. DISTRIBUTION OF LIMB BUD MESENCHYMAL CELLS

   At first, when the somatopleure is still of epithelial
structure, the limb fields are devoid of somite and neural
crest cells. The immigration of myogenic (somite) cells starts
at stage 14 at the wing and at stage 16 at the leg level. Mean-
while the somatopleure has changed into mesenchyme. JACOB et
al. (1978, 1979) have particularly described the migration modalities of the myogenic stem cells.

   After the limb buds are manifest, there is at first no
preferential localization of the somite cells within the proximal part of the limb bud mesenchyme (GUMPEL-PINOT, 1974).
Later on, myogenic cells gather into the dorsal and ventral
intermediate zones of the prospective stylopod and zeugopod,
where they form the premuscular masses, in which a considerable mitotic activity can be found (Fig. 4). The most distal
part of the limb bud mesenchyme is not yet colonized by somitic cells (NEWMAN et al., 1981). Later on the myogenic cells
gradually invade the distally added mesenchyme, proliferate
and undergo differentiation.

   To observe the migration capacity myogenic quail material was grafted to the wing bud of chick embryos. Quail nuclei
were found in the myotubes of the normally developed musculature, not only at the level of the graft, but also distally
from there. However, no chimaeric myotubes can be seen proxi-

mally. It can be concluded that myogenic cells are capable only of migrating in a proximo-distal direction but over considerable distances (WACHTLER et al., 1982).

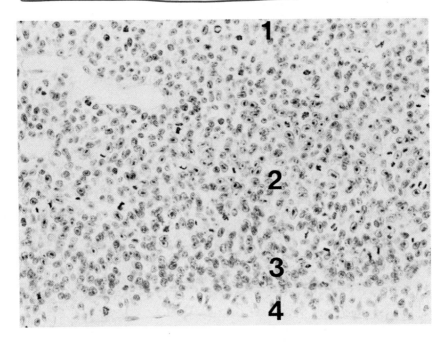

Fig. 4. Ventral premuscular mass of the proximal part of a
leg bud 2 days after replacement of the somites. 1 chondrogenic zone; 2 myogenic zone; 3 dermis; 4 ectoderm.   x 350.

   Concerning the neural crest cells invading the limb bud,
it is well known that during normal development the Schwann
cells migrate along the axons in a proximo-distal direction.
After grafting pieces of quail neural tube into the wing of
chick embryos of a white race, Schwann cells migrate in both,
the proximo-distal and the disto-proximal direction, whereas
the pigment cells as the myogenic cells are only capable to
migrate in one direction (WACHTLER et al. unpublished data).

One can summarize:

1. At first the limb anlage is represented only by mesenchyme
   of somatopleural origin.
2. Already before limb bud outgrowth starts, some of the myogenic cells have invaded the limb.
3. Migration of myogenic cells and their aggregation into pre-

muscular masses takes place gradually in a proximo-distal direction.

4. The occurence of pigment cells is restricted to the epidermis, the dermis and the myogenic zone.

III. DETERMINATION OF LIMB BUD MESENCHYMAL CELLS

The question remains whether the limb bud mesenchymal cells are already determined by their different origin or if their determination depends on their position within the limb.

First the differentiating abilities of somatopleural cells had to be tested. Somatopleural mesoderm from limb levels, which had not yet been invaded by somitic cells, were isolated together with the adjacent ectoderm from quail donors and grafted into the coelomic cavity or onto the chorioallantoic membrane of chick hosts. The grafts, cultivated from 5 to 10 days, exhibit normally arranged skeletal elements and dermis (CHRIST et al., 1977, 1979a). However, these grafts are characterized by a total absence of striated muscles (Fig. 5).

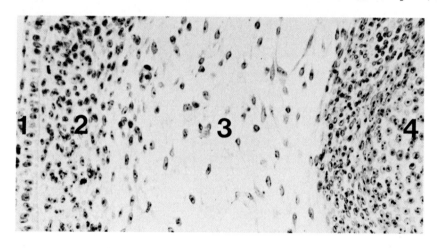

Fig. 5. Quail wing grown (7 days) in the coelomic cavity. Donor stage: 13 (HH). 1 epidermis; 2 dermis; 3 myogenic zone without muscles; 4 cartilage. x 350.

The zone between the cartilage and the dermis contains loosely arranged mesenchymal cells. In the distal part of the muscleless limb tendons can be seen (JACOB and CHRIST, 1980). Grafts cultivated on the chorioallantoic membrane are in all cases devoid of pigment cells. The observation has to be added that after exact destroying the brachial somites by X-irradiation wings develop without muscles (CHEVALLIER et al., 1978; LEWIS et al., 1981).

Because somatopleure has repeatedly been assumed to give rise not only to cartilage and connective tissue,but may also in the absence of somite cells be capable of differentiating into musculature (McLACHLAN and HORNBRUCH, 1979; KIENY, 1980; MAUGER and KIENY, 1980), prospective wing somatopleure of quail embryos (HH-stages 9 - 14) were grafted into the myogenic zone of chick wings (HH-stages 18 - 24). Though the grafts developed in an environment known to permit expression of myogenic potencies, somatopleural cells are never found to differentiate into striated muscle cells in any of the experiments, but they have formed cartilage and soft connective tissue (WACHTLER et al., 1982).

To test the determination state of those mesenchymal cells which belong to the myogenic lineage two-step grafting procedures were performed as described by WACHTLER et al. (1981). Pieces of mesenchyme from chick wing buds containing myogenic cells of quail origin were grafted into chick wing buds, so that myogenic parts of the grafts would be situated in cartilage forming regions of the host. Hosts and donors belonged to any stages between 19 to 25 (HH). Quail cells were found almost exclusively in the blastemata of striated muscles, not within the cartilage (Fig. 6). The quail cells, only incidently observed in soft connective tissue could be of neural crest origin because the somites, grafted at the first experimental step, already could have been invaded by neural crest material.

The results indicate that cells after grafting into regions, which are known to allow the differentiation into cartilage as well as into striated muscle, develop "herkunftsgemäß". Therefore, we consider them to be determined. There is no significance that neural crest cells within the limb bud are able to differentiate into cartilage, muscles or muscle satellite cells as suggested for the head neural crest (LE LIÈVRE and LE DOUARIN, 1975; JOHNSTON et al., 1979). The observation that under extreme experimental conditions cells of one lineage may be able to develop muscle or cartilage phenotypes (KIENY, 1980; NATHANSON and HAY, 1980) are not in all means inconsistent with our findings (for discussion see WACHTLER et al., 1981).

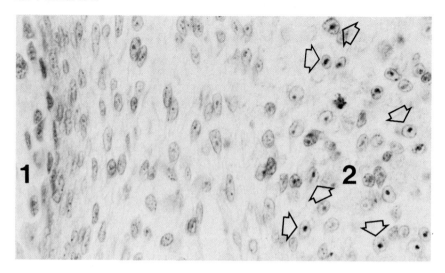

Fig. 6. Zeugopod five days after two-step grafting procedure (see text for details) 1 radius; 2 muscle blastema; arrows: quail nuclei.   x 1000.

Because cartilage and fibroblast progenitor cells both belong to the same lineage of somatopleural cells, the question arises how they become determined with respect to their option on cartilage or non-cartilage development. For better understanding pieces of quail wing mesenchyme were grafted to chick wing buds so, that portions of the prospective cartilage forming regions were replaced by non-cartilage regions and vice versa. Host embryos ranged from HH-stages 19 to 25 and donors from 19 to 28. We found that mesenchyme from the cartilage forming regions is determined to form cartilage from HH-stage 20 on. Mesenchyme from non-cartilage forming regions retains the option to form cartilage or connective tissue at least to HH-stage 26 (WACHTLER et al., 1981). This means that formation of cartilage or non-cartilage phenotype of somatopleural cells depends on their position within the limb bud. Cells possibly would be determined by cellular interactions or by their exposure to certain gradients (SUMMERBELL et al., 1973; GUMPEL-PINOT, 1981).

We summarize:

1. From their origin cells of the distinct lineages which make up the limb bud core are not interchangeable under normal in vivo-conditions.

2. Cells of somatopleural origin differentiate into cartilage or non-cartilage tissue depending on their position within the limb bud.

Conclusions

The mesenchymal cellular diversification during limb development is unavailably bound to the colonization of limb anlagen with the distinct cell lines. Differentiation and arrangement of cartilage and connective tissue should be determined by positional information and cell interactions, in general, by factors within the limb bud.

Acknowledgements

This work was supported by grants of the Deutsche Forschungsgemeinschaft. We wish to thank U. Ritenberg, I. Schaeben-Hamm, I. Burks and H. Hake for excellent technical assistance, A. Boeckmann for helping with the English translation and B. Scharf for typing the manuscript.

Caplan AJ, Koutroupas J (1973). The control of muscle and cartilage development in the chick limb: the role of differential vascularisation. J Embryol Exp Morph 29:571.

Chevallier A, Kieny M, Mauger A (1976). Sur l'origine de la musculature de l'aile chez les Oiseaux. C R Acad Sci Paris Sèrie D 282:309.

Chevallier A, Kieny M, Mauger A (1977). Limb-somite relationship: origin of the limb musculature. J Embryol Exp Morph 41:245.

Chevallier A, Kieny M, Mauger A (1978). Limb-somite relationship: effect of removal of somitic mesoderm on the wing musculature. J Embryol Exp Morph 43:263.

Christ B, Jacob HJ, Jacob M (1974a). Über den Ursprung der Flügelmuskulatur. Experimentelle Untersuchungen mit Wachtel- und Hühnerembryonen. Experientia 30:1446.

Christ B, Jacob HJ, Jacob M (1974b). Experimentelle Untersuchungen zur Entwicklung der Brustwand beim Hühnerembryo. Experientia 30:1449.

Christ B, Jacob HJ, Jacob M (1977). Experimental analysis of the origin of the wing musculature in avian embryos. Anat Embryol 150:171.

Christ B, Jacob HJ, Jacob M (1979a). Differentiating abilities of avian somatopleural mesoderm. Experientia 35:1376.

Christ B, Hirschberg M, Jacob HJ, Jacob M (1979b). Experimentelle Befunde zur Entwicklung der distalen Extremitäten-
muskulatur. Verh Anat Ges 73:519.

Dienstman SR, Biehl J, Holtzer S, Holtzer H (1974). Myogenic
and chondrogenic lineages in developing limb buds grown in
vitro. Develop Biol 39:83.

Grim M (1970). Differentiation of myoblasts and the relationship between somites and the wing bud of the chick embryo.
Z Anat Entwickl Gesch 132:260.

Gumpel-Pinot M (1974). Contribution des mésoderme somitique
a la genèse du membre chez l'embryon d'Oiseau. C R Acad
Sci Paris Série D 279:1305.

Gumpel-Pinot M (1981). Ectoderm-mesoderm interactions in relations to limb-bud chondrogenesis in the chick embryo:
transfilter cultures and ultrastructural studies. J Embryol
Exp Morph 65:73.

Jacob HJ, Christ B (1980). On the formation of muscular pattern in the chick limb. In Merker H-J, Nau H, Neubert D
(eds): "Teratology of the Limbs", Berlin, New York: W de
Gruyter, p 89.

Jacob M, Christ B, Jacob HJ (1978). On the migration of myogenic stem cells into the prospective wing region of chick
embryos. A scanning and transmission electron microscope
study. Anat Embryol 153:179.

Jacob M, Christ B, Jacob HJ (1979). The migration of myogenic
cells from the somites into the leg region of avian embryos.
Anat Embryol 157:291.

Johnston MC, Noden DM, Hazelton RD, Coulombre JL, Coulombre,
AJ (1979). Origins of avian ocular and pericular tissues.
Exp Eye Res 29:27.

Kieny M, Mauger A, Chevallier A, Sengel P (1979). Origine
embryologique des muscles lisses cutanés chez les Oiseaux.
Arch Anat Micr 68:283.

Kieny M (1980). The concept of a myogenic cell line in developing avian limb buds. In Merker H-J, Nau H, Neubert D
(eds): "Teratology of the Limbs", Berlin, New York: W de
Gruyter, p 79.

Le Lièvre C, Le Douarin N (1975). Mesenchymal derivatives of
the neural crest: analysis of chimaeric quail and chick embryos. J Embryol Exp Morph 34:124.

Lewis L, Chevallier A, Kieny M, Wolpert L (1981). Muscle nerve branches do not develop in chick wings devoid of muscle.
J Embryol Exp Morph 64:211.

Mauger A, Kieny M (1980). Migratory and organogenetic capacities of muscle cell in bird embryos. Wilhelm Roux's Arch
Dev Biol 189:123.

McLachlan JC, Hornbruch A (1979). Muscle-forming potential of the non-somitic cells of the early limb bud. J Embryol Exp Morph 54:209.

Nathanson MA, Hay ED (1980). Analysis of cartilage differentiation from skeletal muscle grown on bone matrix. Develop Biol 78:301.

Newman SA (1977). Lineage and pattern in the developing wing bud. In Ede DA, Hinchliffe JR, Balls M (eds): "Vertebrate Limb and Somite Morphogenesis", London: Cambridge University Press, p 181.

Newman SA, Pautou M-P, Kieny M (1981). The distal boundary of myogenic primordia in chimaeric avian limb buds and its relation to an accessible population of cartilage progenitor cells. Develop Biol 84:440.

Rollhäuser-ter-Horst J (1975). Neural crest and early forelimb development in Amphibia. Anat Embryol 147:337.

Summerbell D, Lewis JH, Wolpert L (1973). Positional information in chick limb morphogenesis. Nature (Lond) 244:492.

Wachtler F, Christ B, Jacob HJ (1981). On the determination of mesodermal tissues in the avian embryonic wing bud. Anat Embryol 161:283.

Wachtler F, Christ B, Jacob HJ (1982). Grafting experiments on determination and migratory behaviour of presomitic, somitic and somatopleural cells. Anat Embryol (in press).

Zwilling E (1968). Morphogenetic phases in development. Develop Biol (Suppl) 2:184.

Limb Development and Regeneration
Part B, pages 293–302
© **1983 Alan R. Liss, Inc., 150 Fifth Avenue, New York, NY 10011**

CELL AND TISSUE INTERACTIONS IN THE ORGANOGENESIS OF THE
AVIAN LIMB MUSCULATURE

Madeleine A. Kieny, Docteur d'Etat ès Sciences

ERA 621, Laboratoire de Zoologie
Université Scientifique et Médicale
Grenoble, France 38041

The somitic origin of the skeletal limb myocytes, including satellite cells (Armand et al., 1982), is now well
established. But their arrangement to form separate muscles
is far from being fully understood. The approach followed
in this paper will be to consider the interaction between
muscle cells and presumptive muscle connective tissue cells
in the spatial organization of the musculature in the avian
limb, domain to which our group has been devoted to for the
last years.

The first morphological indication of muscle structure
is the appearance of premuscular masses dorsally and ven-
trally to the chondrogenic core from stage 23 of H.H. on
(Chevallier, 1978). Between stages 27 and 32, repeated fis-
sions of the muscle masses produce the muscle pattern characteristic of each limb segment. Particular attention has
been given to the zeugopod musculature, because it is constituted by numerous muscles. The cleavages start always in
the middle of the segment and from there extend proximally
and distally. In the forearm only binary splittings occur
(Shellswell and Wolpert, 1977). In the tibiotarsal segment,
the splittings of the dorsal mass are binary, those of the
ventral mass are binary, tertiary and binary again (figure
1) (Pautou et al., 1982). When a muscle mass or block splits
in two or in three, a slight decrease in cell density occurs
between the newly formed muscle structures. The mechanisms
whereby in that precise region cell density falls, thus
creating a cleft in the muscular tissue, are not known.

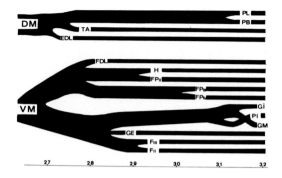

Fig. 1. Normal muscle splitting pattern in the tibiotarsal segment between stages 27 to 32 of H.H. DM, dorsal premuscular mass; VM, ventral premuscular mass. Muscle abbreviation: EDL, M. extensor digitorum longus; F_{II}, M. flexor perforans perforatus digiti II; F_{III}, M. flexor perforans perforatus digiti III; FDF, M. flexor digitorum longus; FP_{II}, M. flexor perforatus digiti II; FP_{III}, M. flexor perforatus digiti III; FP_{IV}, M. flexor perforatus digiti IV; GE, M. gastrocnemius, pars externa; GI, M. gastrocnemius, pars interna; GM, M. gastrocnemius, pars media; H, M. flexor hallucis longus; P or PL, M. plantaris; PB, M. peroneus brevis; PL, M. peroneus longis, TA, M. tibialis anterior.

The dual origin of the muscle tissue, namely the somitic origin of the muscle cells and the somatopleural origin of the muscular connective tissue raises the question of the respective activities of muscle cells and connective tissue cells in the development of the muscle pattern.

There are several experimental results which show that the connective tissue is responsible for the spatial region -specific organization of the muscles. 1) Using heterotopic chick/quail transplantations, previous work (Chevallier et al., 1977; Kieny and Chevallier, 1980) has shown that the pattern of musculature develops appropriately to the limb segment regardless of which group of somites is the source of the myogenic cells. Somitic myogenic precursor cells are equivalent. 2) Even myogenic cells that have already conglomerated into premuscular masses inside the limb bud remain indifferent and participate in a musculature which is in accordance with the limb segment into which they end up (Mauger and Kieny, 1980). 3) By modifying the temporal relationship between myogenic cell invasion and connective tissue cells during early limb bud development new evidence of the morphogenetic role of the connective tissue cells was obtained (Chevallier and Kieny, 1982). Muscle-cell deprived chick wing buds (Chevallier et al., 1978) were allowed to grow up to stages 22 to 27 of H.H., when they received a transplant of quail myogenic cells (somitic mesoderm or limb premuscular mass) into the dorsal aspect of their presumptive upper arm. Results showed that the connective tissue at first controls the proximodistal migratory behaviour of the myogenic cells, and then moulds the muscles in the forearm and hand, provided that the myogenic cells have accumulated before stage 27, which is the stage corresponding to the first cleavage of the forearm premuscular masses.

By what mechanisms do the connective tissue cells organize the pattern of muscles in particular in the tibiotarsal segment ? To further investigate this problem we undertook simultaneously several approaches. At first, a possible involvement of extracellular matrix components in the splitting process was examined. Mauger et al (1982) tried to interfere with collagen synthesis before the muscle splittings start, by administering, in ovo, inhibitors of collagen synthesis : L-azetidine-2-carboxylic acid, α-α'dipyridyl, β-aminopropionitrile. Whatever the drug used, there was no effect on the muscular patterning, although general deleterious effects on the embryos and specific damage on the limb skeleton were detectable. These results of chemical treatments allow us to join Shellswell et al (1980) in their conclusions to reject a primary role of collagen in the cleavage of wing muscle mass. These authors, by immunofluorescence studies on fixed and paraffin-embedded sections showed that collagens were only detectable once the partitions had occurred. As concerns the involvement of other extracellular matrix materials, such as glycosaminoglycans (Mauger et al., 1982) and fibronectin (Chiquet et al., 1982), there is up to now no clearcut implication of these components in the cleavage phenomenon.

Secondly, in collaboration with P.F. Goetinck, the embryology of the muscular dysgenesis in the crooked neck dwarf mutant (cn/cn) chick embryo was undertaken. Sequential histological observations performed on 6- to 12.5-day embryos issued from cn/+ parents showed that the lack of organization of the muscular tissue, reported by Wick and

Allenspach (1978) on 12-day embryos, is not the result of a
primary block of the subdivisions of the premuscular masses.
Indeed, muscle patterning proceeds normally up to the last
splitting (stage 31-32), and secondarily the individuated
muscles fuse together again and become then grouped into an
unpatterned muscle tissue (figures 2 and 3) (Kieny et al.,
1982). Thus, the mutant phenotype is not the result of an
incapability of the connective tissue to organize the muscle cells into individual muscles, but to an inability of

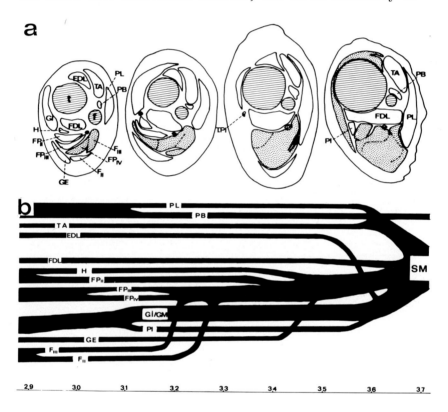

Fig. 2. Muscle patterning in the crooked neck dwarf (cn/cn)
mutant. a, anatomical development of the muscles, from left
to right, at stages 32, 33, 35 and 36 of H.H.; b, diagramma-
tical representation of the individuation and progressive
fusion of the muscles in a single mass (SM), between stages
29 to 37 of H.H. For muscle abbreviations, see figure 1.

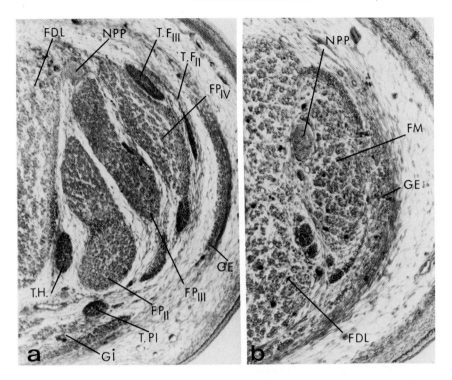

Fig. 3. Zeugopodial posterior muscle structures, at 10 days of incubation in normal (a) and mutant (b) siblings. NPP, peroneus profundus nerve; FM, muscle mass corresponding to coalescent flexor muscles except FDL. For muscle abbreviations, see figure 1.

this organization to become stabilized into definitive structures. The spaces between the muscles are unstable in the mutant and disappear progressively.

In order to gain more insight into the behavior of the connective tissue in the cn/cn mutant, the distribution of several extracellular matrix components was investigated, when the expression of the mutation becomes visible at 7.5 days and once it is further advanced, at 9.5 days. Indirect immunofluorescence technique on frozen sections was used. A comparative study (figure 4) between normal (table) and mutant siblings led to the following features: In the normal embryo, at 7.5 days, there is a clear difference between

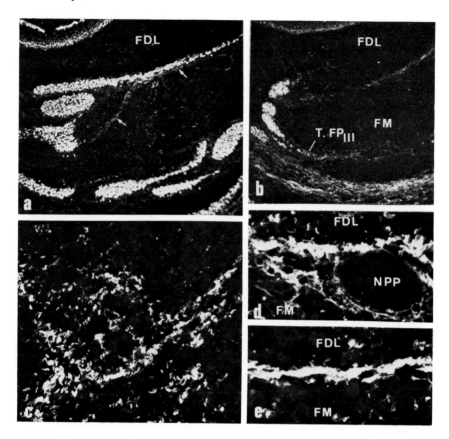

Fig. 4. Immunofluorescent studies of type I collagen in 9.5 day normal ($\underline{a}$) and mutant ($\underline{b}$, $\underline{c}$, $\underline{d}$, $\underline{e}$) posterior muscles. $\underline{a}$, strongly fluorescent areas correspond to tendons. Faint fibrous label decorates intermuscular spaces (arrow). Compare with figure 3a; $\underline{b}$, note the size reduction of tendons and the general weakness of labeling. Compare with figure 3b; $\underline{c}$, dislocation of the tendon of FP_{III} (T. FP_{III}) at a more proximal level as section shown in figure b; $\underline{d}$ and $\underline{e}$, details of figure b. Intermuscular space between $\underline{FDL}$ and the coalescent flexor mass ($\underline{FM}$) is mostly occupied in external ($\underline{d}$) and internal ($\underline{e}$) regions by disrupted tendinous tissue. Note remnants of intermuscular fibrous type I collagen around $\underline{NPP}$ nerve ($\underline{d}$).

Table : Distribution of extracellular matrix components (collagen I, III and V, fibronectin and laminin (I, III, V, F, L) in normal muscle development.

Muscle related structures	7.5 days					9.5 days				
	I	III	V	F¹	L¹	I	III	V	F¹	L¹
Inside muscle bulk	+	±	±	++	+	++	++	+	++	++
Muscle envelope	±	±	–	–	–	+++	++	+	+	–
Intermuscular space	+	±	±	+	–	+	±	+	+	–
Tendon	++	±2	±	++	–	+++	++2	±	++	–
Muscular tunica	+++	+	+	+	–	+++	++	+	+	–

1 around each myofiber; **2** surrounding the tendon only.

anterior and posterior muscles. Only anterior muscles exhibit epimysial envelopes which are stained with anti-type I, and anti-type III collagen antibodies. The epimysial envelopes of the posterior muscles are not clearly defined. But the intermuscular spaces are uniformly occupied by very fine type I-, type III-, and type V- positive fibers or granules, and by fibronectin positive structures.

In the mutant of the same age, the distribution pattern of the extracellular matrix components is grosso modo the same as in the normal, although the structures are more weakly labeled. But in 9.5-day cn/cn embryos, the distribution differs from that in the normal. Labeling of all constituants is weaker. Epimysial envelopes of anterior muscles are still specifically labeled with anti-type I and anti-type III collagen antibodies, but their staining is weak. The architecture of the tunica whereby the whole musculature is surrounded, and which, as in the normal, is constituted, in decreasing order, by type I-, type III-, type V- collagens and fibronectin, is disorganized and disrupted. Moreover, posteriorly there is a characteristic dislocation of the tendinous collagens, particularly so for collagen I. The disrupted collagen blocks are spred inside the muscle tissue of the coalescent muscles or fill the space running between them and the FDL muscle. Apart from this tendinous collagen, the space contains little type I- and type III- fibers. Concerning the distribution of fibronectin, the labeling

with antifibronectin is very much diminished in comparison
with that in normal sibling. The only conspicuous labeling
is carried by the remnants of the tendons and is present
around individual myotubes. In the space fibronectin has
apparently disappeared. It is clear that, in the mutant embryo, there is a loss in types I and III collagens and in
fibronectin, which nevertheless does not seem to be the primary cause of the muscular dysgenesis, but rather an accompanying feature.

Remains an experimental approach to the problem, to determine which of the two component tissues is affected by
the mutation, the somatopleure-originated connective tissue
cells or the somite-originated myoblasts. To answer this
question, somitic mesoderm obtained from a quail embryo, was
grafted in replacement of a piece of somitic mesoderm in putative 2-day mutant chick embryos (Mauger et al, 1982). Histological analysis of the tibiotarsal segment was performed
on the operated and on the non-operated controlateral side,
8 to 11 days later. In the oldest cases, the mutant phenotype of the host was externally discernible, by a filiform
lower leg and shank on the non-operated side. On the controlateral experimental side lower leg and shank were bulky.

The introduction of quail somitic myogenic cells into
a mutant limb, led to a normal patterning of the musculature. All muscles and tendons were individualized and normally distributed in space at all the levels (figure 5).
Exogenous non-mutant myogenic cells were able to restore
completely the normal development and stabilization of the
musculature. From this result it can be concluded that in
the crooked neck dwarf mutant, it is not the connective
tissue but the myoblasts themselves which are primarily defective. It is a defect of the myogenic cell line.

Since we know from the foregoing experiments and observations that the connective tissue is responsible for the
spacing, for the splitting of the muscle masses and for the
region-specific organization and morphogenesis of the musculature, it is clear that the stability of this architecture
is dependent on a constant tissue interaction between the
myogenic cell line and the connective tissue.

What about vessels + nerves?

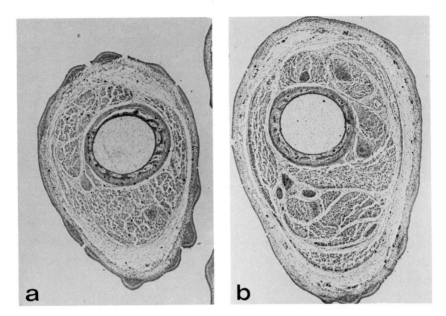

Fig. 5. Restoration of a normal muscle pattern in the crooked neck dwarf mutant by exogenous non-mutant myogenic cells. a, zeugopod of the mutant control hind-limb; b, zeugopod of the controlateral experimental hind limb.

Armand O, Boutineau AM, Mauger A, Kieny M (1982). Origin of satellite cells in avian skeletal muscles. Arch Anat micr Morph exp (submitted).

Chevallier A (1978). Etude de la migration des cellules somitiques dans le mésoderme somatopleural de l'ébauche de l'aile. Wilhelm Roux's Arch 184:57.

Chevallier A, Kieny M (1982). On the role of the connective tissue in the patterning of the chick limb musculature. Wilhelm Roux's Arch (in press).

Chevallier A, Kieny M, Mauger A (1976). Sur l'origine de la musculature de l'aile chez les Oiseaux. C R Acad Sci (Paris) série D 282:309.

Chevallier A, Kieny M, Mauger A (1977). Limb-somite relationship origin of the limb musculature. J. Embryol exp Morph 41:245.

Chevallier A, Kieny M, Mauger A (1978). Limb-somite relationship: effect of removal of somitic mesoderm on the wing musculature. J Embryol exp Morph 43:263.

Chiquet M, Eppenberger HM, Turner DC (1981). Muscle morphogenesis: evidence for an organizing function of exogenous fibronectin. Develop Biol 88:220.

Kieny M, Chevallier A (1979). Autonomy of tendon development in the embryonic chick wing. J Embryol exp Morph 49: 153.

Kieny M, Chevallier A (1980). Existe-t-il une relation spatiale entre le niveau d'origine des cellules somitiques myogènes et leur localisation terminale dans l'aile ? Arch Anat micr Morph exp 69:35.

Kieny M, Mauger A, Hedayat I, Goetinck P (1982). Ontogeny of the leg muscle tissue in the crooked neck dwarf mutant (cn/cn) chick embryo. Arch Anat micr Morph exp (submitted).

Mauger A, Kieny M (1980). Migratory and organogenetic capacities of muscle cells in bird embryos. Wilhelm Roux's Arch 189:123.

Mauger A, Kieny M, Hedayat I (1982). Is the muscular dysgenesis in crooked neck dwarf mutant chick embryo due to deficient muscle cells or to a defective connective environement ? (in preparation).

Pautou MP, Hedayat I, Kieny M (1982). The pattern of muscle development in the chick leg. Arch Anat micr Morph exp (submitted).

Shellswell GB, Wolpert L (1977). The pattern of muscle and tendon development in the chick wing. In Ede DE, Hinchliffe JR, Balls (eds): "Vertebrate Limb and Somite Morphogenesis" p 71.

Shellswell GB, Bailey AJ, Duance VC, Restall DJ (1980). Has collagen a role in muscle pattern formation in the developing chick wing ? I am immunofluorescence study. J Embryol exp Morph 60:245.

Wick RA, Allenspach AL (1978). Histological study of muscular hypoplasia in the crooked neck dwarf mutant cn/cn chick embryo. J Morphol 158:21.

Limb Development and Regeneration
Part B, pages 303–312
© **1983 Alan R. Liss, Inc., 150 Fifth Avenue, New York, NY 10011**

REGIONAL DISTRIBUTION OF MYOGENIC AND CHONDROGENIC PRECURSOR
CELLS IN VERTEBRATE LIMB DEVELOPMENT

Stephen Hauschka and Richard Rutz

Department of Biochemistry
University of Washington
Seattle, WA 98195

One prerequisite for analysis of the mechanisms which
specify pattern formation within the developing limb is to
obtain a precise stage-specific map of the various precursor
cell types. Once embryonic cells express tissue-specific
properties, mapping their location during subsequent developmental stages with histochemical or immunological techniques
is reasonably straightforward. Prior to expression of such
markers, the mapping process is much more complex, yet it is
precisely these early stages that are potentially the most
informative with respect to pattern specification.

Although it is likely that definitive markers for early
stages of the various mesodermal cell lineages will eventually be discovered, we have chosen to approach the mapping
problem by indirect means. Our working assumption is that
when cells are removed from the limb and cultured in vitro,
their previous, in vivo, identity is disclosed by the subsequent behavior of their descendents. Thus, cells whose progeny express muscle or cartilage phenotypes are operation-
ally defined as being myogenic or chondrogenic precursors.
This approach can be used at both the single cell (clonal)
and contiguous cell (explant) level.

The success of an in vitro assay system for identifying
a cell's developmental potential obviously depends upon devising in vitro conditions which permit expression of the
various phenotypes being investigated. The apparent environmental dependence of this approach raises several potential
ambiguities. First, in vitro conditions could conceivably
switch the phenotypic expressions of cells from one pathway

to another (e.g., chondrogenic to myogenic); second, environmental conditions might cause "pluripotential" or "uncommitted" mesenchymal cells (if indeed they exist) to differentiate in a particular direction. Definitive evidence for or against these possibilities has yet to be obtained. What is clear from our studies, however, is: (a) under certain in vitro conditions both phenotypes can be expressed in the same environment; and (b) once the clonal progeny of a single cell have expressed myogenic or chondrogenic phenotypes in environments which permit such expression, their subsequent descendents do not express the "alternative" phenotype when cultured in the alternative environment; instead, they either continue expressing their original phenotype or cease expressing this phenotype until returned to the original permissive culture environment. For this reason the developmental state of cells that fail to express a definitive phenotype in vitro cannot yet be determined; but in those cases where definitive phenotypes are expressed, it seems reasonable to conclude that these are indicative of developmental differences which existed within the precursor cells at the time of their removal from the embryo.

Given these assumptions, we have posed the following general questions: (1) where are myogenic and chondrogenic precursors located during limb development; (2) does their location or frequency within various limb regions change with time; and (3) does the apical ectodermal ridge (which is known to influence patterns of chondrogenic cells within the limb) exert similar effects upon myogenic patterns? In pursuing these questions the bulk of our attention has been focussed on myogenic precursors; but due to the possible existence of "pluripotential" limb mesenchyme cells, and the relevance this would have to pattern specification, we have also attempted to determine the extent to which the myogenic and chondrogenic compartments overlap during stages in which "pluripotentiality" is thought by some to occur (Zwilling, 1968; Searls and Janners, 1969; Caplan, 1981).

We have approached these questions by devising a technique for serially sectioning living chick and human limb buds, and then dissecting small regions from these which are subjected to clonal or explant analysis to determine their content of chondrogenic and myogenic cells (Hauschka and Haney, 1978, Rutz et al., 1982).

METHODS

Briefly, the technique as applied to chick embryos involves removal of leg or wing buds from stage 16 onward, em-
bedding these in a gelatin-culture medium matrix (15% gelatin-
85% medium) and serially sectioning the embedded limbs with
a Vibratome. By approaching the mounted bud from the distal
tip, it is possible to remove grazing sections until the
first 20-30 μ of the distal tip is encountered. By adjust-
ing the Vibratome to the desired section thickness it is
then possible to obtain highly reproducible sections encom-
passing all proximodistal levels of the limb. The resulting
flat sections are easy to dissect, and are sufficiently trans-
lucent to indicate gross tissue boundaries. By photograph-
ing the section and sketching the dissected regions it is
possible to obtain an accurate record of each assayed region
to within plus or minus about 10 cell diameters. Dissected
regions are then either dissociated to the single cell level
for clonal assays or cultured as explants. (Further experimental details are contained in Rutz, et al., 1982).

Muscle colonies were identified by the presence of multinucleated myotubes; and clones derived from early vs. late
MCF cells were distinguished on the basis of their percent
multinuclearity (see Rutz and Hauschka, 1982a). Cartilage
colonies were identified on the basis of positive Alcian
blue staining of a surrounding metachromatic matrix. Explant cultures were observed dialy by phase microscopy and
were scored for muscle and cartilage according to the same
criteria used for clones. MCF: muscle-colony-forming; CCF:
Chondrogenic-colony-forming cells.

RESULTS AND DISCUSSION

LOCATION OF MYOGENIC AND CHONDROGENIC REGIONS

Initial studies were designed to determine the location
of MCF and CCF cells within chick leg buds from stage 21 onward. These studies showed a clear demarcation between myogenic and chondrogenic regions throughout all stages. MCF
cells were restricted to discrete dorsal and ventral regions
(cf. Hauschka et al., 1982, fig. 1 and table 1 for typical
data). Even at early stages in which no morphological indication of myogenic and chondrogenic regions is discernable
(e.g., fig. 1 of this paper), the maximum "overlap" of

myogenic and chondrogenic regions (which we believe is due
wholly to "imperfect" dissections) is 5-10%. Thus even at
stages 21-23, an interval during which "pluripotential" mesenchymal cells have been claimed to exist, there is no strong
indication that colony-forming cells present in the central
(chondrogenic) region will form myogenic clones when cultured
in an environment conducive to muscle differentiation. Similarly, there is no indication that colony-forming cells present in the peripheral (myogenic) regions will form chondrogenic clones when cultured in an environment conducive to
cartilage differentiation.

These results notwithstanding, it is still possible that
"pluripotential" cells do exist, but that they are unable to
form myogenic or chondrogenic colonies in either of the culture environments provided. This possibility was further
investigated by the use of explant cultures. With this technique it was possible to provide an environment which mini-
mized disruption of cell-cell contacts and which did not require extensive cell proliferation prior to the expression
of differentiated phenotypes. In addition, explant cultures
made it feasible to analyze myogenic and chondrogenic compartments at earlier stages in which colony-forming cells
are more difficult to detect (Bonner and Hauschka, 1974).

Explant assays were performed on leg and wing buds between stages 16 and 22. Results from these experiments can
be summarized as follows. (1) Myotube-forming cells are detected in explants from stages as early as 17 in the leg bud
and 16 in the wing bud. (2) During stages 18-22 (when precise
regional dissections of the sections are more feasible) myotube-forming cells are located primarily within the peripheral dorsal and ventral (myogenic) regions; but small numbers
of myotube-forming cells are also detected within explants
from the central core (chondrogenic) region. In contrast,
cartilage nodule-forming cells are restricted primarily to
the central core region, but also appear at low frequency
within explants from peripheral regions. (3) As estimated
from the timing of myotube appearance (which occurs first in
the most proximal explants) and myotube density (which is
greatest in proximal explants), myotube-forming cells are
distributed in a proximodistal gradient which extends to
within 100-150 μ of the distal wing and leg bud tips. This
gradient is presumably the precursor to that observed for MCF
cells at late stages (see below). Throughout all of the explant studies, the relatively infrequent occurrence of

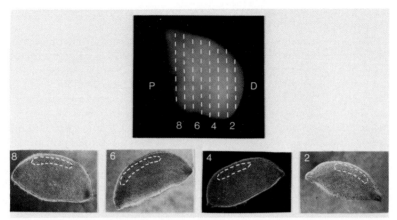

Fig. 1 Transverse sections (100 μm thick) from a stage 23 (3½ to 4-day) chick embryo leg bud. Living sections of reproducible thickness and position are cut (see Materials and Methods), and *prospective* dorsal and ventral myogenic regions are surgically removed for clonal analysis. (The prospective dorsal myogenic regions are indicated by dashed lines on the sections.) Sections are numbered on the whole bud beginning at the distal end. Bud, ×40; sections, ×42.

Table 1

PROXIMODISTAL-DISTAL DISTRIBUTION OF MCF CELLS IN STAGE 23 CHICK LEG BUDS

				Section number (100-μm sections)[a]					
	(Proximal)	8	7	6	5	4	3	2	1	(Distal)
% MCF cells[b]		19[c]	35	30	30	27	11	0	0	
% Late MCF cells[d]		0	0	0	0	0	0	—	—	

[a] Refer to Fig. 1 for section location and depiction of regions dissected for clonal analysis.
[b] Percentage of total colonies that form myotubes. Clonal plating efficiencies ranged from 0.4% in the most distal sections to 10% in the most proximal sections.
[c] Means of four experiments. Standard errors of the means (SEM) for % MCF cells, sections 8 to 1, are 0.6, 3.2, 3.9, 3.5, 3.3, 1.4, 0, 0; SEMs for % late MCF cells are all 0.
[d] Percentage of the total muscle colonies which exhibit the characteristic late muscle clone phenotype (i.e., extensive multinuclearity—see Materials and Methods); stage 23 MCF cells all show the early phenotype.

myotube-forming cells in cultures from the central limb region and of cartilage nodule-forming cells in cultures from
the peripheral region suggests that these minor populations
are due to imperfect dissection rather than to the existence
of a few "pluripotential" cells. Indeed, if such cells did
exist as minor populations within each limb region, it is
not clear why they should express a myogenic phenotype when
present in an explant which is primarily chondrogenic, and
express a chondrogenic phenotype when present in an explant
which is primarily myogenic. (4) Explants from stages 18-20
(which do not contain detectable MCF and CCF cells when
initially cultured) produce such cells after several days in
vitro. Thus cells within the stage 18-20 myogenic and chondrogenic compartments must include the precursors to MCF and
CCF cell types.

PROXIMO-DISTAL GRADIENTS OF MYOGENIC PRECURSOR CELLS

In the process of analyzing the regional distribution of MCF cells within the dorsal and ventral peripheral myogenic compartments, we discovered that MCF cells are distributed in a proximodistal gradient (Hauschka and Haney, 1978; Rutz et al., 1982). As limb development progresses from stage 23 to 31 the proximal myogenic region exhibits an increasingly high plateau level of MCF cells, and the "boundary" between regions containing high and low percentages of MCF cells is located in progressively more distal sections (see % MCF cells, Tables 1-3). In further studies, it was discovered that the gradient consisted of two MCF cell types, and that their distributions changed with development. At stage 23 (fig. 1, table 1) the MCF cell gradient is completely comprised of cells which express the early clonal phenotype (low percent multinuclearity). By stage 25 (data not shown; see Rutz et al., 1982, table 3) the percentage of early MCF

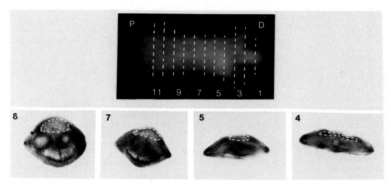

Fig. 2 Transverse sections (200 μm thick) from stage 27 (5-day) lower leg bud. The ventral myogenic region is beginning to split (sections 6 and 8), and distinct chondrogenic elements are appearing. Bud, ×28; sections 4-7, × 29; section 8, ×32.

Table 2

PROXIMODISTAL DISTRIBUTION OF MCF CELLS IN STAGE 27 CHICK LEG BUDS

		Section number (200-μm sections)[a]										
	(Proximal)	11	10	9	8	7	6	5	4	3	2	1	(Distal)
% MCF cells[b]		75[c]	87	81	88	85	84	76	28	0	0	0
% Late MCF cells[d]		65	59	57	45	40	31	14	1	—	—	—

[a] Refer to Fig. 2

[b,d] See footnotes b and d in Table 1. Clonal plating efficiencies ranged from 0.3% in the most distal sections to 21% in the most proximal sections.

[c] Means of five experiments. SEMs for % MCF cells, sections 11 to 1, are 10.9, 3.7, 4.0, 2.1, 4.8, 5.0, 6.0, 9.8, 0, 0, 0; SEM's for % late MCF cells are 9.5, 5.8, 4.6, 6.0, 7.7, 4.4, 4.0, 0.7, —, —, —.

cells within the proximal myogenic regions continues to increase, but in addition, MCF cells which express the late
clonal phenotype (high percent multinuclearity) begin to appear in the more proximal limb regions. It is worth noting
that our results with leg as well as wing buds (Rutz and Hauschka, 1982b) concur to within 200 μ of the distal-most loc-
cation of "muscle" cells observed histologicaly in stage 25
quail-chick chimeric wing buds (Newman et al., 1981). At
stage 27 (fig. 2, table 2) late MCF cells have become the
predominant MCF type within the proximal limb region, whereas in the more distal limb regions early MCF cells still
predominate. This trend, in which late MCF cells supersede
early MCF cells in progressively more distal regions, continues at least through stage 31.

INFLUENCE OF THE APICAL ECTODERMAL RIDGE ON THE FORMATION OF
PROXIMODISTAL MCF CELL GRADIENTS

     Since vertebrate limb outgrowth is known to require the
continuing presence of the apical ectodermal ridge (AER)
(Saunders, 1948; Summerbell, 1974; Rowe and Fallon, 1982), it
seemed of interest to determine whether AER removal at periods prior to establishment of the MCF cell gradient would
affect the myogenic gradient analogously to its well documented effects on the specification of chondrogenic elements.
If AER removal were to have an effect on MCF cell localization, our system seemed to offer several experimental advantages over the chondrogenic system for the analysis of AER-
limb mesoderm interactions. First, the clonal assay is more
readily quantifiable than assessment of partially deleted
chondrogenic elements. Second, the response to AER removal
can be assayed within 1-2 days of the manipulation rather
than waiting 6-8 days for definitive cartilage structures to
develop.

     Because most previous AER studies have focussed on wing
development, this aspect of our study utilized wing buds for
all analyses (Rutz and Hauschka, 1982b). AERs were surgically removed from stage 18 and 20 embryos which were maintained in a modified ex ovo culture system (Dunn et al., 1981).
At stage 23, 25, or 27, operated and contralateral control buds
were removed, serially sectioned, and regions dissected from
the dorsal and ventral myogenic regions were subjected to
clonal analysis. To control for the possibility that distal
myogenic cells were removed along with the AER, AERs were

Table 3

Proximodistal Distribution of MCF Cells in Stage 27 Wing Buds after AER Removal at Stage 18

	Operated Buds (AER removed)							
	Section Number (200 μm sections)							
(Proximal)	7	6	5	4	3	2	1	("Distal")
% MCF Cells[a]	78[b]	71	64	57	46	41	38	
% Late MCF Cells[c]	86	82	87	85	76	76	71	

	Unoperated Buds (Contralateral Normal)									
	Section Number (300 μm sections)									
(Proximal)	10	9	8	7	6	5	4	3	2	1 (Distal)
% MCF Cells[a]	87[d]	78	59	54	54	43	23	7	0	0
% Late MCF Cells[c]	85	89	82	81	74	62	46	4	-	-

[a,c]See footnotes b and d in Table 1. Clonal plating efficiencies ranged from 0.4% in the most distal sections to 23% in the most proximal sections.

[b]Means of 3 experiments. SEM for % MCF cells, sections 7 to 1, are 1.4, 2.1, 3.5, 2.9, 1.8, 2.9, 4.5; SEMs for % Late MCF cells are 1.5, 3.2, 5.5, 0.9, 3.5, 2.7, 4.0.

[d]SEMs for % MCF cells, sections 10 to 1, are 3.5, 4.8, 5.0, 2.8, 2.8, 4.9, 4.3, 0, 0; SEMs for % Late MCF cells are 2.2, 2.1, 4.8, 4.1, 3.7, 4.9, 6.2, 5.8, -, -.

routinely cultured as explants to determine whether myotube-forming cells would appear. No cases of such inadvertant deletions were encountered. As an additional control to ascertain that AER removal according to our procedure caused the expected effects upon chondrogenic elements, some embryos were permitted to develop to day 10 and the operated wings were then analyzed by standard wholemount procedures. In all cases, missing cartilage structures were appropriate for the stage at which the AER had been removed (Summerbell, 1974).

Typical results from the AER deletion experiments are shown in table 4. In this stage 27 specimen it can be seen that the control wing bud exhibits a proximodistal gradient of total and late MCF cells as well as an extensive distal region which lacks MCF cells. By contrast, the operated bud exhibits an abrupt truncation of the myogenic gradient and the MCF cell-free distal region is totally lacking. When AER removal is delayed until stage 20, the MCF cell gradient observed at stages 25 and 27 is again truncated; but since the operated limbs are at least 500 μ longer, the level of truncation is at a more distal portion of the MCF cell gradient. These results suggest that continuous presence of the AER is required for establishment of normal myogenic gradients. However, since the proximodistal morphogenesis of chondrogenic elements is similarly affected by AER removal, it is not clear whether the AER exerts a direct influence on both myogenic and chondrogenic precursor cells, or whether it affects the localization of one precursor cell type and this effect is secondarily exerted on the other type via chondrogenic-myogenic tissue interactions. Although such interactions may occur, evidence from other experimental manipulations suggest that chondrogenic elements can develop in the absence of muscle (Shellswell, 1977), and that myogenic elements can develop in the absence of their corresponding cartilage elements (Javois and Iten, 1982); thus, direct requirements of both mesodermal populations for AER interaction might be anticipated.

To summarize, by utilizing an experimental approach that provides a precise and reproducible method for localizing and assaying myogenic and chondrogenic colony-forming cells or their precursors, it has been possible to delineate the stage-specific locations of these cell types to within about 100 μ between stages 17 and 31 of chick limb development. The data obtained indicates that myogenic and chondrogenic cells are restricted to discrete limb compartments well before their overt differentiation, and that there is no compelling evidence to support the existence of "pluripotential" limb mesenchyme cells. Clonal analysis of serial limb sections from different stages indicates that myogenic cells of all types (MCF cell precursors as well as early and late MCF cells) are distributed in proximodistal gradients which change as development progresses. If the AER is removed from stage 18 and 20 wing buds and the operated buds are assayed at stages 23-27, proximal portions of the MCF cell gradients are seen to persist, whereas distal portions are totally absent. This behavior is consistent with the well documented proximodistal specification of chondrogenic elements by interaction with the AER, and suggests that elaboration of myogenic gradients also requires an AER interaction.

REFERENCES

Bonner PH, Hauschka SD (1974). Clonal analysis of vertebrate myogenesis. I. Early developmental events in the chick limb. Develop. Biol. 37:317.

Caplan AI (1981). The molecular control of muscle and cartilage development. In Subtelny S, Abbott UK (eds): "Levels of Genetic Control in Development," 39th Symposium of the Society for Developmental Biology. New York: A.R. Liss.

Hauschka SD, Haney C (1978). Use of living tissue sections for analysis of positional information during development. J. Cell Biol. 79:24a.

Hauschka SD, Rutz R, Linkhart TA, Clegg CH, Merrill GF, Haney CM, Lim RW (1982). Skeletal muscle development. In Schotland DL (ed): "Disorders of the Motor Unit" New York: John Wiley & Sons, p. 903.

Javois LC, Iten LE (1982). Supernumerary limb structure after juxtaposing dorsal and central chick wing bud cells. Develop. Biol. 90:127.

Newman SA, Pautou M-P, Kieny M (1981). The distal boundary of myogenic primordia in chimeric avian limb buds and its relation to an accessible population of cartilage progenitor cells. Develop. Biol. 84:440.

Rowe D, Fallon JF (1982). The proximodistal determination of skeletal parts in the developing chick leg. J. Embryol. Exp. Morphol. 68:1.

Rutz R, Haney C, Hauschka, S (1982). Spatial analysis of limb bud myogenesis: A proximodistal gradient of muscle colony-forming cells in chick embryo legs buds. Develop. Biol. 90:399.

Rutz R, Hauschka S (1982a). Clonal analysis of vertebrate myogenesis. VII. Heritability of muscle colony type through sequential subclonal passages in vitro. Develop. Biol 91:103.

Rutz, R Hauschka S (1982b). Spatial analysis of limb bud myogenesis: Elaboration of the proximodistal gradient of myoblasts requires the continuing presence of apical ectodermal ridge. Develop. Biol. (submitted).

Searls RL, Janners MY (1969). The stabilization of cartilage properties in the cartilage-forming mesenchyme of the embryonic chick limb. J. Exp. Zool. 170:365.

Shellswell GB, Wolpert L (1977). The pattern of muscle and tendon development in the chick wing. In Ede DA, Hinchliffe JR, Balls M (eds): "Vertebrate Limb and Somite Morphogenesis" Cambridge: Cambridge Press, p. 71.

Summerbell D (1974). A quantitative analysis of the effect of excision of the AER from the chick limb-bud. J. Embryol. Exp. Morphol. 32:651.

Zwilling E (1968). Morphogenetic phases in development. Develop. Biol. 2 (suppl.):184.

Limb Development and Regeneration
Part B, pages 313–322
© **1983 Alan R. Liss, Inc., 150 Fifth Avenue, New York, NY 10011**

ULTRASTRUCTURAL ASPECTS OF MIGRATION MODALITIES OF THE
MYOGENIC LIMB BUD CELLS

Monika Jacob, H.J. Jacob, B. Christ

Institute of Anatomy, Ruhr-University D-4630 Bochum, Federal Republic of Germany

On the basis of the experimental studies of CHRIST et
al. (1974, 1977) and CHEVALLIER et al. (1976, 1977) it is well
established that myogenic stem cells migrate from somites into
prospective limb buds, where later on they differentiate into
muscle cells. Based on earlier studies (JACOB et al., 1978b,
1979) it is the purpose of the present paper to outline our
current view on migration modalities of myogenic limb bud
cells using SEM, TEM and histochemical techniques.

Looking at the wing level opposite the somites 14 - 20
with SEM, we found that in a HH-stage 14 embryo, groups of medio-laterally oriented cells are located lateral to somite 14
(Fig. 1). Between somite 15 and the somatopleure only one cell
is visible, more caudally none. Consequently the migration of
the myogenic stem cells proceeds in a cranio-caudal sequence.
To describe the early events of this movement, we first study the region at the level of somite 15 (Fig. 2). The space
between the somite and the somatopleure, about 30 μm broad, is
filled with a fibrillar network,whose main strands are oriented in a medio-lateral direction. Elongated cells are moving
within this network. On the ventro-lateral border of the somites (Figs. 2 and 3) filopodia of somitic cells extend into
this space and have contact to the fibrillar bundles or, in
the more cranial somites (Fig. 4), to the groups of spindle
shaped transversally oriented cells which also are of somitic
origin.

The leg buds arise opposite to somites 26 - 32. In HH-
stage 16 embryos first signs of cell migration are found at
the just formed somites 26 - 28 (Fig. 5). Single elongated

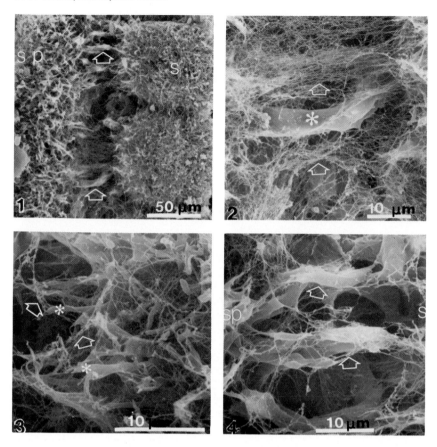

Fig. 1. Dorsal view of the somites 14 and 15 (s) and the opposite somatopleure (s p) of a stage 14 chick embryo. The ectoderm was removed after critical-point drying. Arrows: cells migrating into the prospective wing region.

Fig. 2. Higher magnification of the space between somite 15 and somatopleure. Asterisk: migrating cell with lamellipodium; arrows: strands of microfibrils.

Fig. 3. The ventro-lateral edge of somite 14. Filopodia of the somitic cells (asterisks) are in contact with bundles of fibrils (arrows).

Fig. 4. Groups of medio-laterally oriented cells (arrows) between somite 14 (s) and somatopleure (s p).

and transversally oriented cells reach from the cranial part
of the last somite to the somatopleure (Fig. 6).
These cells are spanned over the whole subectodermal space
and their leading ends touch the somatopleural cells or extracellular fibrils with feeler-like filopodia (Fig. 7).
Their features resemble cells of other migrating systems in
vertebrate embryos (ed. BARD and HAY, 1975; EBENDAL, 1977;
MARKWALD et al., 1977; JACOB et al., 1978a; LÖFBERG and AHL-
FORS, 1978). In stage 17 and 18 chick embryos the number of
cells detaching from the pelvic somites increases, so that
a broad cell stream fills the space between somites and leg
anlagen (Fig. 8).

Corresponding TEM micrographs reveal that the ectodermal
side of the somites is covered by a nearly complete basal lamina (Fig. 9), but the ventro-lateral border shows only fragments of this lamina (Fig. 10), because numerous tapering cell
processes of somite cells break through it.

The enormous elongation of the cell at the ventro-lateral
border of the somites initiates the cell migration. This elongation correlates with the appearance of numerous microtubuli
oriented parallel   to the long axis of the cells (Fig. 11).
This observation is in agreement with the findings of BYERS
and PORTER (1964) and KARFUNKEL (1972) on other elongating
cells. The thin filopodia at the leading ends of the migrating cells contain mainly microfilaments (Fig. 12) which may
enable a contraction of these cell processes (TILNEY and GIB-
BINS, 1969).

The fibrillar network seems to be the substrate for the
migrating cells because they are in contact with the fibrils.
In stage 14 embryo the fibrils are c. 10 nm thick. Such fibrils in the axial region of chick embryo disappear after collagenase treatment (FREDERICKSON and LOW, 1971). From stage
16 on a faint cross-striation of the now c. 15 nm thick fibrils is visible (Fig. 13). We therefore assume that the fibrils are collagenous. By immunoflouresence, VON DER MARK et
al. (1976) have shown collagen in the trunk of 2-day old chick
embryos. The thinner fibrils may be in part fibronectin which
has been identified   around migrating neural crest cells by
MAYER et al. (1981).

Whereas the cells within the migrating strands are sometimes connected by gap junctions (Fig. 14), the main type of
junction we found is characterized by plaques of electron

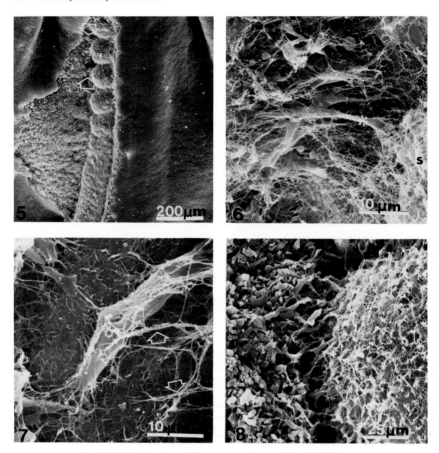

Fig. 5. Leg bud region of a stage 16 chick embryo. Ectoderm
has been removed. Arrows: migrating cells.

Fig. 6. A single elongated cell (asterisk) extending from
the last somite (s) to the somatopleure.

Fig. 7. Migrating cell which branches at its leading end
into filopodia. Arrows: strands of fibrils.

Fig. 8. Numerous cells which emigrate from somite 27
(HH-stage 17).

Fig. 9. Basal lamina (arrows) on the ectodermal side of somite 15 (HH-stage 14).

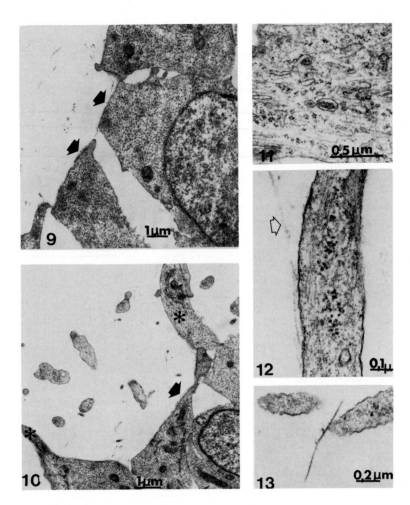

Fig. 10. Ventro-lateral border of the same somite. Asterisks: cell processes of somitic cells. Arrow: fragment of basal lamina.

Fig. 11. Microtubuls in an elongated cell.

Fig. 12. Filopodium with microfilaments. Arrow: extracellular fibrils.

Fig. 13. Filopodium in contact with a fibril being faintly cross-striated.

dense material on the opposing cell membranes (Fig. 15). As
similar cell types are found between other migrating cells
(EBENDAL, 1977; HOUBEN, 1976; HEAYSMAN and PEGRUM, 1973), we
assume that they are built up either at the medial end when
the cell elongates or at the leading end during retraction of
the cell at the trailing end.

TRELSTAD et al. (1967) postulated that the Golgi apparatus is always located in the trailing end of migrating cells.
Therefore, we stained the Golgi apparatus (Fig. 16) and observed that this is also true for migrating myogenic stem cells.
The TEM micrograph shows (Fig. 17) that the Golgi apparatus
consists of many stacks of saccules surrounded by uncoated
and bristle-coated vesicles, indicating the production of extracellular material (MARKWALD et al., 1977).

Some migrating cells as neural crest cells can be distinguished from fixed cells by staining with toluidine blue (DI
VIRGILIO et al., 1967; NICHOLS, 1981). The same results are
obtainable by treatment with alcian blue (Fig. 18). The neural
crest cells are then very osmiophil, whereas the neural tube
cells are pale. It is interesting that some cells of the ventro-lateral somite border at the beginning of migration are,
like the neural crest cells, darkly stained (Fig. 19). This
phenomenon may be due to a higher RNA content (JOHNSTON and
LISTGARTEN, 1972).

To elucidate the role of the extracellular material, histochemical techniques were applied. After treatment with RR
the space between somite and somatopleure is filled with RR-
positive particles often attached to the cell surface or to
the fibrils (Fig. 20). From these particles fine 3 - 5 nm thick
microfibrils radiate. After MORRIS and SOLURSH (1978) the
fibrils consist of hyaluronate to which proteoglycans are
attached. As first observed by TOOLE and TRELSTAD
(1971) hyaluronate is necessary to enable cell migration.

An even cell coat, about 20 μm thick, is found after Con
A-treatment (Fig. 21). This is in agreement with the findings
of MESTRES and HINRICHSEN (1974) that the migrating primary
mesoderm cells exhibit a thicker cell coat than the epiblast
cells.

The characteristics of cell migration are:
1. Cells of the ventro-lateral somitic border, distinguishable
   by their greater affinity to alcian blue, break through
   the basal lamina and elongate.

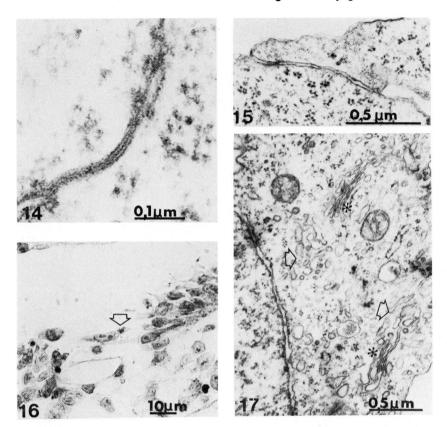

Fig. 14. Gap junction between cells migrating in groups.

Fig. 15. Plaque-like junction.

Fig. 16. Golgi apparatus (arrow) in a migrating cell, stained with osmium.

Fig. 17. Golgi apparatus (asterisks) with vesicles (arrows) in the trailing end of a migrating cell.

2. They move through a matrix consisting of collagenous fibrils, hyaluronate and proteoglycans.
3. The migrating cells have filopodia at their leading end. A distinct cell coat and GAG granules at their surface enable the contact with the fibrillar network. Microfilaments within them indicate the ability of cell contraction.

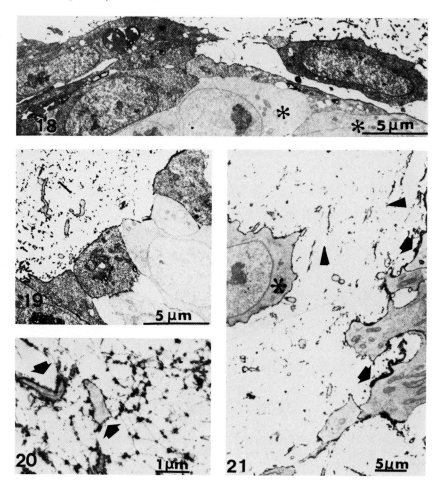

Fig. 18. Darkly stained neural crest cells after treatment by 0,5% alcian blue during fixation. Asterisks: neural tube.

Fig. 19. Darkly stained cells at the ventro-lateral border of the somite (HH-stage 14) after treatment with alcian blue.

Fig. 20. RR-positive granules from which microfibrils radiate. Arrows: cell processes surrounded by granules.

Fig. 21. The space lateral to somite 30 at a stage 17 embryo after Con A-peroxidase reaction. Arrows: basal lamina of the somite; arrowheads: extracellular material;asterisk: migrating cell with a distinct cell coat.

4. Plaque-like contacts are built to fix the trailing end while the cell elongates, or to fix the leading end when the trailing end retracts.
5. The Golgi apparatus is located in the trailing end of the migrating cell.

Acknowledgements

This work was supported by grants of the Deutsche Forschungsgemeinschaft. We wish to thank U. Ritenberg, I. Schaeben-Hamm, I. Burks and H. Hake for excellent technical assistance, A. Boeckmann for helping with the English translation and B. Scharf for typing the manuscript.

Bard JBL, Hay ED (1975). The behavior of fibroblasts from the developing avian cornea. Morphology and movement in situ and in vitro. J Cell Biol 67:400.
Byers B, Porter KR (1964). Oriented microtubules in elongating cells of the developing lens rudiment after induction. Proc Natl Acad Sci USA 52:1091.
Chevallier A, Kieny M, Mauger A (1976). Sur l'origine de la musculature de l'aile chez les Oiseaux. C R Acad Sci Paris 282:309.
Chevallier A, Kieny M, Mauger A (1977). Limb-somite relationship: origin of the limb musculature. J Embryol Exp Morph 41:245.
Christ B, Jacob HJ, Jacob M (1974). Über den Ursprung der Flügelmuskulatur. Experimentelle Untersuchungen an Wachtel- und Hühnerembryonen. Experientia 30:1446.
Christ B, Jacob HJ, Jacob M (1977). Experimental analysis of the origin of the wing musculature in avian embryos. Anat Embryol 150:171.
Di Virgilio G, Lavenda N, Worden JL (1967). Sequence of events in neural tube closure and the formation of neural crest in the chick embryo. Acta Anat 68:127.
Ebendal T (1977). Extracellular matrix fibrils and cell contacts in the chick embryo. Possible roles in orientation of cell migration and axon extension. Cell Tissue Res 175:439.
Frederickson RG, Low FN (1971). The fine structure of perinotochordal microfibrils in control and enzyme-treated chick embryos. Am J Anat 130:347.
Heaysman JEM, Pegrum SU (1973). Early contacts between fibroblasts. An ultrastructural study. Exp Cell Res 78:71.
Houben J-JG (1976). Aspects ultrastructuraux de la migration de cellules somitiques dans les bourgeons de membres postérieurs de souris. Arch Biol (Bruxelles) 87:345.

Jacob HJ, Jacob M, Christ B (1978a). Die Feinstruktur des Wolffschen Ganges bei Hühnerembryonen. Verh Anat Ges 72: 363.

Jacob M, Christ B, Jacob HJ (1978b). On the migration of myogenic stem cells into the prospective wing region of chick embryos. A scanning and transmission electron microscope study. Anat Embryol 153:179.

Jacob M, Christ B, Jacob HJ (1979). The migration of myogenic cells from the somites into the leg region of avian embryos. An ultrastructural study. Anat Embryol 157:291.

Johnston UL, Listgarten UA (1972). The migration, interaction and early differentiation of the oro-facial tissues. In Slavekin HC, Bavetta LA (eds): "Developmental Aspects of Oral Biology", New York, Academic Press, p 53.

Karfunkel P (1972). The activity of microtubules and microfilaments in neurulation in the chick. J Exp Zool 181:289.

Löfberg J, Ahlfors K (1978). Extracellular matrix organization and early neural crest cell migration in the axolotl embryo. Zoon 6:87.

Markwald RR, Fitzharris TP, Manasek FJ (1977). Structural development of endocardial cushions. Am J Anat 148:85.

Mayer BW Jr, Hay ED, Hynes RO (1981). Immunocytochemical localization of fibronectin in embryonic chick trunk and area vasculosa. Develop Biol 82:267.

Mestres P, Hinrichsen K (1974). The cell coat in the early chick embryo. Anat Embryol 146:181.

Morris G, Solursh M (1978). Regional differences in mesenchymal cell morphology and glycosaminoglycans in early neural-fold stage rat embryos. J Embryol Exp Morph 46:37.

Nichols DH (1981). Neural crest formation in the head of the mouse embryo as observed using a new histological technique. J Embryol Exp Morph 64:105.

Tilney LG, Gibbins JR (1969). Microtubules and filaments in the filopodia of the secondary mesenchyme cells of Arbacia punctulata and Echinarachnius parma. J Cell Sci 5:195.

Trelstad RL, Hay Ed, Revel JP (1967). Cell contact during early morphogenesis in the chick embryo. Develop Biol 16: 78.

von der Mark H, von der Mark K, Gay G (1976). Study of differential collagen synthesis during development of the chick embryo by immunofluorescence. I. Preparation of collagen type I and type II specific antibodies and their application to early stages of the chick embryo. Develop Biol 48:237.

Limb Development and Regeneration
Part B, pages 323–332
© 1983 Alan R. Liss, Inc., 150 Fifth Avenue, New York, NY 10011

MUSCLE DIFFERENTIATION IN CULTURES AND CHORIOALLANTOIC
GRAFTS OF PRE-BUD WING TERRITORIES

Terry Kenny-Mobbs and Brian K. Hall

Department of Biology, Dalhousie University,
Halifax, Nova Scotia, Canada, B3H 4J1

Recent studies have shown that cells from the brachial
somites (somites 15-20) migrate into the somatic mesenchyme
of the wing territories and there give rise solely to the
myofibres of the skeletal muscle of the wing. These findings
were derived primarily from experiments using various chick/
quail recombinations of brachial somites and wing lateral
plate (Christ, Jacob and Jacob 1974; 1977; 1979; Chevallier,
Kieny and Mauger 1976; 1977; Chevallier 1978; Kieny and Chevallier 1980; Mauger, Kieny and Chevallier 1980). Morphological
studies on normal embryos, at both light and electron microscopic levels (Christ et al. 1977; Chevallier 1978; Jacob,
Christ and Jacob 1978; Mauger et al. 1980), destruction of
brachial somites by irradiation (Chevallier, Kieny and Mauger
1978; Chevallier 1979) or their replacement by non-somitic
tissue (Chevallier et al. 1977; 1978), and growing isolated
wing territories as coelomic or chorioallantoic grafts (Christ
et al. 1977; 1979) have all corroborated the conclusion that
somitic cells enter wing territories and there differentiate
into skeletal myofibres. The cartilage, bone, smooth muscle
and connective tissues of the wing, including the connective
tissues of the skeletal muscles, develop from the somatopleural
mesenchyme of the lateral plate.

Christ and his colleagues have consistently maintained
that the somitic cells within the wing are the sole source of
wing myofibres. Chevallier's group, on the other hand, concede
that wing mesenchyme of somatopleural origin may give rise to
myofibres, for they found skeletal muscle in wings of all
embryos in which the brachial somites had been replaced by a
piece of gut and in 33% of the embryos in which the somites

were destroyed by irradiation (Chevallier et al. 1978). Furthermore, when quail brachial somites were replaced with those of chick, the resulting quail wings contained some muscles composed mostly, or exclusively of host cells (Chevallier et al. 1977). Experiments by McLachlan and Hornbruch (1979) seem to support the view of Chevallier. Grafts to the chick coelom of quail wing anlagen isolated prior to the reported stages of somitic cell migration contained myotubes of donor origin when recovered seven days later. In similar experiments, however, Christ et al. (1977; 1979) found no myotubes in coelomic grafts of quail wing anlagen of stages HH 14 (Hamburger and Hamilton 1951) and younger.

The important issue to be resolved in these latter studies is the origin of the cells that gave rise to the skeletal muscle. If it can be clearly shown that no somitic cells are present in wings in which myofibres develop, it would indicate that wing mesenchyme of somatopleural origin, like somitic mesenchyme, has the ability to differentiate into skeletal muscle. The approaches used in the experiments described above do not preclude the possibility that somitic cells invaded the wing anlagen subsequent to the experimental manipulations. It is possible that some somitic cells were left behind after the destruction of brachial somites or their replacement with a piece of gut (Chevallier et al. 1978). Because quail tissues grow faster than those of chick (Zacchei 1961), a few host somitic cells left behind after the replacement experiments could result in a disproportionately higher number of quail nuclei in wing myofibres relative to the initial number of somitic cells that entered the wing. Coelomic grafts may also be invaded by somitic cells via their connection to the host somatopleure (Dhouailly and Kieny 1972; Christ and Jacob 1980; Mauger and Kieny 1980; Mauger et al. 1980). Interspecific grafting (McLachlan and Hornbruch 1979) should permit the distinction of cells of host versus donor origin but, because of the dual origin of skeletal muscle tissues, requires careful analysis of the cell distribution within the graft and it cannot distinguish between host cells of somitic versus somatopleural origin.

As long as the wing territories remain accessible to somitic cell invasion, one cannot be confident that the wings are completely devoid of somitic cells. By isolating the wing territories from embryos at stages prior to the migration of somitic cells into the wing and growing them as explants outside the body, this possibility is precluded. This is the approach we have taken to determine if the somatopleural mesenchyme of the wing has myogenic potential when somitic cells are prevented from entering the wing territories.

CULTURES AND CAM GRAFTS OF ISOLATED WING TERRITORIES

The location of the wing territory is presaged by a condensation of the somatopleural mesenchyme of the lateral plate opposite somites 15-20 during stage HH 15. A wing bud first appears between stages HH 16-17. Before the wing bud forms, the migration of somitic cells into the wing regions begins; the first cells migrate at stage HH 13/14 (20-22 somite pairs) and the last during stage HH 18 (30-36 somite pairs) (Christ et al. 1977; Chevallier 1978). For this study we isolated wing territories from White Leghorn chick embryos of two pre-bud stages; stage HH 12/13 (15-19 somite pairs) and stage HH 15 (24-27 somite pairs). Primordia from stage HH 12/13 embryos should contain no somitic cells while stage HH 15 primordia may contain a few since somitic cell migration would have been underway for a couple of hours prior to the isolation of the wing territories.

The dissected wing regions, which included all three germ layers of the lateral plate, extended from (and occasionally included) the intermediate mesoderm to the distal edge of the area pellucida in a proximal/distal direction and from the level of somites 12/13 to somites 20/21 in the anterior/posterior direction. With the clearer definition of the wing region at stage HH 15, the anterior boundary of the explant was usually made at the level of somites 14/15. Wing primordia that included the brachial somites also were dissected from both stages of embryos. The boundaries were as described above with the exception of the medial boundary which extended to the midline of the embryo and included half of the neural tube and portions of the notochord.

The dissected wing primordia were positioned on 2 mm^2 pieces of 0.45 μm Millipore filters with the endodermal surface down. For culturing, these tissue/filter assemblies were placed on wire mesh supports in 35 mm Falcon tissue culture dishes to which 1.5 ml of culture medium was added. The culture medium consisted of BGJ$_b$ medium with 15% horse serum and 5% embryo extract (GIBCO). Three dishes were placed in a 100 mm glass petri dish and transferred to a humidified CO_2 incubator maintained at 37°C and an atmosphere of 5% CO_2 in air. The

medium was changed every two days. Chorioallantoic grafting
of the wing primordia followed the procedure described by
Hall (1978) although the tissue/filter assemblies were usually
placed filter side down on the CAM vessels. Cultures and
grafts were recovered 5 to 6.5 days later and processed for
light microscopic examination.

TISSUE DIFFERENTIATION IN EXPLANTED WING PRIMORDIA

Previous studies have shown that wing primordia isolated
from embryos of stage HH 15 (25 somite pairs) and grown as
coelomic grafts gave rise to limbs of normal morphology
(Hamburger 1938) with a normal pattern of skeletal elements
(Pinot 1970). Prior to stage HH 15 normal wing morphology was
obtained only if axial tissue was included in the explant
(Pinot 1970). To test the abilities of the culture conditions
and the CAM to foster good growth and differentiation of wing
tissues, wing primordia were cultured and grafted with the
brachial somites included in the explants. Neither the cultured

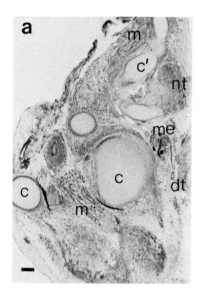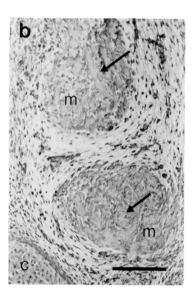

Fig. 1. CAM grafts of HH 12 wing primordia and brachial somites recovered after 6 days. a. Axial (top) and wing (bottom)
tissues show normal differentiation. b. Myotubes (arrows) are
present within distinct muscle bundles. c':vertebral cartilage;
c:wing cartilage; dt:digestive tube; m:muscle; me:mesonephros;
nt:neural tube. Bar = 100 μm.

nor the CAM-grafted explants gave rise to normally shaped wings
for either stage. In culture a slightly elevated, rounded
mound of tissue formed. The CAM grafts formed large mounds
of tissue which ranged from round and featureless to distinctly
wing-like in structure. Both in vitro and on the CAM (fig. 1)
the explants contained histologically normal cartilage, muscle,
neural tube, notochord, mesonephros, digestive tube and, occasionally in older grafts, periosteal bone and feather germs.
Discrete muscle bundles, containing myotubes (fig. 1b), were
readily identifiable. McLachlan and Hornbruch (1979) have
suggested that the CAM might not be able to provoke myogenic
differentiation in wing primordia. While CAM grafts and especially organ cultures of HH 12 and 15 wing primordia do not
result in externally-identifiable wings, both environments
permit normal differentiation of all wing and adjacent axial
tissues to occur.

In the absence of brachial somites, explants of wing
primordia from both stages also exhibited good growth and
normal differentiation of wing tissues but wing-like external
morphologies were even less frequent. There was a marked difference in the amount of muscle and cartilage present in the
explanted tissues which varied both with the stage of the wing
primordium and the conditions under which it was grown (Table
1). Differentiation of skeletal muscle within the explants,
assessed by histological appearance, location and the presence
of myotubes, occurred more frequently in cultured explants
(fig. 2) than in CAM grafts(fig. 3) and more frequently in
stage HH 15 explants than those of stage HH 12. When present,
the muscle primordia were usually small and only one or two
individual bundles were seen. In several explants from both
stages small condensations of cells were observed in areas
of loose connective tissue surrounding the cartilage elements
(fig. 3a), but no myotubes could be detected within these

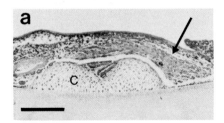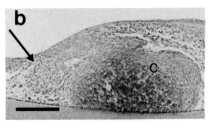

Fig. 2. Wing primordia cultured for 6 days. a. Stage HH 12
b.Stage HH 15. Muscle bundles are present in both explants
(arrows). c:cartilage. Bar = 100 μm.

Table 1. *Tissue differentiation in cultures and CAM grafts of wing territories*

AGE OF WING TERRITORY (HH stage)	CONDITIONS OF GROWTH	NUMBER OF SAMPLES	MUSCLE		CARTILAGE	
			myotubes	condensations only	1-2 elements	$\geqslant$3 elements
HH 12	in vitro	16	25%	38%	69%	31%
	CAM	37	8%	19%	73%	27%
HH 15	in vitro	11	55%	9%	36%	64%
	CAM	13	23%	31%	69%	31%

Table 2. *Tissue differentiation in cultures and CAM grafts of stage HH 12 wing territories*

CONDITIONS OF GROWTH	NUMBER OF SAMPLES	MUSCLE		CARTILAGE	
		myotubes	condensations only	1-2 elements	$\geqslant$3 elements
in vitro	23	0%	52%	96%	4%
CAM	18	0%	11%	72%	28%

condensations.

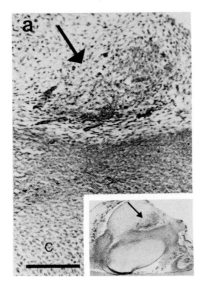 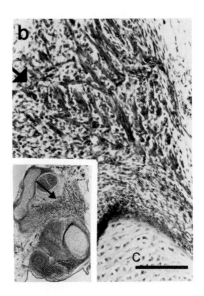

Fig. 3. Myogenic areas of CAM-grafted wing primordia. a. Small condensation (arrow) but no myotubes apparent; HH 15, 5 days on CAM. b. Myotube development within condensation; HH 12, 5.5 days on CAM. Bar = 100 μm.

While the appearance of skeletal muscle in the stage HH 15 explants was not unexpected - since somitic cells should be present in the grafted wing primordia from this stage - its presence in the stage HH 12 explants was not anticipated. In light of these results the experimental procedure was reviewed to determine if somitic cells could have been included inadvertently with the stage HH 12 explanted wing primordia. A possible source of somitic cells was from the lateral plate tissue anterior to the wing territory; lateral plate as anterior as somite level 12 was sometimes included in the explants. Results from other studies suggest that somitic cell migration into the lateral plate occurs in a cephalocaudal sequence (Christ et al. 1977; 1978; Chevallier 1978; Mauger et al. 1980) so it seemed possible that cells from the somites anterior to the brachial somites may have already begun to populate the adjacent lateral plate. Consequently, the experiments on the stage HH 12 embryos were repeated but this time the explanted wing primordia consisted only of the lateral

plate tissues adjacent to somites 14/15 to 20/21. The medial
boundary was made lateral to the intermediate mesoderm and
the block of axial and intermediate tissues remaining after
the dissection was carefully examined to ensure that the somites were intact. The cultures and grafts of these stage HH 12
wing primordia differed from those in which more anterior tissue was included in that no myotubes could be found (Table 2)
although condensations were seen in the connective tissue
adjacent to the cartilage elements (fig.4).

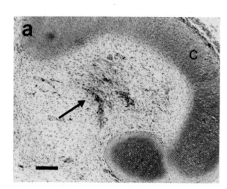
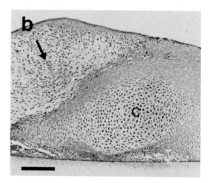

Fig. 4. Stage HH 12 wing primordia grown a. for 6 days on the
CAM and b. for 6.5 days in vitro. Condensations (arrows);
c:cartilage. Bar = 100 μm.

CONCLUSIONS

    The results of this study are consistent with a somitic
origin for the wing musculature: in the absence of the brachial somites, skeletal muscle differentiates only in explants
of wing primordia that already contain somitic cells (stage
HH 15 explants). Wing territories isolated from stage HH 12
embryos, in which migration from the brachial somites has
not yet occurred, do not contain myotubes after being cultured or CAM-grafted for 5-6.5 days. The appearance of
myotubes in the first series of explants of stage HH 12 wing
primordia seems to be attributable to the presence of somitic
cells in the lateral plate tissue anterior to the wing region
which was included in the explants to ensure that all of the
wing primordium was dissected.

    The results also suggest that the somitic cells are the
sole source of myogenic cells for the skeletal muscles of the

wing since no myotubes developed in the wing primordia in the
absence of somitic cells. This finding supports the view of
Christ et al. (1979) but contravenes that of Mauger et al.
(1980) and McLachlan and Hornbruch (1979). It is difficult to
pinpoint the source(s) of the discrepancies in the results
of these latter two groups and those of this study. Two possibilities seem to be 1) that somitic tissue was accidently
included in the grafted wing primordia, especially primordia
from stage HH 11 embryos and younger (see Pinot 1970) or 2)
that it is difficult to clearly distinguish the exact location
of chick and quail nuclei in the chimeric muscles that formed
in their grafts. One of the experiments of Mauger et al. (1980)
in fact supports our conclusion. Quail wing and leg somatopleure, isolated from embryos prior to and during brachial
somite segmentation, and grown as explants between the folds
of the chick neural tube, contained no myotubes, with the
exception of one leg graft which may have contained somitic
cells when grafted since it was from a 27 somite stage embryo.

It is concluded that the somitic cells that migrate into
the wing territory represent a distinct cell lineage and are
the sole source of myogenic cells for the skeletal musculature
of the wing.

This research was supported by the Natural Sciences and
Engineering Research Council of Canada.

REFERENCES

Chevallier A (1978). Etude de la migration des cellules somitiques dans le mésoderm somatopleural de l'ébauche de
  l'aile. W Roux's Arch 184:57.
Chevallier A (1979). Role of the somitic mesoderm in the
  development of the thorax in bird embryos. II. Origin of the
  thoracic and appendicular musculature. J Embryol exp Morphol 49:73.
Chevallier A, Kieny M, Mauger A (1976). Sur l'origine de la
  musculature de l'aile chez les oiseaux. C r hebd Seanc
  Acad Sci Paris D 282:309.
Chevallier A, Kieny M, Mauger A (1977). Limb-somite relationship: origin of the limb musculature. J Embryol exp Morphol 41:245.
Chevallier A, Kieny M, Mauger A (1978). Limb-somite relationship: effect of removal of somitic mesoderm on the wing
  musculature. J Embryol exp Morphol 43:263.

Christ B, Jacob HJ (1980). Origin, distribution and deter-
mination of chick limb mesenchymal cells. In Merker H-J,
Nau H, Neubert D (eds): "Teratology of the Limbs", Berlin:
Walter de Gruyter & Co, p 67.
Christ B, Jacob HJ, Jacob M (1974). Uber den Ursprung der
Flügelmuskulatur. Experimentelle Untersuchungen mit Wachtel-
und Hühnerembryonen. Experientia 30:1446.
Christ B, Jacob HJ, Jacob M (1977). Experimental analysis of
the origin of the wing musculature in avian embryos. Anat
Embryol 150:171.
Christ B, Jacob HJ, Jacob M (1978). Differentiating abilities
of avian somatopleural mesoderm. Experientia 35:1376.
Dhouailly D, Kieny M (1972). The capacity of the flank somatic mesoderm of early bird embryos to participate in limb
development. Develop Biol 28:162.
Hall BK (1978). Grafting of organs and tissues to the chorioallantoic membrane of the embryonic chick. TCA Manual
4:881.
Hamburger V (1938). Morphogenetic and axial self-differen-
tiation of transplanted limb primordia of 2-day chick
embryos. J exp Zool 77:379.
Hamburger V, Hamilton HL (1951). A series of normal stages
in the development of the chick embryo. J Morphol 88:49.
Jacob M, Christ B, Jacob HJ (1978). On the migration of myogenic stem cells into the prospective wing region of chick
embryos. A scanning and transmission electron microscopic
study. Anat Embryol 153:179.
Kieny M, Chevallier A (1980). Existe-t-il une relation spatiale entre le niveau d'origine des cellules somitiques
myogènes et leur localisation terminale dans l'aile? Arch
d'Anat microsc Morphol exp 69:35.
Mauger A, Kieny M (1980). Migratory and organogenetic capacities of muscle cells in bird embryos. W Roux's Arch 189:
123.
Mauger A, Kieny M, Chevallier A (1980). Limb-somite relationship: myogenic potentialities of somatopleural mesoderm.
Arch d'Anat microsc Morphol exp 69:175.
McLachlan JC, Hornbruch A (1979). Muscle-forming potential
of the non-somitic cells of the early avian limb bud. J
Embryol exp Morphol 54:209.
Pinot M (1970). Le rôle du mésoderme somitique dans la mor-
phogenèse précoce des membres de l'embryon de Poulet. J
Embryol exp Morphol 3:109.
Zacchei AM (1961). Lo sviluppo embrionale della quaglia
giapponese (Coturnix coturnix japonica T. e S.). Arch
Ital Anat Embriol 66:36.

Limb Development and Regeneration
Part B, pages 333–341
© **1983 Alan R. Liss, Inc., 150 Fifth Avenue, New York, NY 10011**

PROBLEMS OF MUSCLE PATTERN FORMATION AND OF NEUROMUSCULAR
RELATIONS IN AVIAN LIMB DEVELOPMENT

Heinz Jürgen Jacob, B. Christ, M. Grim

Institute of Anatomy, Ruhr-University D-4630 Bo-
chum, Federal Republic of Germany; Department of
Anatomy, Faculty of Medicine, Charles University
Prague, Czechoslovakia

In the course of avian limb development mesenchymal cells
give rise to various tissues, which become arranged in a typical spatial, species specific pattern.

Concerning muscle development, it is well known that
first of all the dorsal and ventral premuscular masses appear
within the myogenic zone, separated by cartilage condensations.
In temporal accordance with skeletal development and ingrowing
of nerves, the premuscular masses subdivide into the premuscle
blastemata and later-on into individual muscles (ROMER, 1927,
1964; WORTHAM, 1948; SULLIVAN, 1962; SHELLSWELL and WOLPERT,
1977), which are characterized by spatial arrangement, form,
muscle-specific pennation and distribution of motor endplates.

The way the formation of these patterns is controlled is
still under discussion.

Although innervation doesn't play an important role in
muscle separation (HARRISON, 1904; HUNT, 1932; HAMBURGER and
WAUGH, 1940; EASTLICK, 1943; GRIM and CARLSON, 1978; SHELLS-
WELL, 1977; CHRIST et al., 1979; JACOB and CHRIST, 1980), it
is necessary for the maintenance of muscles and the subsequent
differentiation (EALSTLICK and WORTHAM, 1947; ZELENA, 1962).

Also, normal anatomical arrangement of limb muscles was
observed in cases in which the source of the muscle cells was
replaced by non-brachial somites (CHEVALLIER et al., 1977;
CHRIST et al., 1978).

From limb transplantation experiments it is well known

that the factors which determine the peripheral branching pattern and gross distribution of the nerves, are located within
the limb itself (BRAUS, 1905; HARRISON, 1907; DETWILER, 1936;
PIATT, 1942, 1956; STRAZNICKY, 1963; NARAYANAN, 1964; LEWIS
et al., 1981). On the other hand, limbs deprived of muscles
show a normal pattern of nerve trunks (LEWIS et al., 1981).
MORRIS (1978) pointed out that in supernumerary limbs a definitive peripheral innervation pattern was formed.

With regard to the factors controlling the development
of functional motor endplates, different attempts have been
performed reaching from the concept of non-selective developmental mechanisms to the assumption of an initially determined
neuromuscular specifity (for discussion see HOLLYDAY, 1980;
LANDMESSER, 1980; SUMMERBELL and STIRLING, 1981).

On the basis of the species specific differences between
chick and quail wing muscles either in form and pennation or
in the distribution pattern of the motor endplates (GRIM et
al., 1982; GRIM and KOMANCOVA, 1982; KLEPACEK, 1981), we have
analyzed the influence of muscle material and of nerves on the
establishment of the pattern involved.

A) DEVELOPMENT OF SHAPE AND PENNATION IN CHIMAERIC WING MUSCLES

The extensor medius brevis muscle (EMB) in the chick wing
shows a well developed muscle belly, while in the quail the
muscle primordium is only provable up to the 12th day of incubation and later on degenerates into a small rudiment of
connective tissue.
Another species specific difference could be detected in
the gross anatomy of the flexor digitorum superficialis muscle
(FDS), which is a thin long muscle in the chick, peeping from
behind the radial side of the flexor carpi ulnaris muscle.
In the quail wing this muscle is more vigorous, showing two
muscular origination zones and lying fully under the flexor
carpi ulnaris muscle (FCU). In both species the muscle is
bipennate. In combination with the quail chick marker system,
introduced by LE DOUARIN and BARQ, 1969, we used these differences in form and pennation of EMB and FDS as criteria for
the judgement upon our experiments.

Two series of experiments were performed:
1. The heterotopic substitution of wing myogenic material in
    chick hosts by quail somites.
2. Heterotopic substitutions of brachial neural tube in the

chick by quail neural tube.

To prove the origin of the muscles, we excised small muscular tissue blocks and examined them in light microscopic sections after Feulgen-reaction. To document that the muscles in the second experimental series are innervated by heterotopic moto-neurons, they were treated by HRP-injection into the wing muscles of the zeugopodium. The connection of graft nerves with wing muscles was verified by gross anatomical dissection of the thoracal part of the spinal cord and identification of the brachial plexus and afterwards by staining of the spinal cord sections with Feulgen-reaction.

After heterotopical substitution of brachial somites in the chick hosts by quail somites, the chimaeras showed a normally developed chick-like FDS muscle (Fig. 1) and a normally developed chick-like EMB muscle (Fig. 2). The muscle fibres are made up by quail cells (Fig. 3).

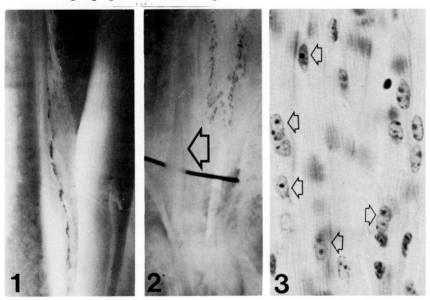

Fig. 1. Chick-like M. flexor digitorum superficialis (FDS) of quail material after replacement of brachial somites.

Fig. 2. Chick-like M. extensor medius brevis (EMB) of quail material (arrows) after replacement of brachial somites.

Fig. 3. Section through the wing shown in Fig. 1. Muscle fibres contain quail nuclei (arrows) stained with Feulgen-reaction.

When the brachial neural tube of the chick hosts was substituted by quail neural tube of the leg level, the chimaeras also showed normally developed wing muscles. After injection of horseradish peroxidase (HRP) the motoneurons of the graft were seen to be stained (Figs. 4 and 5).

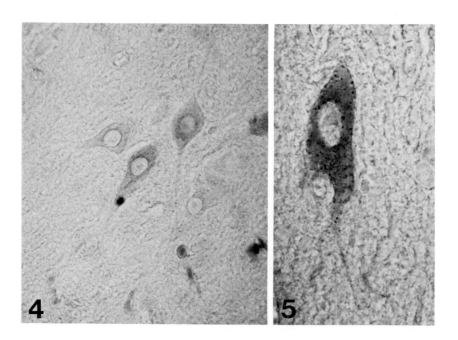

Fig. 4. Group of motoneurons of the grafted neural tube after HRP-treatment.

Fig. 5. Motoneuron at higher magnification.

The results show that neither the myogenic cells nor the nerves determine the muscle pattern of the wing.

B) DEVELOPMENT OF ENDPLATE PATTERN IN CHIMAERIC WING MUSCLES

Recently GRIM et al. (1982) pointed out the distribution of endplates in the wing muscles of chick and quail. The motor endplates were stained in the whole muscles by the indigogenic method for non-specific esterase as modified by LOJDA et al. (1976).

They found the ulnimetacarpalis muscle (UMD) in chick

and quail, having endplates of the "en plaque" and "en grappe" type characteristically distributed. The ventral part shows focally, the dorsal part multiple innervation.

In order to determine, whether the characteristic endplate pattern depends upon the myogenic material or the motoneurons, the brachial neural tube or the brachial somites in the chick were replaced in two experimental series by heterotopic grafts of either somites or neural tube of the quail.

After replacement of the myogenic material, the wing muscles of the chick are made up by quail muscle fibres and are seen to be innervated by normally developed chick host nerves. The endplate pattern analysis exhibits no variations of innervation pattern in limb muscles, especially in the UMD muscle.

After replacements of brachial neural tube by leg level neural tube of quail origin, the developed muscles exhibit a normal wing-like innervation pattern. The UMD shows a multiply innervated dorsal and a locally innervated ventral part. The histological controls reveal muscle fibres of chick origin with nerves accompanied by quail Schwann cells (Figs. 6, 7, 8).

We concluded that the formation of the normal innervation pattern cannot be disturbed by the exchange of myogenic material and not by replacement of the neurons which normally innervate these muscles.

Therefore the determinative factors of the normal innervation pattern of the muscles cannot be situated immediately within the muscles or the ingrowing axons.

We had to concentrate on those possible factors, which might be localized within the developing limb anlage, preferentially the connective tissues. These are considered to be determined in a very early stage of development. Perhaps the neural crest derivates, which come in a close contact with early structural elements of the limbs, belong to those factors.

Futhermore it should be mentioned that not only the fibroblasts or the intercellular material may be important for pattern formation in the wing. Other factors also should be considered, such as the general growth of the wing anlage, interactions between ectoderm and mesenchyme as well as in-

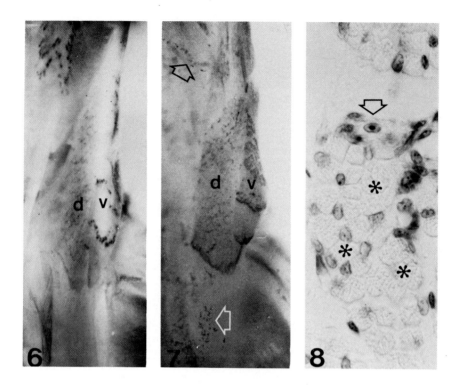

Fig. 6. M. ulnimetacarpalis dorsalis (UMD) of the chick. The ventral part (v) is focally (en plaque), the dorsal part (d) is multiply (en grappe) innervated.

Fig. 7. M. ulnimetacarpalis dorsalis of the just hatched chick after replacement of the brachial neural tube by quail neural tube from the leg level. The pattern of neuro-muscular connections remains unchanged. Ventral part (v), dorsal part (d). Arrows: pigment cells.

Fig. 8. Section through a muscle of the wing shown in Fig. 7. Asterisks: Muscle fibres of the chick. Arrow: Nerve with a Schwann cell of quail origin.

fluences from possible sources of morphogens.

Acknowledgements

This work was supported by grant of the Deutsche Forschungsgemeinschaft (Ch 44/5-1). We wish to thank U. Ritenberg, I. Schaeben-Hamm, I. Burks and H. Hake for excellent
technical assistance, A. Boeckmann for helping with the English translation and B. Scharf for typing the manuscript.

Braus H (1905). Experimentelle Beiträge zur Frage nach der
  Entwicklung peripherer Nerven. Anat Anz 26:433.
Chevallier A, Kieny M, Mauger A (1977). Limb-somite rela-
  tionship: origin of the limb musculature. J Embryol Exp
  Morph 41:245.
Christ B, Jacob HJ, Jacob M (1979). Experimentelle Untersu-
  chungen zum Problem der Muskelindividuation in den Extremi-
  täten von Vogelembryonen. Verh Anat Ges 73:537.
Detwiler SR (1936). Neuroembryology: An Experimental Study.
  New York: Macmillan Co.
Eastlick HL (1943). Studies on transplanted embryonic limbs
  of the chick. I. The development of muscle in nerveless and
  in innervated grafts. J Exp Zool 93:27.
Eastlick HL, Wortham RA (1947). Studies on transplanted em-
  bryonic limbs of the chick. III. The replacement of muscle
  by "adipose tissue". J Morphol 80:369.
Grim M, Carlson BM (1978). Morphogenesis of muscle primordia
  in aneurogenic fore limbs in the axolotl (Ambystoma mexicanum)
  In XIXth Morphol Congr Symp 1976 Klika E, ed, pp 127 Praha:
  Univ. Karlova
Grim M, Komancová A (1982). The morphology and pennation of
  some chick wing muscles and their pattern of motor innervation. Folia Morphol (Prague) in press.
Grim M, Christ B, Klepácek I, Vrabcová M (1982). A comparison
  of motor end-plate distribution and the morphology of some
  wing muscles of the chick and quail. J Histochem (in press).
Hamburger V, Waugh M (1940). The primary development of the
  skeleton in nerveless and poorly innervated limb transplants
  in chick embryos. Physiol Zool 13:367.
Harrison RG (1904). An experimental study of the relation of
  the nervous system to the developing musculature in the embryo of the frog. Am J Anat 3:197.
Harrison RG (1907). Experiments in transplanting limbs and
  their bearing upon the problems of the development of nerves.
  J Exp Zool 4:239.

Hollyday M (1980). Motoneuron histogenesis and the develop-
  ment of limb innervation. Current Topics in Developmental
  Biology 15:181.
Hunt EA (1932). The differentation of chick limb buds in
  chorioallantoic grafts, with special reference to the muscle.
  J Exp Zool 62:57.
Jacob HJ, Christ B (1980). On the formation of muscular pattern
  in the chick limb. In Merker HJ, Nau H, Neubert D (eds):
  "Teratology of the limbs", Berlin, New York, W de Gruyter,
  p 89.
Klepácek I (1981). Anatomy of the zeugopodium and autopodium
  of the wing. II. The distribution of the muscle groups and
  muscles of the autopodium of the chick wing in ontogenesis.
  Folia morphol (Prague) in press.
Landmesser L (1980). The generation of neuromuscular speci-
  fity. Ann Rev Neurosci 3:279.
Le Douarin N, Barq G (1969). Sur l'utilisation des cellules
  de la caille japonaise comme "marqueurs biologiques" en
  embryologie expérimentale. C R hebd Sc Acad Sci Paris D
  269:1543.
Lewis L, Chevallier A, Kieny M, Wolpert L (1981). Muscle nerve
  branches do not develop in chick wings devoid of muscle.
  J Embryol Exp Morph 64:211.
Lojda Z, Gossrau R, Schiebler TH (1976). In "Enzymhistoche-
  mische Methoden". pp 1, Springer, Heidelberg-New York.
Morris DG (1978). Development of functional motor innerva-
  tion in supernumerary hind limbs of the chick embryo.
  J Neurophysiol 41:145o.
Narayanan CH (1964). An experimental analysis of peripheral
  nerve pattern development in the chick. J Exp Zool 156:49.
Piatt J (1942). Transplantation of aneurogenic forelimbs in
  Ambystoma punctatum. J Exp Zool 91:79.
Piatt J (1956). Studies on the problem of nerve pattern. 1.
  Transplantation of the fore limb primordium to ectopic sites
  in Amblystoma. J Exp Zool 131:172.
Romer AS (1927). The development of the thigh musculature of
  the chick. J Morphol Physiol 43:347.
Romer AS (1964). "The vertebrate body". Saunders, Philadelphia, Pennsylvania.
Shellswell GB (1977). The formation of discrete muscles from
  the chick wing dorsal and ventral muscle masses in the absence of nerves. J Embryol Exp Morph 41:269.
Shellswell GB, Wolpert L (1977). In Ede DA, Balls M, Hinchliffe
  (eds) "Proc Symp Vertebrate Limb and Somite Morphogenesis"
  p 71, Brit Soc Dev Biol, Cambridge Univ. Press, London
  and New York.

Straznicky K (1963). Function of heterotopic spinal cord segments investigated in the chick. Acta Biol Acad Sci Hung 14:145.

Sullivan GE (1962). Anatomy and embryology of the wing musculature of the domestic fowl (Gallus). Anst J Zool 10:458.

Summerbell D, Stirling V (1981). The innervation of dorsoventrally reversed chick wings: evidence that motor axons do not seek out their appropriate targets. J Embryol Exp. Morph 61:233.

Wortham RA (1948). The development of the muscles and tendons in the cower leg and foot of chick embryos. J Morphol 83:105.

Zelená J (1962). In Gutmann E (ed) "The Denervated Muscle" p 103, Publ House of the Czechoslovak Academy of Sci, Prague.

Limb Development and Regeneration
Part B, pages 343–348
© 1983 Alan R. Liss, Inc., 150 Fifth Avenue, New York, NY 10011

STABILITY OF MUSCLE CELLS IN THE EMBRYONIC CHICK LIMB

John C. McLachlan

Department of Zoology,
University of Oxford,
South Parks Rd., OXFORD OX1 3PS.

Amata Hornbruch

Department of Biology as Applied
to Medicine,
Middlesex Hospital Medical School,
Cleveland St., LONDON W1P 6DB.

INTRODUCTION

Many people have wished to change cells from one cell
type into another. It has been felt that, if it is possible
to demonstrate that cells can be influenced by their experimental environment in such a way as to alter their type,
this might indicate that environment determines cell type in
the first place. The logic of this, however, is not absol-
utely rigorous. In any case, attempts to do this have met
with many technical difficulties, one of which is the
problem of knowing when you've actually succeeded. If you
wait until a cell has expressed a completely mature cell
type, then, while you will not be in any doubt as to what
that cell intended to turn into, you may find that it is
impossible for trivial reasons to turn it into anything
else. A fused myotube, or an erythrocyte, may be too far
gone to change its mind. But if you act at a more promising
early stage, it is very difficult to be sure that all the
cells you are dealing with were going to turn into one type.
Any uncertainty over this can lead to the accusation that
you are dealing with selection of one cell type by the
environment, rather than conversion of one cell type into
another by the environment.

One possible way round this problem is to use a clone of cells, some of whose members are already expressing a particular cell type. It may then be possible to persuade either the differentiated cells, or failing this their un-differentiated siblings, to turn into something else. This is the approach adopted by Nathanson and his co-workers (Nathanson, Hilfer, & Searls 1978; Nathanson 1979). They grew cells from muscle in dispersed culture, and identified clones as either muscle or fibroblast. Each type was scraped off and pooled separately, and tests of clones failed to reveal contamination.

It is now necessary to find an environment likely to convert such pooled clones to something else. A very prom-ising system is that employed by Marshall Urist and his colleagues over a number of years, and involves the use of demineralised bone matrix (see, for example, Nogami & Urist 1974). When a piece of this is implanted ectopically into a rat, it induces cartilage and bone formation from the surrounding tissue. Similarly, if muscle from neonatal or embryonic animals is placed on bone matrix and maintained in organ culture, cartilage will develop from the explant. Nathanson therefore placed his pooled clones onto Urist bone matrix, and subsequently obtained cartilage, both from rat fibroblast and chick muscle clones. This he interpreted as indicating that cells had been converted from muscle to cartilage.

However, when working with pooled clones, the risk of contamination is always present, particularly since, by the nature of cloning, only very small total samples can be obtained (in Nathanson's case, only one pool of cloned muscle cells could be assembled). It would be necessary to have a marker which was specific to the muscle cells before-hand, and which was subsequently identified on the cartilage cells, before such a conclusion could be drawn with absolute confidence.

Such a marker is now available. It has been shown (Christ, Jacob, & Jacob 1977; Chevallier, Kieny, & Mauger 1977) that if the somites opposite the wing forming region of a chick embryo of about 2 days of incubation, are replaced by equivalent somites from a quail embryo, then in the wing which subseq ently develops opposite the graft, muscle cells, and muscle cells only, are of quail origin. (Quail cells can be identified histologically from chick

cells, as they possess a nucleolus which is visible after
staining by the Feulgen method; see Le Douarin 1973). All
the other tissues of the limb, including even the connective
tissues of the muscles are of chick origin.

We therefore decided to employ this experimental technique to investigate the lability of the muscle cells of
the embryonic chick limb.

MATERIALS AND METHODS

Host eggs were prepared as follows. Cross-bred chick
eggs (Rhode Island Red X Light Sussex, obtained from
Orchards Farm, Great Missenden, Bucks, U.K.) were incubated
for 44 hours at 39°C, and were then windowed and staged.
The embryos were made visible by injecting Indian Ink
diluted in three parts Minimum Essential Medium into the
yolk, via a hole in the blastoderm outside the area vasculosa.
The somites opposite the wing region were removed with an
electrolytically sharpened tungsten needle.

Donor Quail eggs from our own breeding stock were
incubated for 40 hours. The yolks were tipped into a dish
of warm sterile saline, the blastoderm was cut free, and the
embryos were transferred to a dish with a black wax base.
The embryo was pinned out under Calcium- and Magnesium-Free
sterile saline, endoderm uppermost, and the endoderm peeled
away with sharp needles. The embryo was turned over, and
the ectoderm removed in the same way. A piece including the
desired somites was cut out with angled scissors, and transferred to a dish of M.E.M., where the somites were isolated,
prior to being transferred to the host and placed in the
prepared trench. Host eggs were resealed and returned to
the incubator.

After a further 7 days, the host embryos were sacri-
ficed, and the operated limbs removed. The distal tip of
the wing (from the hand) was removed, and embedded in wax
for sectioning and staining by the Feulgen method, to check
that the operation had been successful. The ectoderm was
removed from the remaining part of the wing, and the tendons
and cartilage elements taken out. The soft tissue left
behind was placed onto a hemicylinder of demineralised bone
matrix (kindly given by Dr. Nathanson) which had been
prepared by pre-soaking in fresh chicken serum. An

additional piece of bone matrix was placed on top of the soft tissue to hold it in place, and the whole combination placed on a stainless steel grid, and maintained in organ culture in medium CMRL 1066 for 28 days, with the medium being changed every 2 days. After this period, the bone matrix was removed from culture, and embedded in Araldite resin. 1 μm serial sections were cut, and stained altern- ately with Toluidine Blue, or by the Feulgen method. About half the length of each piece of bone matrix was usually examined. Sections were then examined for presence of cartilage in Toluidine stained sections, or quail cells in sections stained by the Feulgen method.

To ensure that the bone matrix was functioning corr- ectly in our hands, control experiments using pure chick tissue were carried out beforehand. Similarly, experiments were performed with pure quail soft tissue, to ensure that the quail nucleolar marker could survive this period in culture.

RESULTS

In the controls, cartilage was readily observed in almost all preparations. Quail controls also showed good preservation of the quail marker. 61 somite grafts were carried out for the experiment proper, of which 18 survived for a further seven days. On examination, 9 of these were found to show insufficient presence of quail cells in the hand region, and so were discarded. These figures are con- sistent with those obtained by us in other experimental series. 9 experimental organ cultures were therefore estab- lished. 1 of these became infected over the culture period, and was therefore discarded. On sectioning, no sign of cartilage could be found in one of the organ cultures: the other 7 all contained cartilage in varying degrees. In no case were quail cells observed in any of the sections. To ensure that the Schiff's reagent in the staining technique was functioning correctly, some of the control sections of the pure quail preparation were processed with each of the experimental batches.

DISCUSSION

In this experiment, no evidence was found which would

indicate that embryonic muscle cells, or cells which would normally have given rise to muscle, had been converted by their environment to cartilage cells. This is at variance with the findings reported by Nathanson et al. However, some comments must be made. Firstly, a negative finding does not prove that the phenomenon cannot occur. Secondly, our cells had never been exposed to dissociation by any technique, and this process may alter the readiness of cells to divert from one cell type to another. Thirdly, it is not clear how many of our quail cells were already fused into myotubes, and may therefore have been unavailable for conversion into cartilage cells. In this regard, it is worth pointing out that severe selection must have operated on the cells; the numbers placed on the bone matrix were of the order of millions, but the numbers recovered were of the order of thousands. In a recent experiment by Madeleine Kieny and her co-workers (Kieny 1980) somite grafts carried out, and the resulting wing buds were removed, the mesoderm dissociated to single cells, the single cells spun down into pellets which were then shaped into pieces for recovering by ectodermal hulls, and these combinations pinned back to the flanks of chick hosts. The resulting outgrowths contained cartilage elements, and among the cartilage cells were cells of chick origin.

Workers have tended to concentrate in attempting to interconvert muscle and cartilage cells; largely, I think, because of the readily recognisable phenotype of these cells. However, in our experiments, the cartilage we obtained was derived from tissues which would not normally give rise to cartilage, and this may represent a diversion of cells from their fate. It may also conceivably represent the presence of a concealed cartilage stem cell in the connective tissue.

Finally, it is worth commenting that, even if it were unequivocally demonstrated that the cell types in the embryonic chick limb could be interconverted, this would not prove that cell types in the normal developing limb arise under the influence of local environmental factors.

J.McL. was supported by the Medical Research Council.

REFERENCES

Chevallier A, Kieny M, Mauger A (1977). Limb-somite relationship: origin of the limb musculature. J Embryol exp Morph 41: 245.

Christ B, Jacob HJ, Jacob M (1977). Experimental analysis of the origin of the wing musculature in avian embryos. Anat and Embryol 150: 171.

Kieny M (1980). The concept of a myogenic cell line in developing avian limb buds. In Merker HJ, Nau H, Neubert D (eds) "Teratology of the Limbs", Berlin, New York: Walter de Gruyter, p 79.

Le Douarin NM (1974). A biological cell labelling technique and its use in experimental embryology. Devl Biol 30: 217.

Nathanson MA (1979). Skeletal muscle metaplasia: formation of cartilage by differentiated skeletal muscle. In Mauro A (ed) "Muscle Regeneration", New York: Raven Press, p 83.

Nathanson MA, Hilfer SR, Searls RL (1978). Formation of cartilage by non-chondrogenic cell types. Devl Biol 64: 99.

Nogami H, Urist MR (1974). Substrata prepared from bone matrix for chondrogenesis in tissue culture. J Cell Biol 62: 510.

Limb Development and Regeneration
Part B, pages 349–358
© 1983 Alan R. Liss, Inc., 150 Fifth Avenue, New York, NY 10011

THE INVOLVEMENT OF NERVES IN CHICK MYOBLAST DIFFERENTIATION

Philip H. Bonner and Thomas R. Adams

T. H. Morgan School of Biological Sciences
University of Kentucky, Lexington, KY   40506

Previous work has shown that embryonic chick skeletal muscle is not a homogeneous tissue with respect to mononucleated myoblasts; several distinct myoblast classes have been identified in the leg by analysis of the differentiation in tissue culture of clones grown from single myoblasts. Four clonable myoblast types are recognized: CMR-I cells appear in the leg during the third day of egg incubation (E3) and nearly disappear by day E5; CMR-II is another transient myoblast type, being present at high levels between approximately E4 and E9; CMR-III appears during E8 and remains through the rest of embryonic development; FMS myoblasts are seen after E5 (White et al., 1975; Bonner and Hauschka, 1974; Bonner, 1980).

Normal neuromuscular relationships in the legs of embryos have been interrupted permanently by spinal cord cautery or transiently by treatment with neuromuscular blocking agents such as curare and bungarotoxin (Bonner, 1978). Such functional denervation performed on embryos of different ages has developmentally restricted effects on the composition of the total myoblast population (Bonner, 1980). Transient denervation during E3 and E4 or after E6 has no discernible effect on clonable myoblast classes. Denervation during E5 and E6, however, prevents the later appearance of CMR-III myoblasts during E8 and reduces the proportion of FMS myoblasts to about 60% of normal. The proportions of CMR-II and CMR-I are significantly increased. This window of denervation sensitivity corresponds roughly to the period during which initial innervation of the leg musculature occurs (Fouvet, 1973; Landmesser and Morris, 1975).

A normal E10 leg contains very low levels of CMR-II and CMR-I myoblasts; most of the clonable cells are CMR-III or FMS. Denervation during the E5-E6 window changes these proportions dramatically so that by E10, in denervated tissue, CMR-II is the predominant type. These shifts are not detectable by clonal analysis until E8. Thus there is some nerve-sensitive event or events which occur during the window but which are not expressed for another 2 or 3 days.

Attempts to sort out possible cell interactions during and immediately after the window using whole embryos as an experimental system have proved fruitless due to the limited number of approaches which can be taken. To circumvent many of these problems a completely in vitro system which allows interaction between nerve and muscle cells has been developed.

METHODS

Nerve cell cultures are initiated and, after neuron differentiation is well under way, muscle cells are added so that cellular interaction can occur. Briefly, embryonic spinal cords free of meningeal tissues are enzymatically reduced to single cell suspensions and the cells added to gelatin-coated 35 mm plastic Petri dishes at 1.2×10^6 cells per dish. Cytosine arabinoside (Ara-C; $2\mu g/ml$) is added at 24 hr of culture, after cell attachment and initial neuron differentiation, to kill dividing non-neuronal cells. The cultures are washed to remove Ara-C during the third day, fed, and on the 4th or 5th day myogenic cells are added at 2×10^5 cells/dish. Myogenic cells from the same preparations used for the nerve-muscle cocultures are also cultured alone as controls. After 48 hr of culture all cells from the two types of culture are enzymatically removed from the dishes and replated at low density on fresh plates for clonal analysis of myoblast populations. Because of the Ara-C treatment essentially the only clonable cells retrieved from the primary cocultures are the myogenic cells. Complete details of procedures can be found in Bonner and Adams (1982).

Fresh medium sufficient (FMS) myoblasts are defined by their ability to grow and form differentiated clones in fresh medium (FM; Ham's F-10 with 15% horse serum and 5% chick embryo extract). Conditioned medium requiring (CMR) myoblasts grow in FM but do not differentiate; clonal differentiation occurs only in conditioned medium (CM; FM which has been exposed to cultures of confluent muscle cells). CMR types I, II, and III are distinguished by clonal morphology and the stability of their differentiated states.

RESULTS

Our initial objective was to determine whether or not cultured spinal cord cells could affect myoblast populations of normally developing embryo legs. In an extensive series of experiments cells from prewindow, window, and postwindow muscle tissue were cocultured for 48 hr with spinal nerve cells or alone as controls. Prewindow cocultures and controls showed no difference of percentage muscle clones (%MC, figure 1). Cells from postwindow embryos similarly are unaffected by nerve coculture. As in the embryo it was only with E5 and E6 window cells that a nerve effect on clonable myoblasts could be demonstrated. The effect is a substantial increase in the proportion of myoblast clones grown in either medium.

This effect on window cells increases linearly from a few hours after beginning coculture to a maximum between 24 and 48 hr. The generation time of both control and cocultured cells is constant at about 20 hr, cell morphologies are identical under the two conditions, and the clonal plating efficiencies of cells from either type of primary culture are the same. Myogenic cell density is kept low enough so that fusion does not occur during primary culture, thus ensuring not only that myoblasts are not removed from the clonable pool by entrance into myotubes but also that the nerve effect is not transmitted to clonable mononucleated cells by way of myotubes. In most of the experiments of figure 1 the nerve cells were from spinal cords of E5 and E6 embryos, ages chosen for ease of dissection rather than for effectiveness; cords of both older and younger embryos also increase %MC of window cells in coculture.

Spinal cord cells could be interacting with myogenic cells in a variety of ways. Neurons or non-neuronal glia and fibroblasts could be conditioning either the culture medium or the plate surface by release of trophic substances; on the other hand, contact between cells may be required. In the case of neurons, contact between neurites and myogenic cells is frequent and obvious (figure 2).

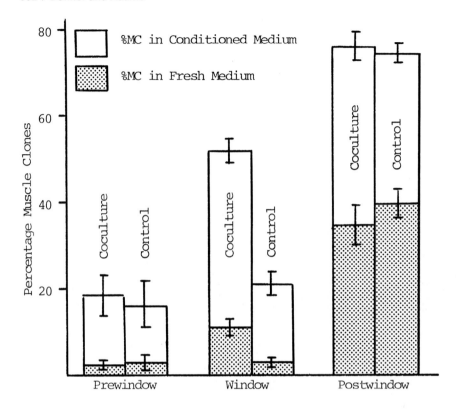

Figure 1. Mean percentage muscle clones ± SEM of n separate experiments. Prewindow: Clonal analysis in FM and CM of primary cocultures and control, solitary cultures of myogenic cells from stage 22 through stage 24 legs; n = 13 for both media. Window: stages 26 through 30, n = 28 for FM clonal analysis, n = 36 for CM. Postwindow: stages 31 through 38, n = 15 for FM, 18 for CM. To determine significance of difference, the proportions from which the means were calculated were subjected to an arcsine transformation and the transformed means and standard deviations of cocultures and controls compared by a two-tailed t test. For prewindow, P < 0.5 in both FM and CM; window, P < 0.02 in FM and P < 0.001 in CM; postwindow, P < 0.4 in FM and P < 0.5 in CM.

A series of experiments was performed using different primary culture methods to get a clearer picture of how the effect could be transmitted to susceptible cells.

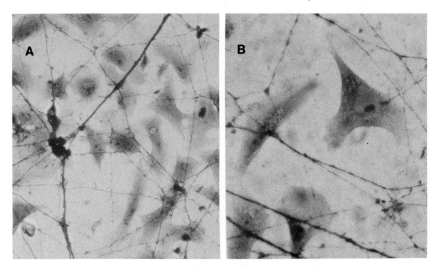

Figure 2. Spinal cord and window myogenic cell cocultures fixed and stained during the second day of culture.
Typical neurite networks and the relationships between
neurites and myogenic cells are shown in A (200X). A few
dark staining neurons are present. Note that essentially all
non-neuronal cells are contacted by neurites. The cells
shown in B are from an identical coculture and show at higher
magnification (400X) the sorts of contacts made. The left-
center cell is crossed by at least three neurites while the
right-center cell has a well-defined neurite terminal near
its nucleus.

Cultures were set up in which myogenic and nerve cells
were on the same plates but were physically separated from
each other. Glass cylinders of 9 mm diameter were sealed to
Petri plates and spinal cord cells added either inside or
outside the cylinders. After Ara-C treatment and neuron
differentiation myogenic cells were added only to the cylinders. The cylinders were removed after myogenic cell attachment (6 hr) and the plates washed to remove unattached cells.
In one case the myogenic and nerve cells were together inside
a corral (coculture), in another the myogenic cells were inside the corral but separated from the nerve cells outside
by a 0.5 to 1 mm wide ring of bare plate surface left by
removal of the cylinder, and, as a control, myogenic cells
were in corrals with no nerve cells present anywhere on the

plates. In some experiments all plates were continuously rocked back and forth so that corraled muscle cells would be exposed to nerve-conditioned medium in case diffusion should prove to be insufficient. After 48 hr of culture, cells were removed for clonal analysis (Table 1). Clearly separation of nerve and muscle cells abolishes the increase of %MC and it can be concluded that soluble substances released from nerve cells do not act on myogenic cells over relatively long distances.

It is possible that such putative neurotrophic factors may act over only short distances. They may be so unstable or dilute that active components never reached the muscle cell corrals. To examine this point standard cocultures and controls were continuously perfused with culture medium at high rates--10 ml per hour across a plate containing 5 ml-- to wash out or dilute short-range unstable factors. Perfusion did not reduce the coculture effect. In fact, the %MC of cocultures was increased by perfusion from 39% to 53% and of controls from 15% to 25%. This perfusion dependent increase is presumed to be due to improved metabolic health of the cells. In any event, rapid perfusion does not abolish the effect and strongly implies that medium conditioning is not the only effect.

Nerve cells could be conditioning the plate surface, an effect which cannot be ruled out by the corral or perfusion experiments. Substrate-attached-material (SAM) generated by cultured spinal cord cells was tested for its ability to affect clonable myoblast populations of window cells. Four to five day old nerve cultures were treated with EDTA in Ca^{++}-Mg^{++}-free saline to remove cells without removing SAM (Sieber-Blum and Cohen, 1980). Window cells cultured on these SAM plates or on fresh, control plates showed no differences of myoblast content by clonal analysis: $28 \pm 8.6\%$ (SEM) muscle clones from SAM-cultured cells and $25 \pm 8.2\%$ from controls.

At this point all signs point toward actual physical contact between effector cells and myogenic cells as being the mediator. The question now is which cells are acting on myogenic cells, neurons or other cell types? If neurons are in fact the responsible cells then the nerve effect in culture should be sensitive to curare if it is at all similar to the observed in vivo effects, which certainly are curare sensitive. We have attempted to interfere with transmission

TABLE 1

Separation of Nerve and Muscle Cells
Abolishes the Nerve Effect

Primary Culture Conditions	% Muscle Clones[a]
Nerve and myogenic cells inside corrals	43 ± 6.6 n = 6
Myogenic cells inside, nerve cells outside corrals	17 ± 5.9 n = 7
Myogenic cells inside corrals, no nerve cells	18 ± 4.1 n = 8

[a]Cells were grown 48 hr under the indicated conditions and then cloned in conditioned medium. Mean percentage muscle clones ± SEM of n experiments. Line 1 is significantly different from both lines 2 and 3 at $P < 0.01$.

of the effect by adding curare to cocultures and controls at concentrations from 0.05 to 0.5 mg/ml and, because curare is somewhat unstable in culture medium, the curare was replenished by partial refeeding 3 to 7 times during the 48 hr of primary culture. Controls were unaffected by all concentrations but the %MC increase of cocultures in 0.5 mg/ml curare was only 48% as large as cocultures with no curare ($P < 0.01$). There was no significant effect at 0.1 mg/ml curare or less. The highest concentration of curare used is close to the maximum non-toxic levels tolerated by nerve and muscle cells; 1.0 mg/ml is toxic to both cell types.

The nerve-dependent increase in %MC is only reduced, not abolished, by curare. These results imply that neurons are indeed exerting an effect on myogenic cells and that the effect is mediated in part by acetylcholine receptors. The inability to completely inhibit the effect, however, suggests other mechanisms may also be operable. To this end we have tested cells from a few non-neuronal embryonic tissues for their ability to interact functionally with window muscle cells. Cells from E6 liver and heart were grown under the same conditions as nerve cells and cocultured 48 hr with myogenic cells. The number of heart or liver cells used to begin the cultures (50,000 cells per 35 mm dish) gave, after Ara-C treatment, a cell density similar to that of the non-neuronal cells in spinal cord cultures. Clonal analysis in conditioned medium of cocultured and control myogenic cells showed no effect by liver cells (22% and 18% MC, respectively) but with heart cells there was a definite, reproducible, and statistically significant (P < 0.02) difference of %MC between cocultures and controls (42 ± 4.2% SEM and 22 ± 5.3%, respectively). Corral experiments showed that intimate coculture is required -- separation of heart and leg cells allowed no increase in %MC -- and Petri plates with heart-produced SAM on them likewise did not affect window cells. Cultures were also subjected to continuous fresh medium perfusion before clonal analysis and in this situation the stimulatory effect of heart cells was abolished; the %MC in conditioned medium of perfused heart cocultures was 38 ± 13.6% and of perfused controls 32 ± 16.1% (SEM). These numbers are not significantly different. Prewindow myogenic cells were also cocultured with heart cells without perfusion and, as with nerve cultures, there were no differences between cocultures and controls by clonal analysis. Heart cells then, but not liver cells, in an undisturbed coculture do cause demonstrable changes in the clonable myoblast population of window cells. Because perfusion abolishes these changes it is most likely that some sort of short-range, soluble heart cell products are responsible; long range stable factors are unlikely in view of the corral results.

Our conclusion from all of these experiments is that in cocultures of spinal cord and window myogenic cells there are two kinds of effects, both of which tend to increase the clonable myoblast proportion. One is a curare-sensitive interaction between neurons and myogenic cells which involves contact between cells; the other is mediated over short distances by components produced by cells and is a form of medium conditioning. The curare-insensitive increase of %MC in spinal cord cocultures is considered to be of this latter type and analogous to the effects of heart cells. Whether neurons or non-neuronal cells produce the conditioning factors is not known.

DISCUSSION

Coculture of nerve and muscle cells closely mimics events occurring in the embryo -- prewindow and postwindow muscle cells are unaffected while the clonable myoblast populations of window cells are altered. It is also known that in vitro coculture can induce the differentiation of CMR-III myoblasts from a neuron-sensitive precursor (Bonner and Adams, 1982). CMR-III myoblasts were induced from leg myogenic cells of 10-12 day old embryos which had been permanently denervated by cautery before the window. The cells had passed through the window without being affected by the nervous system and these denervated legs had accumulated abnormally large numbers of nerve-sensitive precursor cells which could be induced to become CMR-III by coculture. The same sorts of control experiments described in this paper were performed and it was concluded that induction of CMR-III from precursor requires contact with a neuron and that soluble and insoluble trophic substances are not involved at all. While the identity of the precursor remains obscure, the CMR-II myoblast, or a cell with similar clonal characteristics, is the most likely candidate because its proportion in denervated 10 day legs is indeed increased to abnormally high levels (Bonner, 1978, 1980). The increase in %MC observed in the present work on window cells is due almost entirely to production of cells which display all the culture characteristics of CMR-II myoblasts, as defined in White et al. (1975).

In summary, it appears that cultured nerve cells produce CMR-II-like cells from normal window muscle cells, induce CMR-III myoblasts from a CMR-II-like precursor, and, in the embryo, that blockage of nerve action during the window abolishes the later appearance of CMR-III. In the normally innervated embryo a 2-3 day maturation period after the window is required before CMR-III myoblasts appear. We propose that, during the window, nerves cause the appearance of a CMR-II-like cell population which will become CMR-III after nerve-independent maturation.

When embryos are permanently denervated before the window an abnormally large population of true CMR-II myoblasts is present at day 10. These cells, when cocultured with neurons, can either be directly induced to clonable CMR-III or else they can undergo their maturation phase in culture. In some of the window coculture experiments, appearance of CMR-III myoblast clones was specifically examined and it was found that low levels (5-10% of clones) were produced in about half of the experiments and then only from the cocultures and not from the controls. This suggests

that the maturation phase of CMR-III induction can occur in
culture as an infrequent event when window cells are used
but as a highly probable, frequent event when denervated
postwindow (10 day) cells are used. It is not surprising
that some cells, especially younger ones, have relatively
unstable differentiated states and cannot carry out their
complete developmental program in culture. In fact, if
window cells could complete their programs in culture then
CMR-III myoblasts should be clearly apparent after clonal
analysis of normally innervated window myogenic cells, a
situation which does not occur (Bonner, 1980).

Bonner, PH and Hauschka SD (1974). Clonal analysis of verte-
    brate myogenesis. I. Early developmental events in the
    chick limb. Develop Biol 37:317.
Bonner, PH (1978). Nerve-dependent changes in clonable myo-
    blast populations. Develop Biol 66:207.
Bonner, PH (1980). Differentiation of chick embryo myo-
    blasts is transiently sensitive to functional denervation.
    Develop Biol 76:79.
Bonner PH and Adams TR (1982). Neural induction of chick
    myoblast differentiation in culture. Develop Biol 90:175.
Fouvet B (1973). Innervation et morphogenese de la patte
    chez l'embryon de poulet. I. Mise en place de l'inner-
    vation normale. Arch Anat Microsc Morphol Exp 62:69.
Landmesser L and Morris DG (1975). The development of func-
    tional innervation in the hind limb of the chick embryo.
    J. Physiol (London) 249:301.
Sieber-Blum M and Cohen AM (1980). Clonal analysis of quail
    neural crest cells: they are pluripotent and differen-
    tiate in vitro in the absence of noncrest cells. Develop
    Biol 80:96.
White, NK, Bonner PH, Nelson DR, and Hauschka SD (1975).
    Clonal analysis of vertebrate myogenesis. IV. Medium-
    dependent classification of colony forming cells. De-
    velop Biol 44:346.

Limb Development and Regeneration
Part B, pages 359–368
© **1983 Alan R. Liss, Inc., 150 Fifth Avenue, New York, NY 10011**

CELLULAR ORIGIN OF EXTRACELLULAR MATRIX COMPONENTS DURING
MUSCLE MORPHOGENESIS REVEALED BY MONOCLONAL ANTIBODIES

Matthias Chiquet and Douglas M. Fambrough

Carnegie Institution of Washington
115 West University Parkway
Baltimore, Maryland 21210

INTRODUCTION

During muscle morphogenesis in the chick embryo, interactions between muscle and connective tissue cells are
thought to be important in several processes. Limb bud
mesenchymal cells seem to provide "positional information"
for the muscle precursor cells, which migrate into the limb
from the myotomes of the somites (Chevallier et al., 1977).
The same mesenchymal cells might be involved in the splitting
of the muscle anlagen into individual muscles (Shellswell
and Wolpert, 1977). Finally, the extracellular matrix of
the muscle, including the basement membrane surrounding each
muscle fiber, probably arises in cooperation of muscle and
connective tissue cells (Lipton et al., 1977; Chevallier et
al., 1977). Therefore, it has been postulated (Shellswell
et al., 1980; Chiquet et al., 1981) that limb bud nonmuscle
cells might form an extracellular framework populated by
muscle precursor cells; such a framework might serve both
to transfer morphogenetic information from nonmuscle to
muscle cells and to anchor the arising muscle fibers in their
functional position. Indirect evidence for such an hypothesis
stems from work on the distribution and function of the extracellular matrix components, collagens (Bailey et al., 1979;
Shellswell et al., 1980) and fibronectin (Chiquet et al.,
1979; 1981). Fibronectin might act as an information carrier
since fibroblasts produce five times more of it than myogenic
cells (Chiquet et al., 1981; Gardner and Fambrough, 1982).
Myogenic cells are highly responsive to exogenous fibronectin:
it promotes their attachment, elongation, and alignment
(Chiquet et al., 1979; 1981).

Inherent to all studies on the function of extracellular matrix components in morphogenesis is the problem of their origin, both in cell and organ cultures and especially in vivo. This problem needs to be solved in order to determine the direction (i.e., source and target) of morphogenetic information transferred between different cells. To solve this problem one must be able to distinguish between the molecules produced by the different cell types involved. We therefore decided to isolate non-crossreacting monoclonal antibodies against homologous extracellular matrix molecules of different species; such antibodies can be used to determine the cellular origin of a given component in mixed interspecies cultures (e.g., chick myoblasts with rat fibroblasts) and eventually in interspecies limb bud grafts. With the aid of two species-specific anti-fibronectin antibodies, we are already able, in mixed chick/rat cultures, to determine the respective contribution of both cell types to the overall fibronectin production. Other chick-specific antibodies against extracellular matrix components are available in our laboratory (Fambrough et al., 1981), one of them (called M1) being described here. Using the mixed culture system and the species-specific antibodies, we hope to study the interactions of myogenic and fibrogenic cells in synthesizing and assembling an extracellular matrix during muscle formation.

METHODS AND RESULTS

Fibronectin Production in Mixed Chick/Rat Cultures Determined by Species-Specific Anti-Fibronectin Antibodies

A monoclonal antibody against chick fibroblast surface fibronectin was isolated earlier (Gardner and Fambrough, 1982) and was kindly provided by Dr. J. Gardner. This antibody, called B3, is an IgG and reacts with chick cellular and plasma fibronectins; it crossreacts with quail, but not rat, fibronectins. Fibronectin from rat serum (GIBCO) was isolated as described (Chiquet et al., 1979) and injected into mice (200 µg/mouse). The fusion protocol is published elsewhere (Fambrough et al., 1981). Hybridoma supernatants were screened by a microtiter assay with rat serum fibronectin as a bound antigen and ^{125}I-labeled rabbit anti mouse Ig as second antibody. Of the eleven positive hybridomas, four were cloned and the antibodies isolated from ascites fluids. One of them, an IgG called M9, was used in the further experiments and was shown to be an anti-fibronectin

by the following criteria: immunofluorescence staining of
rat fibroblast cultures with M9 codistributed with the
fibrillar fibronectin pattern revealed by a (crossreacting)
rabbit anti human fibronectin antiserum (Chiquet et al.,
1979). Furthermore, M9 precipitated a M_r = 230,000 polypeptide from medium conditioned by ^{35}S-methionine labeled
rat fibroblasts; this precipitation was inhibited by the
anti-fibronectin antiserum (data not shown).

To demonstrate the species specificity of B3 and M9
antibodies and their general usefulness for the detection
of the respective fibronectins in mixed rat/chick cultures,
we set up the following experiment. A chick fibroblast, a
rat fibroblast, and a mixed chick/rat fibroblast culture
were labeled with 100 µCi/ml ^{35}S-methionine for 24 hr. One
hundred microliter aliquots of the three conditioned media
were incubated with 10 µg of either B3 or M9 antibody or
with a monoclonal mouse IgG of unknown specificity (MOPC 21)
as a control. Goat anti mouse IgG was used as a second

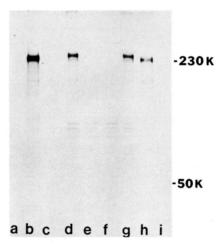

Fig. 1. Fibronectin production in mixed chick/rat cultures
revealed by species-specific anti fibronectin antibodies.
Immunoprecipitations were made from medium conditioned by
chick fibroblasts (a-c), rat fibroblasts (d-f), and mixed rat/
chick cultures (g-i) with anti rat fibronectin M9 (a,d,g),
anti chick fibronectin B3 (b,e,h), and mouse control IgG
(c,f,i). 5-15% polyacrylamide-SDS gel. Other explanations
are given in the text.

antibody. Immunoprecipitates were collected, washed, subjected to electrophoresis on SDS gels and the gels fluorographed as described (Chiquet et al., 1981). As seen in Fig. 1, only M9, but not B3 antibody precipitated labeled fibronectin from rat conditioned medium, whereas both antibodies precipitated the respective antigens (which can be identified by their slightly different migration rates on 5-15% acrylamide-SDS gradient gels) from the conditioned medium of the mixed culture. For immunoprecipitations both with B3 and M9, cross-contamination with fibronectin of the unrelated species was 5% or less (not shown). The two antibodies can thus be used to reveal the contribution of chick myogenic and rat fibrogenic cells (or vice versa) to fibronectin synthesis and deposition in a common extracellular matrix.

M1-Antigen, a Novel Chick Extracellular Matrix Component?

Different collagen types have distinct locations in muscle connective tissue, suggesting different functions and/ or different origin during development (Bailey et al., 1979; Shellswell et al., 1980). The most interesting collagen in muscle might be Type V, which is reported to be located exclusively in the endomysium surrounding single muscle fibers (Bailey et al., 1979) and to be synthesized by chick muscle cells (Sasse et al., 1981). In an attempt to raise monoclonal anti chick Type V collagen antibodies, we injected mice with material kindly provided by Drs. H. and K. von der Mark (Munich). For the boosts and screenings, more Type V collagen was isolated from 16 d chick embryo according to published methods (von der Mark et al., 1979); on SDS gels, the preparation revealed α_1 (V) and α_2 (V) chains in the ratio 2:1 as well as a minor contaminant of $M_r = 60,000$ (not shown). Fusion and screening of hybridomas were done as described above. Unfortunately, none of the 40 supernatants positive in microtiter assays could be shown to precipitate ^{125}I-labeled Type V collagen. Six hybridomas were nevertheless cloned and antibodies isolated; one of them, called M1, was further characterized.

Immunofluorescence staining of adult chick ALD muscle cryosections with M1 antibody was restricted to the endomysium (Fig. 2) and to capillaries; epi- and perimysium were not labeled at all. This pattern thus corresponds to that reported for an anti Type V collagen antiserum (Bailey et al.,

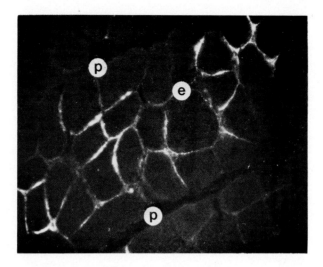

Fig. 2. Labeling of adult chick ALD muscle cryosections
with M1 antibody (7.5 µg/ml) and fluorescein labeled second
antibody. e, endomysium; p, perimysium.

1979). Whereas in chick myogenic cultures M1 antigen formed
a very fine, patchy or fibrillar network, it was partially
codistributed with fibronectin fibrils in chick fibroblast
cultures; rat cultures were not stained (data not shown).
From conditioned media of both chick (but not rat) fibroblast
and primary myogenic cultures labeled with ^{35}S-methionine,
M1 antibody precipitated three polypeptides of M_r = 220,000,
200,000, and 190,000 (Fig. 3) which were always found in
approximate ratios of 2:1:1. A fourth band of M_r = 150,000
(Fig. 3) showed variable yields. The relationship between
the different polypeptides is not yet known; to which of
them M1 antibody binds remains to be determined.

Although the M_r = 220,000 component migrates only
slightly faster than secreted chick fibronectin on SDS gels
(Fig. 3), M1 antigen is not related to fibronectin, since
M1 antibody precipitated the same bands (and in the same
amounts) from conditioned medium pretreated with monoclonal
anti-fibronectin, B3, and second antibody (not shown). In
conditioned medium treated with pepsin, M1 antibody recognized
one single peptide band of M_r = 80,000 (Fig. 3), suggesting

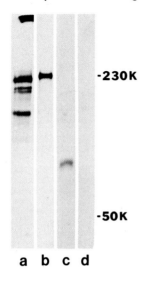

-230K

-50K

a b c d

Fig. 3. Polypeptides precipitated by Ml antibody from chick
conditioned medium labeled with
^{35}S-methionine. Immunoprecipitations were made from fresh
(a,b) or pepsin treated (2 mg/ml
for 24 hr) (c,d) conditioned
medium with antibodies Ml (a,c)
or B3 (b,d). 5-15% polyacrylamide
gel containing SDS and mercaptoethanol. Other explanations are
given in the text.

a large collagen-like domain in one (or all?) of the native
polypeptides precipitated. No bands related to the Type V
collagen chains were found.

Ml antigen is a major secretory component of both chick
fibroblast and primary myogenic cultures, the amounts found
in conditioned medium being comparable even to released
fibronectin (see below). Attempts to solubilize matrix-
bound Ml antigen in a form compatible with immunoprecipitation
have not been successful.

The Effect of Rat Fibroblasts on Ml and Fibronectin Release
by Chick Myogenic Cells

Because of its abundance and interesting in vivo distribution, Ml antigen was chosen together with fibronectin for
a pilot study on the influence of rat fibroblasts upon
secretion of these components by chick primary myogenic cells.
In a first experiment, chicken breast myoblasts were either
plated directly into gelatinized 35 mm culture dishes or
mixed in a 3:1 ratio with rat secondary fibroblasts and then
plated. After two days, both sets of cultures were labeled
with ^{35}S-methionine (100 µCi/ml) for 24 hrs. Conditioned

media were collected and aliquots taken for immunoprecipitations with the antibodies M1, B3, M9, and mouse control
IgG as described above. After electrophoresis, radioactivity
in the respective bands was determined by liquid scintillation
counting. Nonspecific backgrounds found with control IgG
were subtracted from specific counts which were normalized
on a per cell basis by counting nuclei in parallel cultures.

Plated Cells	250,000 Chick Mb	200,000 Chick Mb 50,000 Rat Fb
Total nuclei at 72 hr x 10^{-5}	6.8	10.2
Mt nuclei at 72 hr x 10^{-5}	4.4	3.3
M9 (cpm/10^5 cells)		7360
M1 (cpm/10^5 cells)	130	218
B3 (cpm/10^5 cells)	706	213
ratio M1/B3	0.18	1.02

Table 1. Effect of rat fibroblasts on M1 antigen and
fibronectin release by chick myogenic cells. Mb, myoblasts;
Fb, fibroblasts; Mt, myotubes. M9, anti rat fibronectin;
M1, M1 antibody; B3, anti chick fibronectin. Other explanations are given in the text. Values are the mean of two
immunoprecipitation experiments.

As seen in Table 1, chick fibronectin release per cell
was found to be repressed in the chick myogenic cultures
mixed with rat fibroblasts as compared to the control. (Of
course, total fibronectin was much higher due to the rat
fibroblasts). On the contrary, release of M1 antigen seemed
to be stimulated. This is most clearly seen in the ratio of
M1 antigen to chick fibronectin release which is increased
from 0.2 to 1.0. This result was confirmed in a second
experiment where a lower plating density (200,000 cells/dish)
and a shorter labeling period (6 hr) were chosen. In this
case, chick myoblasts were mixed with rat fibroblasts in
ratios of 20:0, 19:1, 18:2, and 15:5. The percentage of

nuclei in myotubes was 67, 60, 50, and 42% at 72 hr, and
final cell densities ranged from 4.7 to 6.1 x 10^5 cells
per dish. With increasing proportion of rat fibroblasts in
the chick myogenic culture, the ratio of M1 antigen to chick
fibronectin release increased from 0.71 (without rat fibroblasts) to 1.07, 1.53, and 1.65.

Control experiments were performed to exclude some
trivial explanations for this phenomenon. First, both chick
fibroblasts and myoblasts plated at different densities
revealed no obvious dependence on cell density of the M1/
chick fibronectin release ratio, which was always smaller
than one. Second, this effect cannot be due solely to the
chick fibroblasts (about 10-15%) contaminating the primary
myogenic cells. This was shown by mixing chick fibroblasts
with rat fibroblasts and determining the M1/chick fibronectin
release ratios which, although increased compared to the
controls, were smaller than one. Therefore, the contaminating
chick fibroblasts are at least not overproportionally
affected as compared to the chick myogenic cells. This
point however remains to be confirmed by adding rat fibroblasts to pure chick myotube cultures where the chick fibroblasts have previously been killed by a cytotoxic monoclonal
antibody (Fambrough et al., 1981).

It is not yet known if the described effect actually
reveals changes in fibronectin and M1 synthesis in the mixed
cultures, or if only the ratios of cell-bound vs. released
molecules are changed. Experiments have so far been
hampered by the insolubility of cell bound M1 antigen. The
cause for this change therefore remains completely unknown.

CONCLUSION

The significance of the observed shift in M1 antigen and
chick fibronectin release in mixed chick myoblast/rat
fibroblast cultures is unclear. The experiments however show
that, with the aid of species-specific monoclonal antibodies,
it is possible in principle to detect minor changes in the
synthesis of fibronectin and other extracellular matrix
components by myogenic cells against a huge background of
extracellular matrix produced by fibroblasts of another
species in the same culture. With more antibodies available,
we hope to get a more complete picture on how myogenic and
fibrogenic cells interact to form the extracellular matrix

of the muscle.

It would, of course, be worthwhile to use species-specific antibodies as probes in developmental studies on interspecies limb bud grafts. We are aware of the potency of the chick/quail transplantation system (Le Douarin, 1973), unfortunately, each of our anti chick monoclonal antibodies crossreacts with the corresponding quail antigen. We are currently setting up fusions solely with the aim to get chick and quail specific anti fibronectin antibodies, respectively.

REFERENCES

Bailey AJ, Shellswell GB, Duance VC (1979). Identification and change of collagen types in differentiating myoblasts and developing chick muscle. Nature 278:67.

Chevallier A, Kieny M, Mauger A, Sengel P (1977). Developmental fate of the somitic mesoderm in the chick embryo. In Ede DA, Hinchliffe JR, Balls M (eds): "Vertebrate Limb and Somite Morphogenesis", Cambridge: Cambridge Univ. Press, p 421.

Chiquet M, Puri EC, Turner DC (1979). Fibronectin mediates attachment of chicken myoblasts to a gelatin-coated substratum. J Biol Chem 254:5475.

Chiquet M, Eppenberger HM, Turner DC (1981). Muscle morphogenesis: Evidence for an organizing function of exogenous fibronectin. Develop Biol 88:220.

Fambrough DM, Bayne EK, Gardner JM, Anderson MJ, Wakshull E, Rotundo R (1982). Monoclonal antibodies to skeletal muscle cell surfaces. In Brockes J (ed): "Neuroimmunology", New York: Plenum, in press.

Gardner JM, Fambrough DM (1982). The biosynthesis and expression of fibronectin during myogenesis in vitro. Submitted.

LeDouarin N (1973). A biological cell labeling technique and its use in experimental biology. Develop Biol 30:217.

Lipton BH (1977). Collagen synthesis by normal and bromo-deoxyuridine-modulated cells in myogenic culture. Develop Biol 61:153.

Sasse J, von der Mark H, Kuehl U, Dessau W, von der Mark K (1981). Origin of collagen Types I, III, and V in cultures of avian skeletal muscle. Develop Biol 83:79.

Shellswell GB, Wolpert L (1977). The pattern of muscle and tendon development in the chick wing. In Ede DA, Hinchliffe JR, Balls M (eds): "Vertebrate Limb and Somite

Morphogenesis", Cambridge: Cambridge Univ.Press, p.81.

Shellswell GB, Bailey AJ, Duance VC, Restall DJ (1980). Has collagen a role in muscle pattern formation in the developing chick wing? 1. An immunofluorescence study. J. Embryol Exp Morph 60:245.

Von der Mark H, von der Mark K (1979). Isolation and characterization of collagen A and B chains from chick embryos. FEBS Letters 99:101.

Limb Development and Regeneration
Part B, pages 369–378
© **1983 Alan R. Liss, Inc., 150 Fifth Avenue, New York, NY 10011**

COLLAGEN TYPES IV AND V IN EMBRYONIC SKELETAL MUSCLE:
LOCALIZATION WITH MONOCLONAL ANTIBODIES

Thomas F. Linsenmayer*, John M. Fitch*, and
Richard Mayne+.
Developmental Biology Laboratory*, Depts.
Medicine & Anatomy, Harvard Medical School and
Mass. Gen. Hosp., Boston, Mass. 02166; Dept.
Anatomy+, Univ. Alabama at Birmingham,
Univ. Station, Birmingham, Ala. 35294.

During embryogenesis, skeletal muscle develops in
association with several different types of connective
tissue cells and the extracellular matrices they elaborate.
These connective tissue components, in addition to
performing structural and supportive roles in muscle, have
also been implicated in controlling its normal development
(Konigsberg, Hauschka 1965; Hauschka, White 1970).

Extracellular matrices fall into two general
categories, the basement membranes and the stromas. In
skeletal muscle, basement membranes are found surrounding
each myofibril, and associated with nerves, capillaries and
larger blood vessels. Stromal matrices surround groups of
myofibrils and entire muscle bundles. The collagenous
components of these matrices constitute a group of at least
five well characterized molecules (collagen types I-V) but
probably many more (for reviews see Bornstein, Sage 1980;
Linsenmayer 1982a). General agreement exists that collagen
types I-III are structurally related and occur chiefly in
stromal matrices (for several exceptions in embryos see
Linsenmayer 1982a,b). There is also agreement that type IV
collagen is a component solely of basement membranes
(Kefalides 1978). With type V ,however, conflicting
results have been the rule rather than the exception.
Several immunocytochemical studies using conventional
antibodies have reported a basement membrane location for
this molecule in which it co-distributes with type IV (Roll
et al 1980, Bailey et al 1979); others have reported a
stromal matrix location (Poschl and von der Mark 1980), and
still others have suggested that the molecule is

pericellular (Gay et al 1981). We felt that some of the ambiguities about the distribution of type V collagen and its spatial relationship, if any, to that of type IV collagen might be resolved by immunocytochemical analyses using monoclonal antibodies specific for these molecules.

Type V collagen, as originally isolated, contained two different chains termed α A and α B, now designated $\alpha2(V)$ and $\alpha1(V)$, respectively (for reviews see Bornstein, Sage 1980; Linsenmayer 1982a). Since the chains were present in a ratio of 1:2 it was suggested that the major molecular form is $[\alpha1(V)]2\alpha2(V)$. Subsequent studies have generally verified this conclusion, but in addition have shown that some tissues may have a small portion of their molecules containing $\alpha1(V)$ chains only. Still other tissues may have molecules containing a third chain, the $\alpha3(V)$ chain.

Type IV collagen molecules have both triple-helical regions (collagenous domains) and regions in which the chains are not folded into a helix (non-collagenous domains). A model for the supramolecular organization of type IV proposes that one non-helical end of a type IV molecule becomes associated with the same end of several others, forming a highly crosslinked region, termed 7S (Timpl et al 1981). Extending from 7S are long, triple-helical domains which, when isolated from the chick contain at least two collagenous fragments termed F3 and $(F1)2(F2)$ (Mayne and Zettergren, 1980).

In this report we describe two monoclonal antibodies: one specific for the $(F1)_2F2$ domain of type IV collagen (Fitch et al 1982) and the other specific for type V molecules, probably of the chain composition $[\alpha1(V)]_2\alpha(2)$ (Linsenmayer et al 1982). By immunocytochemical staining with these, we show that type IV collagen is indeed localized in basement membranes, while type V, conversely, is located within dense stromal matrices.

METHODS USED IN THE PRODUCTION OF MONOCLONAL ANTIBODIES:

Chick collagen types III, IV and V were obtained by pepsin extraction of adult chicken gizzard; type II was from a similar extract of sterna, and type I from an acid extract of lathyritic chick skin. All collagens were separated by fractional salt precipitations; some were further purified by ion exchange chromatography on CM-cellulose or DEAE-cellulose or by gel filtration on Agarose A-15M, all performed under non-denaturing conditions (Mayne, Zettegren 1981; Reese, Mayne 1981). The preparations were characterized by SDS-PAGE, CM-cellulose chromatography and amino acid analysis.

For hybridoma production, the general methodology we employ has been described (Linsenmayer, Hendrix 1982), as well as the specific details for the production of those making antibodies against collagen types IV (Fitch et al 1982) and V (Linsenmayer et al 1982). The characterization of the antibodies themselves was done largely by inhibition ELISA (Rennard et al 1980) using Immulon microtiter plates (Dynatech) coated with 2ug of collagen per well and B-galactosidase-conjugated sheep anti-mouse IgG as the enzyme-linked secondary antibody (BRL). Aliquots of inhibitor collagens were preincubated with the culture medium antibody and then were transferred to the collagen-coated wells to allow free antibody to bind. Bound antibody was determined by ELISA.

Fluorescence immunocytochemistry was performed on cryostat sections from 18-day chick embryos. Sections were incubated with hybridoma antibodies followed by rhodamine-conjugated goat anti-mouse IgG and mounted in glycerol. Before application of antibodies, some slides were pretreated for 0.5 hr at room temperature with either pepsin (0.1 mg/ml) in 0.5M HAc or the HAc solution itself.

MONOCLONAL ANTIBODIES AND THEIR CHARACTERIZATION:

The hybridomas were produced by fusions of NS1 myeloma cells with splenocytes from mice that had been immunized with either type IV or type V collagen. In the fusion with splenocytes from the type V-immunized mouse (Linsenmayer et al 1982), the antibody-positive wells were selected by passive hemmagglutination with type V collagen-coated erythrocytes. In the one using splenocytes from a mouse immunized with type IV (Fitch et al 1982), positive wells were selected by fluorescence immuno-cytochemistry. The two antibodies used in this present report are IV-IA8 which is specific for the $(F1)_2F2$ region of type IV collagen, and

V-DH2 which is specific for type V, probably in the molecular form $[\alpha 1(V)]_2\alpha 2(V)$.

By inhibition ELISA, both antibodies were completely specific for their respective collagen types. As can be seen in figure 1, antibody IV-IA8 is strongly inhibited by type IV and none of the other native collagen molecules. It is also inhibited by the $(F1)_2 F2$ native, helical fragment of type IV, purified by gel filtration on Agarose A-15M (see figure insert and legend), but not the F3 or 7S regions.

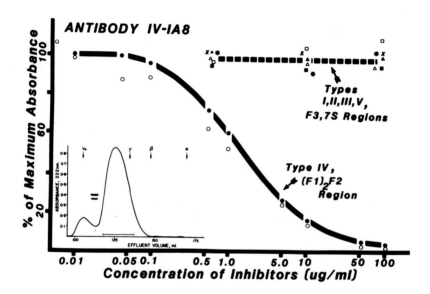

Fig. 1 Inhibition ELISA of antibody IV-IA8. Wells coated with type IV collagen. Inhibition (solid line) was obtained with type IV (●) and the $(F1)_2 F2$ region (O). No inhibition (dashed line) was obtained with types I (▲), II (△), III (■) and V (□), nor with the F3 (✱) and 7S (✗) regions. Insert shows the isolation of native $(F1)_2 F2$ by column chromatography of type IV collagen on Agarose A-15 M run under non-denaturing conditions. The major peak consists of native $(F1)_2 F2$, as deduced by SDS-Page (left of major peak) which shows exclusively F1 and F2 bands in a 2:1 ratio.

Antibody V-DH2 is equally specific for native type V collagen, being inhibited only by this molecule and none of the other known collagens (Fig.2). Among those which do not react are the "minor cartilage collagens" (unpublished results), whose component chains (1^α, 2^α, 3^α) closely resemble the chains of type V (see Mayne et al, this volume). The type V collagen used in the present study was predominantly, if not exclusively $[\alpha 1(V)]_2\alpha 2(V)$ molecules. This can be seen in the CM-cellulose chromatogram and the SDS-gel shown in figure 2 (insert). Both of these show,the $\alpha 1(V)$ chain (chromatographic Peak B and upper gel band) and the $\alpha 2(V)$ chain (chromatographic Peak A and lower gel band) to be present in a 2:1 ratio. The small CM-cellulose chromatographic peak labeled "post B" seems to be similar, if not identical, to the $\alpha 1(V)$ chain by all criteria including migration on SDS-PAGE (Mayne, unpublished data).

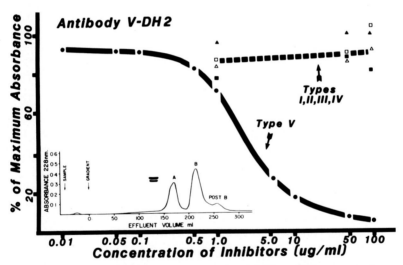

Fig. 2 Inhibition ELISA of antibody V-DH2. Wells coated with type V collagen. Inhibition was obtained with native type V (●) (solid line), and native type V which had undergone additional purifications by alkaline-urea precipitation (Miller, Rhodes 1982) and DEAE-cellulose chromatography (not shown). No other collagen showed detectable inhibition (dashed line), including types I (□), II (△), III (■) and IV (▲). The insert is a CM-cellulose chromatogram and an SDS-PAGE gel showing the chain composition of the type V. The bands of the SDS-gel appear to the left of peak A.

The antigenic sites against which both antibodies are directed are conformational dependent. That is, they require an intact triple helix for their generation and upon denaturation of the collagen they cease to exist (for reviews see Timpl 1976; Linsenmayer 1982b). Conformational dependence was examined by inhibition ELISA, using as inhibitors collagen samples that had been heated to progressively higher tempratures. These data were then compared to thermal denaturation curves of the helical structure of the collagens obtained by optical rotation measurements made in a spectropolarimeter. The results (see Fitch et al 1982; Linsenmayer et al 1982) showed that the curves of helicity were very similar to those of the temperature-dependent loss of binding of each antibody.

IMMUNOCYTOCHEMICAL ANALYSIS OF EMBRYONIC SKELETAL MUSCLE:

IV-IA8 and V-DH2 were used as the primary antibodies for indirect immunofluorescence staining of sections of embryonic skeletal muscle. The anti-type IV antibody (Fig. 3A) stained most basement membranes including those surrounding individual muscle cells, and those associated with capillaries, nerves, and the smooth muscle layers of arteries and veins. When the anti-type V antibody was used on identical sections of skeletal muscle, no staining could be seen (Fig.3B). One possibe explanation for this was masking of antigenic sites. Several methods of unmasking were tried, including pretreatment with pepsin dissolved in dilute HAc to try to remove non-collagenous proteins, and with the HAc solution itself as a control. Surprisingly, both pretreatments resulted in dramatic new staining of the dense connective tissue layers of the muscle (Fig. 3C) but no detectable staining of basement membranes. In fact, in this tissue, as well as in others we have examined (Linsenmayer et al 1982), the staining produced by the anti-type V antibodies frequently appears to be the reciprocal of that produced by the anti-type IV antibodies (compare A and C). The unmasking by acid pretreatment is specific for type V antibodies and is not due to a generalized increase in IgG binding. Identically pretreated sections stained with anti-type IV antibodies show no change in the pattern of structures normally stained by these.

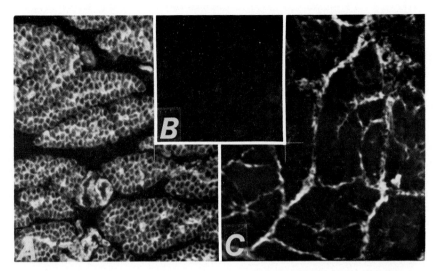

Fig. 3 Fluorescence micrographs of 18-day embryonic chick skeletal muscle stained with monoclonal antibodies IV-1A8 (A), and V-DH2 (B&C) followed by rhodamine-conjugated goat anti-mouse IgG. A and B are fresh cryostat sections. C was pretreated with 0.1M HAc for 0.5hr at room temperature.

DISCUSSION:

The specificity of the monoclonal antibodies we have produced against collagen types IV and V extends even to individual regions within the molecules. This is evidenced most clearly by IV-1A8, which binds to the $(F1)_2 F2$ region of type IV collagen and to no other domain. Additional anti-type IV monoclonals we have produced are equally specific for binding to the 7S or F3 regions of this molecule (unpublished data). In type V collagen such a diversity of molecular domains has not yet been detected (but see Fessler et al 1981). Even so, we have been able to determine that at least two of the different monoclonal antibodies we have produced against type V collagen are against different sites within the molecule. These have the same general properties as antibody V-DH2 described here, including their behavior when used for immunocytochemical staining. By competition ELISA, however, they can be demonstrated to bind to a different site within the

molecule (Linsenmayer et al 1982). One possible future use
for these antibodies, then, is for molecular dissection of
various regions within collagen molecules, providing, for
example, a means to determine whether various regions of
collagen molecules interact differently with other matrix
components and cells.

The immunocytochemical data show that type IV
collagen, as expected, is present exclusively in basement
membranes (but perhaps not in all, see Fitch et al 1982).
This type V molecule, alternatively, seems to be present in
stromal connective tissue and not in basement membranes. In
addition to skeletal muscle, we have obtained strong
staining in the dense connective tissues of a number of
other organs. This has been confirmed using several
different anti-type V monoclonal antibodies, all of which
require acid pretreatment for staining (Linsenmayer et al
1982).

The high degree of masking of the antigenic
determinants in type V collagen was quite surprising, and
equally surprising was the fact that acid pretreatment
alone effected unmasking. Early in these experiments we
detected hybridoma wells with antibodies that gave
extensive staining of connective tissues in nonpretreated
sections. These antibodies, however, upon analysis by ELISA
did not give a detectable reaction with type V collagen.
Thus, they were either directed against a minor component
of type V collagen, or against a contaminant. Although this
constituent of the type V preparation must be
quantitatively very minor, it is antigenically potent,
possibly explaining some of the inconsistent results that
have been obtained with conventional antibodies against
type V (for review on this point see Linsenmayer 1982b;
Timpl 1976).

We do not yet know the exact mechanism of unmasking of
the antigenic sites by pretreatment with dilute acid. Our
current hypothesis is that it may be due to swelling,
either of the entire tissue or more probably of a
supramolecular structure in which the type V collagen
molecules are assembled. The same mechanism of unmasking
could work, for example, if type V collagen molecules were
arranged in, or complexed with, fibrils composed largely of
another collagen, such as type I.

Acknowledgements: This is publication number 896 of the Robert W. Lovett Group for the Study of Diseases Causing Deformities. Supported by NIH Grants EY02261 & AM3560 (TFL); and HL21665 & AM30418 (RM).

Bailey AJ, Shellswell GB, Duance, VC (1979). Identification and change of collagen types in differentiating myoblasts and developing chick muscle. Nature (Lond.) 278:67.

Bornstein P, Sage H (1980). Structurally distinct collagen types. Ann. Rev. Biochem. 49:957.

Fessler LI, Kumamoto CA, Meis ME, Fessler JH (1981). Assembly and processing of procollagen V (AB) in chick blood vessels and other tissues. J. Biol. Chem. 256:9640.

Fitch JM, Gibney E, Sanderson RD, Mayne R, Linsenmayer TF (1982). Domain and basement membrane specificity of an anti-type IV collagen monoclonal antibody. (Submitted).

Gay S, Martinez-Hernandez A, Rhodes RK, Miller EJ (1981). The collagenous exocytoskeleton of smooth muscle cells Coll. and Rel. Res. 1:377.

Hauschka SD, White NK (1970). Studies of myogenesis in vitro. In Baker QB et al (eds): "Research in Muscle Development and the Muscle Spindle", Amsterdam, Excerpta. Med. Found., p 53.

Kefalides NA (1978). Biology and Chemistry of Basement Membranes, New York, Academic Press.

Konigsberg IR, Hauschka SD (1965). Cell and tissue interactions in the reproduction of cell type. In Locke M (ed): "Reproduction: Molecular, Subcellular, and Cellular", New York, Academic Press, p243.

Linsenmayer TF (1982a). Collagen. In Hay ED (ed): "Cell Biology of the Extracellular Matrix", New York, Plenum Press, p 5.

Linsenmayer TF (1982b). Immunology of purified collagens and their use in localization of collagen types in tissues. In Jayson M, Weiss J (eds): "Collagen in Health and Disease", Edinburgh, Churchill Livingston, p 244.

Linsenmayer TF, Hendrix MJC (1982). Production of monoclonal antibodies to collagens and their immunofluorescence localization in embryonic cornea. In Furthmayr H (ed): "Immunochemistry of the Extracellular Matrix", CRC Press, chapter 12 (in press).

Linsenmayer TF, Fitch JM, Schmid TM, Zak NB, Gibney E, Sanderson R, Mayne R (1982). Monoclonal antibodies against chick type V collagen: production, specificity, and use for immunocytochemical localization in embryonic cornea and other organs. (submitted).

Mayne R, Zettergren JG (1980). Type IV collagen from chicken muscular tissues. Isolation and characterization of the pepsin-resistant fragments. Biochemistry 19:4065.

Miller EJ, Rhodes RK (1982). Preparation and characterization of the different types of collagen. Methods in Enzymology 82:33.

Poschl A, von der Mark K (1980). Synthesis of type V collagen by chick corneal fibroblasts in vivo and in vitro. FEBS Letters 115:100.

Reese CA, Mayne R (1981). Minor collagens of chick hyaline cartilage. Biochemistry 20:5443.

Rennard SI, Berg R, Martin GR, Foidart JM, Robey PG (1980). Enzyme-linked immunoassay (ELISA) for connective tissue components. Anal. Biochem. 104:205.

Roll FJ, Madri JA, Albert J, Furthmayr H (1980). Codistribution of collagen types IV and AB_2 in basement membranes and mesangium of the kidney. An immunoferritin study of ultrathin frozen sections. J. Cell. Biol. 85:597.

Timpl R (1976). Immunological studies on collagen. In Ramachandran GN, Reddi AH (eds): "Biochemistry of Collagen", New York, Plenum, p 319.

Timpl R, Weidemann H, Van Delden V, Furthmayr H, Kuhn K (1981). A network model for the organization of type IV collagen molecules in basement membranes. Eur. J. Biochem. 120:203.

Limb Development and Regeneration
Part B, pages 379–389
© 1983 Alan R. Liss, Inc., 150 Fifth Avenue, New York, NY 10011

PROTEOGLYCANS PRODUCED BY SKELETAL MUSCLE IN VITRO AND IN VIVO

David A. Carrino and Arnold I. Caplan

Department of Biology
Case Western Reserve University
Cleveland, Ohio 44106

Proteoglycan analysis from this and other laboratories has provided information concerning the characterization of the proteoglycans synthesized by chick limb bud chondrocytes in culture and the possible role of cartilage proteoglycan structure in chondrocyte phenotypic expression. These studies have demonstrated that chick limb bud chondrocytes in culture produce proteoglycans that are very similar to the proteoglycans isolated from animal hyaline cartilages such as bovine nasal septum. The proteoglycans synthesized by Day 8 limb bud chondrocyte cultures have a monomer molecular weight of 2-2.5 million daltons, chondroitin sulfate chains of 20,000 daltons, keratan sulfate chains of 9,500 daltons, 58% 6-sulfated chondroitin, two classes of oligosaccharides, and the ability to aggregate with hyaluronic acid (Hascall et al. 1976).

In contrast to cartilage, very little is known about the proteoglycans synthesized by limb skeletal muscle. Some reports on the glycosaminoglycans produced by skeletal muscle have appeared (Ahrens et al. 1977; Pacifici, Molinaro 1980), but little is known about their parental proteoglycans. Also unlike cartilage, in which the functional properties of the tissue are due in large measure to the extracellular matrix components, the predominant functional properties of skeletal muscle are primarily due to the intracellular contractile apparatus. Nevertheless, muscle development does take place in a complex extracellular environment composed of products of diverse cell types. Moreover, mature skeletal muscle arises from the fusion of precursor myoblasts, a process which very likely

involves the cell surface and may also involve the extracellular matrix since extracellular matrix components have
been shown to affect myogenesis and to change during myogenesis. Hence, the proteoglycans synthesized by skeletal
muscle and related fibrous connective tissue elements may
play a key role in muscle differentiation, morphogenesis,
and/or function, and consequently a characterization of
muscle and muscle fibroblast proteoglycans could lend
insight into these various aspects of muscle physiology.

Chick leg muscle and leg muscle-fibroblast cultures
were prepared by a vortex procedure as previously developed
in this laboratory (Caplan 1976). For the muscle cultures,
Day 1-Day 2 represents the period of myoblast and
muscle-fibroblast proliferation; Day 2-Day 3 represents the
period of major myoblast fusion, so that by Day 3 the
cultures consist primarily of freshly fused myotubes and a
small number of muscle-fibroblasts. After Day 3, the
multinucleated muscle cells mature into cross-striated,
spontaneously contracting muscle straps, by Day 5 or Day 6;
the muscle-fibroblasts continue to proliferate, and on Day 6
are densely packed between and on top of the contracting
muscle straps. In addition, cultures can be treated for
18-24 h from Day 3 overnight to Day 4 with cytosine
arabinoside, an inhibitor of DNA synthesis. Such exposure
is toxic to dividing cells and thus eliminates the majority
of the muscle-fibroblasts without adversely affecting the
multinucleated muscle cells which cease proliferation upon
fusion. Like the untreated cultures, these myotubes mature
by Day 6 into contracting muscle straps. Moreover, cultures
of muscle-fibroblasts with no multinucleated muscle cells
can be obtained from untreated Day 6 muscle cultures. By
choosing the time of radioisotopic labeling, four different
cell samples were investigated: freshly fused myotubes with
a small number of muscle-fibroblasts (Day 3), mature muscle
straps with a large number of muscle-fibroblasts (Day 6),
mature muscle straps with a small number of, if any,
muscle-fibroblasts, which were labeled on Day 6 (ara C), and
muscle-fibroblasts, which were labeled on the first day
after such cultures were established (i.e., 7 days out of
the embryo).

The proteoglycans were isolated and analyzed by
standard procedures for the isolation and characterization
of cartilage proteoglycans (Hascall et al. 1976). At the
end of the labeling period, the cell layers with associated

matrix were washed twice with cold Tyrode's balanced salt solution and scraped into 4 M guanidinium chloride containing protease inhibitors. The cell layers were extracted by stirring overnight at 4°C in the 4 M guanidinium chloride plus protease inhibitors. Monomer fractions (D1) were prepared directly from the clarified extracts by dissociative CsCl equilibrium density gradient centrifugation (initial density, 1.50 gm/ml; 35,000 rpm; 46 h; 10°C). The gradients were cut into four approximately equal fractions (termed D1, D2, D3, and D4 from densest to lightest), and the small pellicles (D4 gels) routinely found at the top of such gradients were solubilized in the same solution used for extraction of the cell layers. The gradient fractions were dialyzed against water and lyophilized. The CsCl gradient profiles for the various cell samples are shown in Table I.

CsCl Equilibrium Density Gradient Distribution of Incorporated [^{35}S]Sulfate

for Muscle and Muscle Fibroblast Cultures

| Gradient Fraction | Density (gm/ml) | Muscle | | | Muscle-Fibroblast | Cartilage |
| | | Day 3 | Day 6 | Ara C | Day 1 | Day 8 |
		%	%	%	%	%
D1	1.630-1.641	26.2%	13.9%	23.9%	18.8%	86.4%
D2	1.519-1.532	27.5%	20.1%	22.6%	26.6%	6.6%
D3	1.464-1.472	18.7%	26.5%	26.1%	18.0%	4.6%
D4	1.389-1.402	9.0%	22.4%	11.0%	9.9%	1.6%
D4 gel	---	18.6%	17.1%	16.4%	26.7%	0.8%

From the distributions of the incorporated [^{35}S]sulfate, it is apparent that the relative buoyant densities of the sulfated proteoglycans made by cultures of muscle and muscle-fibroblasts are much different than that of cartilage culture proteoglycans. Cartilage proteoglycans isolated from Day 8 high density chick limb bud cultures partition on dissociative gradients such that 80-90% is recovered in the D1 or densest gradient fraction. In contrast, as shown in Table I, a much smaller proportion of the newly synthesized muscle and muscle-fibroblast proteoglycans, in some cases less than 20%, is found in the D1 fraction.

The D1 or densest gradient fractions from each of the cell samples were chromatographed on a column of Sepharose CL-2B in order to estimate the overall monomer size of the proteoglycans synthesized by the various cultures. The column profiles for the muscle cultures are shown in Figure 1. As can be seen in Figure 1A, the majority of the incorporated [^{35}S]sulfate recovered in the D1 fraction prepared from skeletal muscle cultures labeled on Day 3 is in a peak which elutes with a K_{av} of 0.17. This indicates that the average hydrodynamic size of these molecules is substantially larger than that of Day 8 cartilage culture proteoglycans which have a K_{av} on Sepharose CL-2B of 0.25 and whose elution position is indicated by the solid vertical line. In addition, the D1 fraction of Day 3 muscle cultures contains some [^{35}S]sulfate-labeled molecules which elute on Sepharose CL-2B at smaller hydrodynamic sizes, although these are present in smaller amounts than the large proteoglycan species.

FIGURE 1

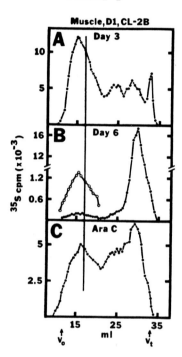

In contrast, on Day 6 (Figure 1B), when the cultures contain a large number of mono-nucleated cells with a fibro-blastic phenotype, the predomin-ant sulfate-containing molecules which are synthesized are of small average hydrodynamic size and have a K_{av} of 0.81, similar to that of ubiquitous or fibro-blastic proteoglycan. However, the D1 fraction of Day 6 muscle cultures does contain a small proportion of a large proteo-glycan which has a K_{av} equal to that of the predominant species synthesized on Day 3. (This peak is plotted with open symbols on an expanded scale.) Likewise, as shown in Figure 2, the principal molecular species synthesized by muscle-fibroblasts and recovered in the D1 fraction is the small

FIGURE 2

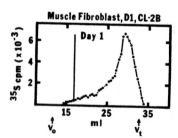

molecular weight class with none of the large proteoglycan observed in the Sepharose CL-2B profile. Moreover, muscle cultures were treated on Day 3-Day 4 with cytosine arabinoside, an inhibitor of DNA synthesis, to eliminate most of the muscle-fibroblasts and then radiolabeled on Day 6. As shown in Figure 1C, the D1 fraction of such cultures contains much less of the smaller species relative to the larger proteoglycan. Thus, it appears that skeletal muscle cultures synthesize two principal size classes of sulfate-containing molecules: a large proteoglycan which is synthesized by muscle cells and a smaller proteoglycan which is synthesized by the muscle-fibroblasts.

The two different size classes of molecules from the various cell samples were treated with alkaline sodium borohydride in order to liberate the glycosaminoglycan chains. The size distributions of the glycosaminoglycan chains were then determined by chromatography on a Sepharose CL-6B column. As shown in Figure 3, the large, muscle-specific proteoglycans, whether from Day 3 cultures or Day 6 cytosine arabinoside-treated cultures, contain unusually large glycosaminoglycan chains. The K_{av} (0.26) indicates that the average molecular weight of these glycosaminoglycan chains is greater than 50,000 daltons. This is much larger than the glycosaminoglycan chains from Day 8 cartilage culture proteoglycans which have an average molecular weight of 20,000 daltons and whose elution position is indicated by the solid vertical line.

FIGURE 3

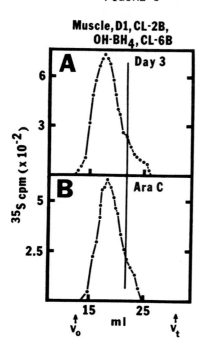

In contrast, Figure 4, on the following page, shows that glycosaminoglycan chains liberated from the small molecular weight proteoglycan, either from untreated Day 6 muscle cultures or from muscle-fibroblast cultures, are considerably smaller than the glycosaminoglycans obtained from the large, muscle-specific proteoglycans. Their K_{av} on Sepharose CL-6B (0.45) indicates that their average molecular weight is 25,000 daltons. This is somewhat larger than the glycosaminoglycans of Day 8 cartilage culture proteoglycans. The small peaks of radioactivity at V_t in Figure 4 perhaps represent degradation products of sulfated glycoproteins which may co-elute on Sepharose CL-2B with the small proteoglycans and may be hydrolized during the alkaline borohydride treatment. In addition, with or without prior alkaline borohydride treatment, the small, fibroblast-specific proteoglycans chromatograph at an identical position on Sepharose CL-6B.

FIGURE 4

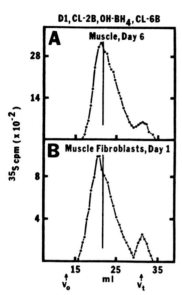

D1, CL-2B, OH-BH$_4$, CL-6B

A Muscle, Day 6

B Muscle Fibroblasts, Day 1

35 S cpm (x 10^{-2})

15   25   35
ml
V_o V_t

Disaccharide analysis was performed on the glycosaminoglycans of the large molecular weight material produced by Day 3 and ara C muscle cultures and the small molecular weight material produced in Day 6 muscle cultures. The various glycosaminoglycans were treated with chondroitinase ABC and chondroitinase AC in order to determine the proportions of 6-sulfated, 4-sulfated, and unsulfated chondroitin and also of dermatan sulfate. These data are shown in Table II.

Proportion of Sulfated Material Present in Glycosaminoglycans

Produced in Muscle Cultures

	Day 3		Ara C		Day 6		Day 8 Cartilage	
	ABC	AC	ABC	AC	ABC	AC	ABC	AC
% resistant	8.8	14.2	9.7	22.8	36.1	81.2	13.3	13.7
% C6S	81.1	76.2	74.8	67.6	18.9	4.9	54.2	53.7
% C4S	10.1	9.6	15.4	9.6	45.0	13.9	32.5	32.6

As can be seen, the glycosaminoglycans of the large proteoglycan synthesized at the two different times are similar to

each other and distinct from both those of the small
molecular weight material and those of Day 8 high density
culture cartilage proteoglycan. For both the Day 3 and
cytosine arabinoside-treated Day 6 glycosaminoglycans, about
9% of the material is resistant to chondroitinase ABC.
Also, as shown by the chondroitinase AC digestion, a large
proportion of the material, as much as 76%, is chondroitin-
6-sulfate. In addition, the muscle glycosaminoglycans contain undetectable levels of unsulfated chondroitin as determined by analysis of samples labeled with [^{3}H]glucosamine
and [^{35}S]sulfate (data not shown). Hence, if the chon-
droitinase-resistant material is disregarded, almost 90% of
the chondroitin in the muscle glycosaminoglycans is sulfated
at the 6 position. In contrast, only 54-58% of the glycosaminoglycans in Day 8 high density culture cartilage proteoglycan is 6-sulfated chondroitin. Moreover, as shown in
Table II, chondroitinase AC digestion of the muscle glycosaminoglycans indicates the presence of a small proportion of
dermatan sulfate. Comparison of the two aliquots treated
with chondroitinase ABC and AC shows that 5% of the Day 3
and 13% of the ara C glycosaminoglycans are dermatan
sulfate. However, this small amount of dermatan sulfate may
very likely arise from streaming of the smaller molecular
weight material into the muscle proteoglycan peak in the
Sepharose CL-2B elution profiles (see Figure 2 and below).

The small molecular weight material produced in Day 6
muscle cultures has a distinctive glycosaminoglycan composition. The data in Table II show that these fibroblast-
associated glycosaminoglycans consist of 36% chondroitinase
ABC-resistant material, which is nearly four times greater
than that of the muscle glycosaminoglycans. Also, only 19%
of the material is chondroitin-6-sulfate, and, if the chondroitinase AC-resistant material is disregarded, only 26% of
the chondroitin is sulfated at the 6 position; this is substantially different from both Day 8 high density culture
cartilage and muscle glycosaminoglycans. In addition, as
measured by comparison of separate aliquots digested with
chondroitinase ABC and AC, the small molecular weight
species synthesized by muscle-fibroblasts contains 45% dermatan sulfate, which is much more than the amount detected
in muscle-specific glycosaminoglycans. Thus, the glycosaminoglycan compositions suggest that the small component is
a distinct species from the large proteoglycan and does not
arise by proteolytic cleavage of the large molecules and,
moreover, probably is synthesized by non-muscle connective

tissue cells.

   In an attempt to analyze the proteoglycans synthesized in ovo, on Day 11 of incubation, fertilized chick eggs were labeled with [^{35}S]sulfate and the lower hind limb muscles removed and extracted as for the culture samples. Day 11 is prior to the major period of myoblast fusion. The [^{35}S]-sulfate-labeled material extracted from Day 11 muscle partitions on a CsCl gradient with 33% in the D1 or densest gradient fraction, which is less than that in the D1 fraction of cartilage extracts, but more than that in the extracts of muscle cultures (see Table I). The remainder of the incorporated [^{35}S]sulfate distributes with 26% in the D2, 16% in the D3, 9% in the D4, and 16% in the D4 gel.

   The D1 fraction was chromatographed on Sepharose CL-2B, and the column profile is shown in Figure 5. The majority of the incorporated [^{35}S]-sulfate elutes in a peak with a K_{av} of 0.17, which is identical to the K_{av} of the muscle-specific proteoglycan extracted from the muscle cultures (dashed vertical line). In addition, there is a small amount of labeled material which elutes at smaller hydrodynamic sizes. Thus, it appears that chick leg muscle in ovo synthesizes a proteoglycan monomer which has an average hydrodynamic size that is the same as that produced by muscle in culture.

FIGURE 5

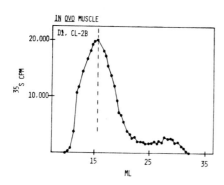

   Analysis of the high density of D1 gradient fractions indicates that chick limb skeletal muscles in ovo and chick limb skeletal muscle cultures produce principally two high buoyant density proteoglycans. One of these, which appears to be synthesized by the muscle fibroblasts, is relatively small in size. Its glycosaminoglycan chains, however, are larger than those of Day 8 chick limb bud culture cartilage proteoglycans. Because the entire molecule has the same average hydrodynamic size as the free glycosaminoglycan chains, these molecules may contain only a single or perhaps two closely spaced glycosaminoglycan chains or may, in fact, be sulfated glycosaminoglycan that is not linked to protein which, to our knowledge, has not been described and seems unlikely. These structural characteristics of the muscle-fibroblast molecules are the same as those of sulfated molecules found by us in cultures on non-chondrocytic cells derived from Day 8 high density chick limb bud cultures (Carrino et al. 1982) and cultures of chick skin fibroblasts (unpublished observations). These molecules may, therefore, represent a class of connective tissue elements that is, as the names generally applied to them imply, ubiquitous, fibroblastic proteoglycans. Further characterization, such as glycosaminoglycan disaccharide analysis, may reveal tissue differences and is currently in progress.

The other major proteoglycans extracted from skeletal muscle cultures appear to be synthesized by the skeletal muscle cells themselves. These proteoglycans, as discussed above, are structurally distinct from cartilage proteoglycans and the muscle-fibroblast molecules. One aspect of their structure which correlates with previous reports of chick skeletal muscle glycosaminoglycans is the proportion of 6-sulfated chondroitin. Two previous reports of total in vitro chick breast muscle myotube glycosaminoglycans prepared by extensive pronase digestion indicate that 80-90% of the sulfated chondroitin is 6-sulfated (Ahrens et al. 1977; Pacifici, Molinaro 1980), which is in good agreement with the results in Table II. Thus, there is the possibility, albeit unlikely, that all of the chondroitin sulfate produced by chick skeletal muscle is contained in the D1 fraction; alternatively, the sulfation pattern of any chondroitin sulfate in the lighter buoyant density gradient fractions may be similar to that of the chondroitin sulfate in the D1 fraction. The material in the other gradient fractions is currently under investigation.

The exact function of the muscle-specific proteoglycans remains to be elucidated. In this regard, several hypothetical roles for skeletal muscle glycosaminoglycan have been postulated. A role of skeletal muscle chondroitin-6-sulfate in muscle cell proliferation has been suggested because of the observation of high levels of chondroitin-6-sulfate in fetal bovine muscle and other tissues (Dietrich, Sampaio 1981) which are diminished in later development and adulthood. These observations are consistent with our own, since with time, the

muscle-fibroblast material, which contains only a small
amount of chondroitin-6-sulfate, becomes predominant. In
contrast, the proposed role for chondroitin-6-sulfate does
not correlate with our results and those of others, because
the chondroitin-6-sulfate-rich muscle proteoglycan (Fig. 1A)
and chondroitin-6-sulfate-rich glycosaminoglycans (Ahrens et
al. 1977; Pacifici, Molinaro 1980) are synthesized in
abundance even after fusion when myogenic cells have ceased
proliferation. Alternatively, L6 skeletal muscle myoblast
cell line glycosaminoglycans have been proposed to be
involved in adhesion of muscle cells to the substrate in
vitro (Schubert, LaCorbiere 1980). Such a role for the
skeletal muscle proteoglycan described in this report is
possible, although no data are available to support a role
for the muscle-specific proteoglycan in this or any other
function. Investigations are currently in progress to
ascertain the role of the muscle proteoglycan in the
development and physiology of muscle. Nevertheless, this
report demonstrates that chick limb skeletal muscle
synthesizes a unique proteoglycan which may be essential in
early developmental events.

Ahrens PB, Solursh M, Meier S (1977). The synthesis and
    localization of glycosaminoglycans in striated muscle
    differentiating in cell culture. J Exp Zool 202:375.
Caplan AI (1976). Simplified procedure for preparing
    myogenic cells for culture. J Embryol Exp Morph.
    36:175.
Carrino DA, Lennon DP, Caplan AI (1982). The role of extra-
    cellular matrix in phenotypic expression: proteoglycns
    synthesized by replated chondrocytes and non-chondro-
    cytes. Submitted to Develop. Biol.
Dietrich CP, Sampaio LO (1981). Changes of acidic mucopoly-
    saccharides and mucopolysaccharidases during histo-
    genesis and fetal development: possible involvement of
    chondroitin sulfate C and hyaluronidase with the
    processes of differentiation and cell division. In
    Yamakawa T, Osawa T, Handa S (eds): "Glycoconjugates:
    Proceedings of the Sixth International Symposium on
    Glycoconjugates," Toyko: Japan Scientific Societies
    Press, p 407.
Hascall VC, Oegema TR, Brown M., Caplan AI (1976). Isolation
    and characterization of proteoglycans from chick limb
    bud chondrocytes grown in vitro. J Biol Chem 251:3511.
Pacifici M, Molinaro M (1980). Developmental changes in
    glycosaminoglycans during skeletal muscle cell differen-
    tiation in culture. Exp Cell Res 126:143.

Schubert D, LaCorbiere M (1980). A role of secreted glycosaminoglycans in cell-substratum adhesion. J Biol Chem 255:11564.

Limb Development and Regeneration
Part B, pages 391–400
© 1983 Alan R. Liss, Inc., 150 Fifth Avenue, New York, NY 10011

STRUCTURE AND REGULATION OF MUSCLE-SPECIFIC GENES DURING
CHICK LIMB DEVELOPMENT

Charles P. Ordahl, Gerald Kuncio, Isaac Peng,
Thomas Cooper
Department of Anatomy
Temple University School of Medicine
3400 N. Broad Street, Philadelphia, PA 19140

TRANSCRIPTIONAL CHANGES DURING LIMB MYOGENESIS

Muscle differentiation during embryonic limb development
involves a complex series of biochemical and morphological
changes which respond to both intrinsic and extrinsic cues.
While the nature and mechanism of the extrinsic cues remain
largely elusive, it is possible to analyze the molecular
mechanisms of one of the intrinsic programs of muscle differentiation, the differential expression of genes during
myogenesis. Gene expression during embryonic development
poses two distinct but related problems for molecular
biology; first how does the embryonic cell 'select' which
genes to express as a differentiated cell? and second, how
does the cell regulate the expression of those genes it has
selected to express?

Muscle differentiation in the embryonic chick limb is a
useful model system for the study of embryonic gene regulation. In addition to expressing "housekeeping" genes which
are expressed in all cells, embryonic muscle cells regulate
a large number of muscle-specific genes encoding muscle-
specific contractile and enzymatic proteins. Work from a
number of different laboratories has indicated that the
regulation of these genes occurs in a coordinate fashion
suggesting the possibility that common, or integrated,
mechanisms operate to effect their co-expression (Devlin and
Emerson, 1978, 1979; Ordahl et al, 1982). Interestingly,
however, when mononucleate myoblasts fuse to form multi-
nucleate myotubes a large number of genes are repressed
concomitant with the turning-on of the muscle-specific

genes (Ordahl and Caplan, 1976; Ordahl et al, 1980a). We
define genes turned-on at the time of muscle fusion as
"class A" genes while those turned-off at fusion, "class B"
genes. To study the molecular mechanisms by which class A
and class B genes are regulated we used recombinant DNA to
clone representative class A and class B mRNAs from embryonic
muscle (for technical details see, Ordahl et al, 1980a,b;
Ordahl et al, 1982).

Figure 1 shows four hybridization experiments which
illustrate the patterns of appearance of four specific mRNAs
during myogenesis. In these experiments, RNA from the tissues
indicated is fractionated on agarose gels and then transferred in situ to a filter paper to which it is covalently
bound. This filter paper is then incubated with a radiolabeled, cloned DNA probe which will hybridize to its
complementary RNA affixed to the paper. Hybridization is
detected by autoradiographing the filter with x-ray film.
Figure 1 (upper left) shows that when this experiment is
performed using a clone derived from a housekeeping mRNA,
hybridization is observed to an RNA species present in all
tissues, at all stages of development indicating that this
mRNA is always present and therefore un-regulated. A clone
derived from a class A mRNA, however, gives a different
pattern as shown in the upper right-hand panel of Figure 1.
Here hybridization is only seen to an RNA species present in
RNA from muscle tissue, and the strongest hybridization is
seen for adult muscle. This RNA species is not detected in
RNA from stage 24 limb buds nor in RNA from either brain or
liver. The mRNA being hybridized in this experiment encodes
troponin I. Identical results have been obtained for clones
of the mRNAs for alpha-actin (Ordahl et al, 1980b), muscle
creatine kinase (Ordahl et al, 1982 and Ordahl et al, submitted), and myosin light chain 2 (Ordahl, unpublished
observations).

Class B sequences, those turned-off during muscle fusion,
can be subdivided into at least 2 sub-categories based upon
their appearance during muscle development and in non-muscle
tissues. The mRNA in the first sub-category (Figure 1,
lower left) resembles a housekeeping mRNA in that it is
present and un-regulated in non-muscle tissue RNA and in RNA
from stage 24 limb buds. During muscle development this
sequence is present until fusion is essentially complete
(embryonic day 18), at which time its concentration dimin-
ishes substantially. There are examples of housekeeping
gene products which are specifically turned-off in muscle.

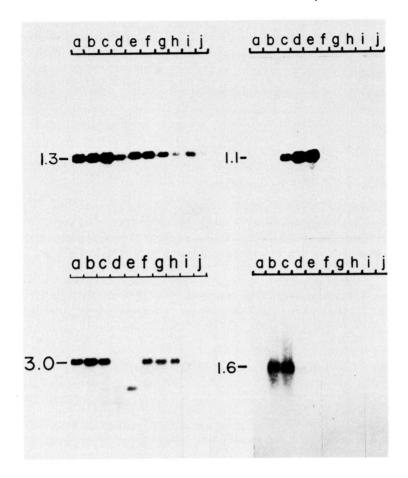

Figure 1.   Class A and class B mRNAs in myogenesis.
Each lane contained 20μg total poly(A) RNA from:
a-stage 24 limb bud; b-embryonic day (ed) 10 thigh muscle;
c-ed 14 thigh muscle; d-ed 18 thigh muscle; e-adult thigh
muscle; f-ed 10 brain; g-ed 18 brain; h-adult brain;
i-ed 18 liver; j-adult liver.   Embryonic thigh muscle is
predominately composed of myoblasts at day 10 but by day 18
fusion is essentially complete.   Conditions for electrophoresis and hybridization are as described elsewhere
(Ordahl et al, 1980a&b).   RNA size is given in thousands of
nucleotides.   The upper right autoradiograph is a 4 hour
exposure, while the other three were exposed for 24 hours.

A good example is the brain isoform of creatine kinase which is a ubiquitous enzyme but which in muscle is turned-off at the same time that the specialized muscle isoform is expressed (Caravatti et al, 1979). Another example is the switch from beta- and gamma-actin gene expression to alpha-actin gene expression (see below and Ordahl et al, 1980b). It is our prediction that the mRNA shown in Figure 1, bottom left also encodes a housekeeping product for which there is a muscle-specific counterpart. Moreover, since we have identified cDNA clones for 2 such sequences, we would further predict that the occurance of such muscle-specific/house-keeping pairs is not unusual.

A second class B clone gives a different hybridization pattern indicating that it is both muscle-specifice and stage-specific (Figure 1, lower right). Therefore, this indicates the gene encoding this mRNA is turned-on in proliferating myoblasts and then turned-off in fused myo-tubes. Some gene products are only transiently expressed during specific stages of myogenesis. For example, Garrels (1979) showed that the heart form of tropomyosin was expressed during proliferation of cultured skeletal muscle cells but disappeared after fusion. Such findings are intriguing because they may indicate that assembly of a myotube involves modeling using transient proteins in a manner similar to the transient modeling of skeletal struc-tures in the growing limb. On the other hand, the transient appearance of certain gene products during development may not reflect a structured, integrated developmental program but may be obligatory remnants of earlier ontological processes similar to the transient appearance of the gill slits in bird and mammalian embryos. In either case, molecular probes for such transiently expressed substances may offer a means to gain insight into the molecular processes which govern muscle differentiation.

STRUCTURAL ANALYSIS OF GENES REGULATED DURING LIMB MYOGENESIS

Cloned cDNA probes from various class A and class B mRNAs permits a determination of how individual mRNA levels change during the course of limb muscle development. How-ever, this does not indicate what molecular mechanisms control the intracellular levels of mRNA because these may operate at a number of levels in the cell. First, as dis-cussed briefly above, a gene may be in an addressable or non-addressable configuration allowing it to be, or not be, accessable to transcription by RNA polymerase. Second, an addressable gene may be modulated either on or off, such as the case with the response of the ovalbumin gene to estrogen challenge in the oviduct. Finally, the amount of transcript from a given active gene may be regulated by post-transcriptional mechanisms. The question of how limb mesenchyme and muscle cells regulate class A and B genes can only be addressed if we can determine the level at which regulation is operating and whether that regulatory control changes at different stages of development. In order to develop strategies to test these possible control mechanisms we are analyzing the structure of the genes which encode the class A and B mRNAs for which we have cDNA clones. For these types of experiments, the segment of genomic DNA encoding each is isolated from a phage library containing fragments of the chick genome. The position and structure of the gene within this cloned genomic fragment is then determined by restriction endonuclease mapping, hybridization analysis and finally, DNA sequencing.

Our greatest progress to date has been with the chick alpha actin gene for which we have determined the entire nucleotide sequence (Fornwald et al, 1982). Figure 2 shows a restriction endonuclease digestion map of the segment of DNA containing the gene and a diagrammatic representation of the organization and position of the coding blocks and intervening sequences. There are 7 coding blocks (I-VII) interrupted by 6 intervening sequences (A-F). The first coding block (I) is only 61 nucleotides in length and does not code for any amino acids but rather for part of the 5' untranslated portion of the mRNA. The first 12 nucleotides of coding block II are also untranslated. The remainder of coding block II encodes amino acids 1-41 (numbered as per Elizinga et al 1973) as well as a met-cys dipeptide at the N-terminus which is not found on mature actin. Intervening sequences B-F interrupt the amino acid coding region of the gene at the following codon positions: B-41/42; C-150; D-204; E-267; F-327/328. Coding block VII encodes amino acids #328-375 plus 272 nucleotides of untranslated region at the 3' end of the mRNA. Also shown in Figure 2 is the position of the translational initiation codon (ATG) and termination codon (TAA) and the position of nucleotide sequences implicated in transcriptional initiation (CAAAT, ATA) and in post-transcriptional addition of poly (A) tracts to the primary transcript of the gene (poly (A) addition

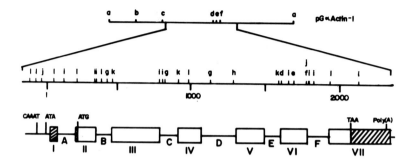

Figure 2. Structure of the chick alpha-actin gene.
Upper portion-Restriction endonuclease cleavage map of the
6200 base pair segment of chick DNA within which the
alpha-actin gene is embedded. This segment was subcloned
into the Hind III site of pBR322 (clone designation
pGaActin-1, see Ordahl et al, 1980b). The expanded region
is the 2500 base pair segment which contains the entire
coding portions of the gene. The scale is in nucleotides
with nucleotide #1 being the first nucleotide of the alpha-
actin mRNA. The symbols for the restriction endonuclease
cleavage sites are: a-Hind III, b-Sac I, c-Sma I, d-BamHI,
e-Kpn I, f-Xho I, g-Pst I, h-Pvu II, i-Ava II, j-Taq I,
k-Hinf I, l-Hha I.
Lower portion-Diagrammatic representation of the position
and size of the coding blocks and intervening sequences in
the alpha-actin gene. Open blocks I-VII represent coding
blocks with cross-hatched regions representing untranslated
regions. Thin lines A-F represent intervening sequences.
The position of signal sequences is noted (see text). The
scale of this diagram corresponds to the restriction map
above. For details see Fornwald et al (1982).

signal, ATTAAA).

   The alpha-actin gene is one member of a large actin
multigene family. Protein sequence data indicates that
birds and mammals contain at least 6 different actin genes
(Vanderkerchkove and Weber, 1978). Hybridization analyses
show that there is substantial nucleotide sequence homology

between the alpha-actin gene and the beta- and gamma-actin genes. Figure 3 shows that a labeled probe derived from coding block VI of the alpha-actin gene hybridizes to the beta- and gamma-actin mRNAs as well as to the alpha-actin mRNA. However, a DNA probe derived from the 3' untranslated segment of the alpha-actin gene hybridizes only to the alpha-actin message (Figure 3). Region-specific probes permit discrimination between different, related mRNAs. This experiment also demonstrates that the regions which encode amino acids in the actin gene are under stronger evolutionary constraints than untranslated regions. The constraint upon protein coding regions of actins is so strong that vertebrate and non-vertebrate actin genes efficiently cross-hybridize.

The structure of the chick alpha-actin gene is very similar to that of the rat alpha-actin gene (D. Yaffe, personal communication) in that: 1. the number and position of the intervening sequences is identical; and 2. the 5' and 3' untranslated regions are of similar size. The alpha-actin genes differ, however, from their beta-actin counterparts in that many intervening sequences do not coincide (Reviewed in Fornwald et al, 1982) and the size of the 3' untranslated region of Beta-actin mRNA is substantially longer. It is too early to tell which features of actin gene structure might play a role in regulation and which might be fixed evolutionary remnants without definite function. However, with this basic structural information in hand it will eventually be possible to determine the functional significance of the various components.

Another approach to understanding the role of gene structural components in regulation can be made by comparing the structure of 2 or more genes which are co-regulated but which are not evolutionarily related. In this case muscle offers an excellent system for study because of the great diversity of regulated genes. We are currently analyzing the structure of several class A and class B genes with a goal towards identifying common structural components which might serve as coordinating regulatory elements.

These efforts to analyze gene structure as a means to understand embryonic gene regulation are based upon 2 hypotheses as to the molecular basis for cell differentiation. First, that phenotypic commitment or determination involves a qualitative change in the addressability of

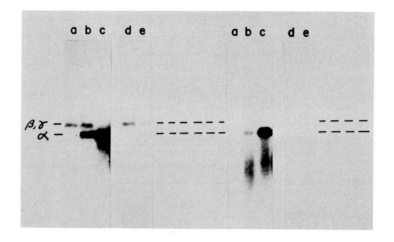

Figure 3. Hybridization using probes from different regions
          of the alpha-actin gene.
Left panel-The hybridization probe was derived from coding
block VI of the alpha-actin gene (see Figure 2). Right
panel-Hybridization probe derived from the 3' untranslated
region of the alpha-actin gene (cross-hatched region of
coding block VII in Figure 2). The hybridization conditions
and other details are as in Figure 1. Poly(A) RNA samples
(20 µg/lane) were from: a-stage 24 limb bud, b-ed 10 thigh
muscle, c-ed 14 thigh muscle, d-ed 18 brain, e-adult brain.
The position of migration of the beta- and gamma-actin mRNAs
and the alpha-actin mRNA is indicated.

specific genes (i.e. some genes become permanently addressable and some become permanently non-addressable). The
second hypothesis is that the ability of the cell to
recognize and regulate such genes derives from structural
components within the genes themselves. Using chick limb
development it is possible to obtain mesenchyme cells prior
to the onset of muscle differentiation in vivo and determine
the status of the class A and B genes in these cells before
and after differentiation. This should allow testing of the
role of structural components implicated by analysis of
cloned gene segments. Ultimately, however, it will be
important to determine how gene regulation is involved in
the process of phenotypic commitment. In order to do this
it is necessary to assay the status of genes in the pluri-
potent cell prior to the time it "decides" its final

phenotype. It is towards this goal that the present studies
are ultimately directed.

Acknowledgements

Supported by Grants to C.P.O. from NIH (GM-25400) and
The Muscular Dystrophy Association. C.P.O. is a recipient
of an NIH Research Career Development Award.

REFERENCES

Devlin RB and Emerson CP (1978). Coordinate Regulation of
  Contractile Protein Synthesis during Myoblast Differ-
  entiation. Cell 13:599-611.
Devlin RB and Emerson CP (1979). Coordinate Accumulation
  of Contractile Protein mRNAs during Myoblast Differentiation. Develop Biol 69:202-216.
Elzinga M, Collins JH, Kuehl WM and Adelstein RS (1973).
  Complete Amino-Acid Sequence of Actin of Rabbit Skeletal
  Muscle. Proc Natl Acad Sci USA 70:2687-2691.
Garrels JI (1979). Changes in protein synthesis during
  myogenesis in a clonal cell line. Develop Biol 73:134-152.
Fornwald JA, Kuncio G, Peng I, and Ordahl, CP (1982). The
  complete nucleotide sequence of the chick a-actin gene
  and its evolutionary relationship to the actin gene
  family. Nucleic Acids Res. in press.
Ordahl CP and Caplan AI (1976). Transcriptional Diversity
  in Myogenesis. Develop Biol 54:61-72.
Ordahl CP and Caplan AI (1978). Diversity in the Poly-
  adenylate RNA Populations of Embryonic Myoblasts.
  J Biol Chem 253:7683-7691.
Ordahl CP, Kioussis D, Tilghman SM, Ovitt C and Fornwald J
  (1980a). Molecular cloning of developmentally regulated,
  low-abundance mRNA sequences from embryonic muscle. Proc
  Natl Acad Sci USA 77:4519-4523.
Ordahl CP, Tilghman SM, Ovitt C, Fornwald J and Largen MT
  (1980b). Structure and developmental expression of the
  chick a-actin gene. Nucleic Acids Res 8:4989-5005.
Ordahl CP, Cooper T, Ovitt CE, Fornwald JA, Caplan AI and
  Calman AF (1982). Molecular cloning and analysis of
  genes regulated during embryonic muscle differentiation.
  In Pearson Mark (ed): "Molecular and Cellular Control of
  Muscle Development," Cold Spring Harbor, New York: Cold
  Spring Harbor Press.
Ordahl CP,Cooper TA, Fornwald JA and Ovitt C. The nucleo-
  tide sequence encoding the active site region of creatine

kinase. Submitted for publication
Vanderkerckhove J and Weber K (1978). At least six
  different actins are expressed in a higher mammal: an
  analysis based on the amino acid sequence of the amino-
  terminal tryptic peptide. J Mol Biol 126:783-802.

Limb Development and Regeneration
Part B, pages 401–408
© **1983 Alan R. Liss, Inc., 150 Fifth Avenue, New York, NY 10011**

BIOCHEMICAL EVIDENCE FOR TWO TYPES OF MYOBLASTS DURING AVIAN
EMBRYONIC DEVELOPMENT

Marc Y. Fiszman[1], Madeleine Toutant[2] and Didier
Montarras[1]
[1]Department of Molecular Biology - Pasteur Institute - Paris
[2]Groupe de Recherches de Biologie et de Pathologie neuromusculaires. CNRS ER 107 - Paris

Skeletal muscle cells in the developing chick leg are
derived from lateral plate mesoderm appearing during the
second day of incubation (Rosenquist, 1971). By transplantation and chorioallantoic membrane grafting experiments, it
was possible to show the presence of myoblastic precursors
in the limb buds of stage 16 embryos (Hunt, 1932 ; Hamburger,
1938 ; Eastlick, 1943). However, one has to wait stage 17-
21 to demonstrate the presence of these myoblastic precursors by *in vitro* experiment (Dienstman *et al.*, 1974 ; Bonner
and Hauschka, 1974). Clonal analysis of muscle cells have
shown that the percentage of muscle colony forming (MCF)
cells increases during limb development and that the morphological appearance of muscle colonies and their nutritional
requirements allow to distinguish four discrete classes of
MCF cells. One class is fresh medium sufficient (FMS) while
the three other are conditioned medium requiring (CMR)(White
et al., 1975). Each class appears at specific times during
development and their temporal order of appearance strongly
suggests a precursor product relationship among them. Lastly
it was found that spinal cord cauterization which interferes
with normal innervation of the limb bud also interferes with
the appearance of one of the class of CMR cells (Bonner,
1978 ; Bonner, 1980). This class of CMR myoblasts which
fail to appear in the absence of innervation can be rescued
in vitro by cocultivation of spinal cord explants with myoblasts from a denervated embryo (Bonner and Adams, 1981).
However, none of these studies have unambiguously shown that
the different CMR-myoblasts do indeed correspond to different classes of muscle cells which upon differentiation can
synthesize different sets of muscle specific proteins. The

purpose of this communication is to present evidence for the existence of at least two types of myoblasts, one "early embryo" and a "late embryo" type, which can be distinguished by the kind of myosin molecule that they synthesize following differentiation. Furthermore, we will also show that the appearance of the late embryo myoblasts is dependent upon innervation and also that these late embryo myoblasts can be converted into early embryo myoblasts by a prolonged subculturing process. The experiments which will be presented have been performed with either quail or chick embryos since both materials were shown to give essentially the same results.

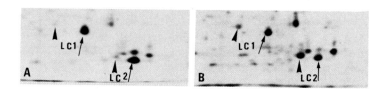

Fig. 1. Myosin light chains synthesized by differentiated primary (A) and subcultured myoblasts (B) derived from pectoralis muscle. Well fused cultures were labeled with 100 μCi/ml ^{35}S-methionine, cell extracts were-prepared as previously described (Montarras $et\ al.$, 1982). and analyzed by two dimensional gel electrophoresis. Only the area corresponding to myosin light chains, as determined by coelectrophoresis of purified markers, is presented.
      ⟶ indicates fast muscle isozymes
      ► indicates slow muscle isozymes

We have observed that myoblasts isolated from 10 day old quail embryos can be subcultured for many generations without losing their ability to fuse into differentiated myotubes. However, when the proteins synthesized by primary myotubes and myotubes derived from subcultured myoblasts are compared, obvious differences are readily observed. This is shown in Fig. 1. Cultures of well fused myotubes were pulse labeled with ^{35}S-methionine and total cell extracts were analyzed by two dimensional gel electrophoresis. Fig. 1A shows an autoradiogram of the proteins synthesized by primary myotubes and proteins synthesized by myotubes derived from subcultured myoblasts are presented in Fig. 1B. When one compares the two autoradiograms it is clear that the two cultures synthesize different myosin light chains. Indeed, myotubes derived from subcultured myoblasts synthesize all types of myosin light chains with the exception of LC3 while primary myotubes synthesize predominantly the myosin light chains specific of fast muscle with the exception of LC3.

To rule out the possibility that the observed synthesis of fast and slow muscle isozyme was related either to the presence of two cell types or to the frequent trypsinisations used during the subculturing process, we have grown primary myoblasts under clonal conditions. Myogenic clones were labeled with ^{35}S methionine and total cell extracts were analyzed by 2D gel electrophoresis. It was observed that all clones tested synthesize slow and fast isozymes of myosin light chains in the same proportions which were found to be synthesized by differentiated cultures obtained from subcultured myoblasts.

We then checked whether the origin of the muscle has any influence on the ability of the myoblasts to be subcultured. We isolated myoblasts from embryonic ALD (anterior latissimus dorsi) and subcultured them according to the same

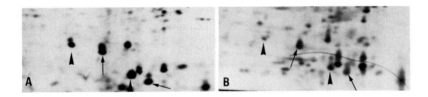

Fig. 2. Myosin light chains synthesized by differentiated subcultured (A) and primary myoblasts (B) derived from ALD The symbols are the one used for Fig. 1.

protocol which was used for myoblasts isolated from either thigh or pectoralis muscle. It was observed that these ALD myoblasts can be subcultured as efficiently as thigh muscle myoblasts and furthermore that myotubes derived from these subcultured ALD myoblasts synthesize a mixture of fast and slow myosin light chains identical to what we have just described (Fig. 2A). However, it has to be noted that primary ALD myotubes already synthesize fast and slow myosin light chains but with a lower proportion of the slow isozymes (Fig. 2B).

In summary, we have shown that myoblasts isolated from either embryonic fast or slow muscles can be maintained undifferentiated for long period of time. However, following this subculturing process progressive changes are detected at the level of the synthesis of myosin light chains. Myotubes derived from subcultured myoblasts synthesize an unbalanced mixture of fast and slow myosin light chains such that LC1F, LC2F and LC2S are present in roughly equimolar amounts and account for 90 % of all the light chains while LC1S accounts for the remaining 10 %.

Could it be that the phenotype is purely artificial and the consequence of our culture conditions or is it possible to observe it in a developing embryo ? Some indications that this later possibility is correct was found in the work by Stockdale and his colleagues who have shown that young embryonic muscle explants synthesize both slow and fast myosin light chains independent of their anatomic location and in proportions very similar to those which we have described for subcultured muscle cells (Stockdale *et al.*, 1981). We therefore analyzed the myosin light chains present during the development of both a fast (PLD : posterior latis simus dorsi) and a slow (ALD) muscle. As shown in Fig. 3 it can be seen that both types of muscle accumulate this unbalanced mixture of light chains when isolated from 9 day old embryos.

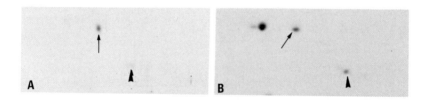

Fig. 3. Two dimensional analysis of myosin light chains present in the PLD (A) and the ALD (B) of a 9 day old chick embryo.
The symbols have been described for Fig. 1.

This result therefore raised the possibility that the phenotype expressed by differentiated subcultured myoblasts could represent a phenotype expressed by myoblasts younger than the ones which were subcultured. To test this

possibility we isolated myoblasts from limb buds of 5 to 7
day old chick embryos and analyzed the myosin light chains
synthesized by differentiated cultures. As shown in Fig. 4
these cells synthesize all four myosin light chains with
LC1F and LC2S as the major constituants and LC1S and LC2F
as the minor forms. It is only when isolated from embryos
older than 8 days that differentiated myoblasts synthesize
predominantly the fast myosin light chains.

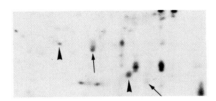

Fig. 4. Myosin light chains synthesized by differentiated
myoblasts isolated from the limb bud of a 6 day old chick
embryo.
The symbols have been described for Fig. 1.

This result therefore suggests that it is possible to
distinguish two classes of myoblasts based on the types of
proteins which are synthesized by the differentiated cultures. The first one, early embryo myoblast, is present in
the limb bud, it forms myotubes which synthesize a myosin
whose light chains are predominantly LC1F and LC2S. The
second one, late embryo myoblast, starts to accumulate after
day 7 *in ovo* and forms myotubes which synthesize a myosin
whose light chains are LC1F and LC2F. The temporal relationship between these early and late embryo myoblasts is
very much reminiscent of the relationship which has been described between the CMR II and CMR III myoblasts (White *et al.*,
1975 ; Bonner, 1980). Since it was shown that the appearance of CMR III myoblast is dependant upon functional innervation (Bonner, 1978 ; Bonner, 1980) we have tested whether
a similar dependance could be demonstrated for the appearance of the late embryo myoblasts. Two and 1/2 day old
chicken embryos (15 HH) were denervated by surgical excision
of the spinal cord from the 22nd somite to the tip of the
tail. These denervated embryos were further incubated. Myoblasts were obtained from the thigh muscle of 12 day old denervated embryos. As a control, myoblasts were isolated

from the pectoralis muscle of the same embryos. Both cultures were allowed to differentiate, labeled with ³⁵Smethionine and total cell extracts were analyzed by two dimensional gel electrophoresis. As shown in Fig. 5A, the culture derived from the leg myoblasts synthesize predominantly LC1F and LC2S while the culture derived from the pectoralis myoblasts synthesize LC1F and LC2F (Fig. 5B). It can therefore be concluded that denervation has prevented the appearance of the late embryo myoblasts which is a strong indication that they are identical to the CMR III class of myoblasts.

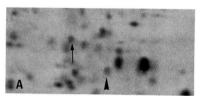 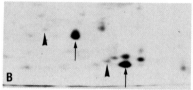

Fig. 5. Myosin light chains synthesized by differentiated myoblasts isolated from the thigh muscle (A) and the pectoralis muscle (B) of a 12 day old chick embryo which had been denervated at 2 1/2 day of incubation *in ovo*.
The symbols have been described for Fig. 1.

In conclusion, we have presented evidence for the existence of two classes of myoblasts during muscle development. The first class, early embryo myoblast, is predominant in the limb bud from day 5 to day 7 while the second, late embryo myoblast, starts to accumulate after day 7 and is the major population of muscle cells in the developing muscles. The two classes can be distinguished according to the type of myosin which is synthesized by the differentiated cells. Differentiated early embryo myoblasts synthesize a myosin whose light chains are predominantly LC1F and LC2S while differentiated late embryo myoblasts synthesize a myosin whose light chains are essentially LC1F and LC2F. At the present time nothing is known about the myosin heavy chain present in differentiated early myoblasts while it has been shown that in late embryo myoblasts the myosin heavy chain is an embryonic isozyme different from both fast and slow myosin (Bandman *et al.*, 1981, Benfield *et al.*, 1981). Our results also indicates that the appearance of late embryo myoblasts is dependent upon functional innervation but it is not clear yet whether the early embryo myoblasts are the precursors of the late embryo myoblasts or whether these latter arise from a precursor cell different from the early embryo myoblasts. Our results would also suggest that during muscle development the early fibers which are made derive from early embryo myoblasts since in both 8 day ALD and PLD the myosin light chains are essentially LC1F and LC2S. Lastly our results also indicate that late embryo myoblasts can revert to a cell type which ressembles early embryo myoblast. This is achieved by repeated subculturing. It is therefore templing to postulate that late embryo myoblasts arise from the interaction of motoneurones with early embryo myoblasts. In other words, a late embryo myoblast could be considered equivalent to an early embryo myoblast imprinted by a nerve signal. This signal is not permanent and in the absence of the neurone it will be lost by dilution.

This work was supported by the Centre National de la Recherche Scientifique, the Institut National de la Santé et de la Recherche Médicale, the Fondation pour la Recherche Médicale Française, the Ligue Nationale Française contre le Cancer and the Muscular Dystrophy Association.

Bandman E, Matsuda R, Micou-Eastwood J, Strohman R (1981) *In vitro* translation of RNA from embryonic and from adult chicken pectoralis muscles produces different myosin heavy chains. Febs Letters 136:301.

Benfield PA, Lowey S, Leblanc DD (1981). Fractionation and characterization of myosins from embronic chicken pectoralis muscle. Biophys J, 33:243a (Abstr.).

Bonner PH (1978). Nerve dependant changes in clonable myoblast populations. Develop Biol 66:207.

Bonner PH (1980) Differentiation of chick embryo myoblasts is transiently sensitive to functional denervation. Develop Biol 76:79.

Bonner PH, Adams TR (1981) Neural induction of chick myoblast differentiation in culture. Develop Biol 90:175.

Bonner PH, Hauschka SD (1974). Clonal analysis of vertebrate myogenesis I early developmental events in the chick limb Develop Biol 37:317.

Dienstman SR, Biehl J, Holtzer S, Holtzer H (1974) Myogenic and chondrogenic lineage in developing limb buds grown *in vitro*. Develop Biol 39:83.

Eastlick HL (1943). Studies on transplanted embryonic limbs of the chick. J Exp Zool 93:27.

Hamburger V (1938). Morphogenetic and axial self differentia
tion of transplanted limb primordia of chick embryo. J Exp
Zool 77:379.
Hunt EA (1932). The differentiation of chick limb buds in
chorio-allantoic grafts, with special reference to the
muscle. J Exp Zool 62:57.
Montarras D, Fiszman MY, Gros F (1982). Changes in tropomyosin during development of chick embryonic skeletal muscles
in vivo and during differentiation of chick muscle cells
in vitro. J Biol Chem 257:545.
Stockdale FE, Raman N, Baden H (1981). Myosin light chains
and the developmental origin of fast muscle. Proc.Nat Acad
Sci 78:931.
White NK, Bonner PH, Nelson DR, Hauschka SD (1975) Clonal
analysis of vertebrate myogenesis IV medium dependent
classification of colony forming cells. Develop Biol 44:
346.

Limb Development and Regeneration
Part B, pages 409–416
© **1983 Alan R. Liss, Inc., 150 Fifth Avenue, New York, NY 10011**

THE ROLE OF SMALL CYTOPLASMIC RNAs IN MUSCLE DIFFERENTIATION

Thomas L. McCarthy, Barbara Mroczkowski, and
Stuart M. Heywood
Genetics and Cell Biology Section
Biological Sciences Group
The University of Connecticut, Storrs, CT 06268

Cellular differentiation is a process which is poorly
understood at the molecular level, but one that is receiving
an ever increasing amount of attention as new methods are
developed to analyze in detail the mechanisms of gene
expression. It is unlikely that there is a single universal
developmental control mechanism that regulates both the
qualitative and quantitative aspects of the expression of
specific genes. Each developing system likely uses a
variety of regulatory mechanisms (transcriptional control,
mRNA processing and transport, differential mRNA utilization,
differential mRNA half-life, protein modification, and
protein turnover). Therefore, muscle differentiation will
use its own blend of control mechanisms to synthesize,
assemble, and maintain the contractile apparatus.

While a number of laboratories have sought to determine which of the regulatory mechanisms likely play the
more strategic roles, we have undertaken to attempt to
establish the mechanisms utilized in translational control
(controls operating at the level of mRNA, ribosomes, or
factors resulting in differential utilization of mRNAs).
We have suggested that during muscle differentiation there
are factors which enhance the translation of specific mRNAs
(Gette, Heywood, 1979; O'Loughlin et. al., 1981), as well as
factors involved in specifically repressing the utilization
of mRNAs (Bester et. al., 1975). In this latter case mRNAs
are sequestered in an inactive form (mRNPs) until they are
utilized, Heywood et. al. (1975a).   Geoghegan et. al.
(1979) have demonstrated that cells likely contain two
forms of mRNAs, a) those which are not actively repressed

but are poor competitors for initiation factors, and b)
those which are actively repressed. In this report we are
concerned only with the latter.

Myosin heavy chain mRNPs (MHC-mRNPs) have been isolated
and purified free of ribosomes and other mRNPs from embryonic
chick muscle as well as primary cell cultures obtained from
embryonic chick muscle. These mRNPs have been shown to have
associated with them, in addition to MHC-mRNA, a number of
small RNAs (Havaranis, Heywood, 1981). Among these small
RNAs is a uridine rich RNA which binds oligo d(A)-cellulose
which we have previously described and termed translational
control RNA (tcRNA). Although originally thought to be
smaller, the advent of reliable sizing gels and sequencing
techniques now show this RNA to be 102 nucleotides in length
(tcRNA$_{102}$. A number of other reports (Bag, Sells, 1981;
Mukherjee, Sarkar, 1981) have demonstrated the presence of
small cytoplasmic RNAs in muscle; however, the authors claim
they are not related to tcRNAs due to different properties
and function. Nucleotide sequence analysis must be performed
to support this claim. The sequence of tcRNA$_{102}$ is:

$$5'UCGGUGAGAE\overset{A}{U}GAAUGUGUUGCUGGUUGUUGAUUGUUGGGUUGUGCGUGNAGUUAA$$
$$\overset{G\quad A}{U}UGUGU\overset{}{G}AGUGUAUAUAUAUAUUGUAGAGGUUGA\overset{GG\ G}{U}\overset{CGG\ GG}{U}GAUUAAACAA3'.\quad In$$

several places the precise nucleotide is unclear but the
choice is never greater than the two indicated. It is not
known if this is due to a small amount of contamination or if
the RNA represents closely related gene products.

We have previously demonstrated that the low molecular
weight, oligo d(A)-binding, RNA isolated from MHC-mRNPs
inhibits MHC synthesis but is not effective in inhibiting
the synthesis of heterologous mRNA (globin mRNA) when present
in stoichiometric amounts (Heywood, Kennedy, 1976). These
experiments have been repeated using the acrylamide gel
purified tcRNA$_{102}$ isolated from MHC-mRNPs which have been
purified on metrizimide buoyant density gradients (Havaranis,
Heywood, 1981). When stoichiometric amounts of MHC-mRNA and
tcRNA$_{102}$ are mixed prior to addition to a cell-free system,
MHC synthesis is completely inhibited while having no effect
on endogenous protein synthesis. In similar experiments
using globin mRNA, VSV mRNA, ovalbumin mRNA, or histone mRNA,
no inhibition of translation of these mRNAs is observed.
However, when muscle mRNAs are isolated from mRNPs of less
than 40S and incubated with tcRNA$_{102}$ from MHC-mRNP the translation of these muscle mRNAs is also inhibited (see below).

These results confirm our previous findings suggesting that the oligo d(A)-binding, $tcRNA_{102}$ isolated from MHC-mRNPs has a degree of specificity for the mRNAs with which it interacts, i.e., it inhibits the translation of muscle mRNAs found in mRNPs, but it has no effect on the translation of those heterologous mRNAs so far tested.

We have previously demonstrated that hybridization of tcRNA to MHC-mRNA would increase the RNase resistance of MHC-mRNA from 5% to approximately 20% (Heywood, Kennedy, 1976). These results suggested that the interaction of MHC-mRNA with tcRNA resulted in a substantial change in the structure of the mRNA. Recently the interaction between $tcRNA_{102}$ with mRNA has been examined using liquid hybridization, and complex formation determined by sucrose density gradient centrifugation. Under the conditions used, $tcRNA_{102}$ remains at the top of the gradient while MHC-mRNA (6700 nucleotides) sediments about three-fourths down the distance of the gradient. When the two RNA species are allowed to hybridize prior to layering on the sucrose density gradients, a complex is formed which appears near the bottom of the gradient giving a positive indication of an interaction between the molecules (Siegel, 1981). Similar results are also obtained when muscle mRNAs obtained from less than 40S mRNPs were hybridized to $MHC\text{-}mRNA\text{-}tcRNA_{102}$ (Mroczkowski, 1982). This again gave a positive indication of an interaction between these molecules which supports the fact that $tcRNA_{102}$ inhibits the translation of these mRNAs by forming complexes with them. Previous results have indicated that the poly A tail of mRNA may be involved with this $tcRNA_{102}$-mRNA interaction (Heywood, et. al., 1975b). This has been verified by liquid hybridization and sucrose density gradient analysis which indicate that $tcRNA_{102}$ can interact and alter the sedimentation of ^3H-poly A (Mroczkowski, 1982).

Since heterologous mRNA translation was not inhibited by the addition of $tcRNA_{102}$ to the in vitro translation system, complex formation experiments were also performed on a variety of other mRNAs. Neither 9S rabbit globin mRNA, TMV RNA, sea urchin histone mRNA, rRNA, or tRNA were found to alter the sedimentation of $tcRNA_{102}$, i.e., form complexes with $tcRNA_{102}$ after being mixed under hybridization conditions. Therefore, the negative assays demonstrating the inhibition of translation combined with the positive assays showing complex formation, strongly suggest a degree of specificity exists allowing $tcRNA_{102}$ to interact only with certain mRNA species.

Considerable attention has recently been focused on both nuclear and cytoplasmic small RNAs associated with RNPs which react with Lupus Erythrmatosus antibodies (Lerner, Steitz, 1979). It was, therefore, of interest to determine if the isolated MHC-mRNPs containing tcRNA would interact with these antibodies. The results indicate that neither antibodies specific for the nuclear or cytoplasmic RNP complexes (anti-Sm or anti-Ro respectively) interact with MHC-mRNP to a greater extent than control anti-sera. In addition, the sequence of $tcRNA_{102}$ is not similar to the sequence of the snRNAs, although some homologous regions do occur (see below).

Using the Sumex Aim facility at Stanford University (Brutlag, et. al., 1982), we have compared the nucleotide sequence of $tcRNA_{102}$ to known small RNA sequences. While no sequence homology to 4.5S RNA exists, some interesting homologies to UlsnRNA are apparent. For example, the following are some examples:

$tcRNA_{102}$	2	CGGUGAGACU GAAUG	**16**
UlsnRNA	115	CGG GAAACUCGACUG	129
$tcRNA_{102}$	32	AUU GUUGGGUUGUGGG	47
UlsnRNA	134	AUUUGU GG UAGUGGG	148
$tcRNA_{102}$	54	UAAUU GUGUGAGUG	67
UlsnRNA	132	UAAUUUGUGGGAGUG	146

The functional role of the homologous sequences in either UlsnRNA or $tcRNA_{102}$ in mRNA metabolism is not known. However, it is clear that, because these homologous sequences are not sequentially positioned, $tcRNA_{102}$ is not a breakdown product of this snRNA.

We have previously suggested a model by which tcRNA may interact with mRNA in an mRNP particle (Bester, et. al., 1975). From the nucleotide sequence data now available as well as the translation and hybridization studies mentioned above we suggest the following model by which $tcRNA_{102}$-mRNA interaction may occur.

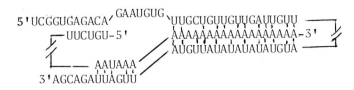

The poly A of mRNA can form a strong hybrid with tcRNA by both Hoogstein and Watson-Crick base pairing (Bina, et. al., 1980). In addition, $tcRNA_{102}$ has a complementary sequence common to most mRNAs near the poly A tail. Since the interaction of MHC mRNA with $tcRNA_{102}$ results in an increase in RNase resistance and a substantial shift in the sedimentation characteristics, we have included in the model base pairing between the first two codons (met, ser) with a region near the 5' end of $tcRNA_{102}$. Whether, in fact, this model is correct will require determining the nucleotide sequence of the muscle mRNAs involved as well as cross-linking studies to determine if the complexes exist in vivo. Preliminary studies using ultra-violet light and amino-methyl trioxalen to cross-link neighboring molecules in vivo, suggest that $tcRNA_{102}$ and MHC-mRNA are indeed associated. However, further nucleotide sequence data is necessary to explain the specificity of $tcRNA_{102}$-mRNA interaction. Nevertheless, at least one function of $tcRNA_{102}$ appears to be involved in the cytoplasmic utilization of mRNA.

The less than 40S mRNPs contain mRNAs which upon trans-lation comigrate with myosin light chains, tropomyosin, tropomin, desmin and vimentin on two dimensional gels. Interestingly we have been unable to detect any species of actin mRNA in the mRNP fraction suggesting that the synthesis of this protein may be under a different regulatory mode. Associated with these small mRNPs is also a small RNA of identical nucleotide sequence to the $tcRNA_{102}$ (described above) which therefore has the identical properties and functions. The specificity of $tcRNA_{102}$ for these mRNAs as well as MHC mRNA, suggests that these mRNAs have a common sequence toward the 5' end of the molecule that is not pre-sent on the heterologous mRNAs tested or rRNAs.

In addition to $tcRNA_{102}$, a number of other small RNAs are found associated with muscle mRNPs. These include a 89 nucleotide RNA ($tcRNA_{89}$) found in MHC-mRNPs and a 90 nucleotide RNA ($tcRNA_{90}$) found in the less than 40S muscle mRNPs. Another RNA(s) migrating at approximately 70-75 nucleotide is found in both sets of mRNPs. The MHC-mRNP-tcRNA$_{89}$ has the following sequence: 5'AAGAGAAGUAGAGAGUUUGG UAUGUGUGUUUGUGUGUGUGUGUUGAGUAUUAGGACAAUAUUCCAAGUAUUAACCG AAACUCAGACA-3'; and the less than 40S mRNP-tcRNA$_{90}$ sequence is: 5'UUGAAUGGUAAGGAGGGUGAGAGCGUGAAGUGUGAGAUGUAUCACAGUUGUG CGGUUCUGUAGGAAUCUAAUGGUGUUAGCUUGCAGCA-3'. The only common feature between the tcRNAs 102, 90, and 89 is their very low cytidine content. A point which should be emphasized from this data is the need for sequence analysis of the RNAs to determine identity. As can be seen from the above sequence analysis the RNA isolated from the polyacrylamide sizing gel at 89-90 nucleotides yields different RNAs depending on the mRNP fraction from which they are isolated. Given this fact, it is puzzling that while MHC-mRNP contains a single tcRNA$_{89}$ and contains only MHC-mRNAs, the 10S-40S mRNP (containing a large group of mRNAs) also has associated with them only one species of this size class tcRNA.

Several studies have been performed similar to those described above for tcRNA$_{102}$ in order to identify the possible function of both tcRNA$_{89}$ and tcRNA$_{90}$. Neither of these tcRNAs inhibit muscle mRNP-mRNA translation in a cell-free system under the same conditions as tcRNA$_{102}$ was found to inhibit translation. When either tcRNA$_{89}$ or tcRNA$_{90}$ was hybridized to their respective mRNAs and analyzed by sucrose density gradient centrifugation, the tcRNAs were found to sediment with the mRNAs, i.e., tcRNA$_{89}$ sedimented at 26S with MHC-mRNA while tcRNA$_{90}$ sedimented at 9-20S with the less than 40S mRNP-mRNAs. These tcRNAs did not hybridize to rRNA, tRNA, globin mRNA, TMV-mRNA or poly A. Although a degree of specificity is observed as previously shown for tcRNA$_{102}$, some important differences are apparent. First, neither tcRNA$_{89}$ nor tcRNA$_{90}$ hybridize to poly A suggesting that these RNAs are not interacting with muscle mRNAs by virtue of the poly A tails as suggested for tcRNA$_{102}$. Secondly, although tcRNA$_{89}$ and tcRNA$_{90}$ hybridize to their respective mRNAs they do not alter the sedimentation of the mRNAs as does tcRNA$_{102}$. This indicates that these RNAs, while binding, do not drastically alter the structure of the mRNAs. This inability to alter the structure may in turn explain the inability of these RNAs to inhibit the translation of the mRNAs. Therefore, the function of these tcRNAs is presently unknown. Further studies involving hybridization sites, structural analysis, and sequence determination of mRNAs are required to determine this. In addition cross-linking studies must be performed to ascertain the association of these molecules in vivo.

Although we have stressed the role of small RNAs in translational repression of cytoplasmic mRNA, we do not wish to exclude the role of mRNP proteins in this function. Indeed, it is likely that the mRNP proteins are important in stabilizing or destabilizing RNA-RNA interactions and that much of the nucleotide sequence of the small RNAs is for specific protein recognition.

The precise role of mRNP as a storage form of mRNA has continually been open to question. Indeed, as with the stored maternal histone mRNA in sea urchins (Gordon, Infante, 1981), the metabolic role of MHC-mRNP in myosin synthesis after myoblast cell fusion is open to question. In both cases the appearance of newly synthesized transcripts occurs immediately after cell fusion or fertilization. Possibly, mRNPs are not only a storehouse of needed mRNAs but also a storage compartment for other macromolecules. These macromolecules may be required for signaling the nucleus to synthesize; process, or transport additional specific mRNAs. In this manner, the tcRNAs may be involved with nuclear packaging and/or transport of newly made mRNA and serve secondarily as regulators of translation once in the cytoplasm. Obviously much more information is required before a complete understanding of the role of small cytoplasmic RNAs in mRNA metabolism is achieved.

Dr. T. McCarthy is supported by Muscular Dystrophy Association of America. This research is supported by NIH grant #5-RO1-HD03316 and Cancer Grant #20882.

Bibliography

Bag, J., Sells, B. H. (1981). Cytoplasmic nonpolysomal ribonucleoprotein complexes and translational control. Mol. and Cell. Biochem., 40:129.

Bester, A. J., Kennedy, D. S., Heywood, S. M. (1975). Two classes of translational control RNA: their role in the regulation of protein synthesis. Proc. Nat. Acad. Sci. U.S., 72:1523

Bina, M., Feldman, R., Deeley, R. (1980). Could poly (A) align the splicing sites of messenger RNA precursors? Proc. Nat. Acad. Sci., U.S., 77:1278.

Brutlag, D., Clayton, J., Frieland, P., Kedes, L. H. (1982). Nucleic Acids Res., 10:279.

Geoghegan, T., Cereghini, S., Brawaman, G. (1979). Inactive mRNA-protein complexes from mouse sarcoma-180 ascites cells. Proc. Nat. Acad. Sci. U.S., 76:5587.

Gette, W. R., Heywood, S. M. (1979). Translation of myosin heavy chain messenger ribonucleic acid in an eukaryotic initiation factor 3- and messenger-dependent muscle cell-free system. J. Biol. Chem. 254:9879.

Gordon, K., Infante, A. (1981). Maternal and newly synthesized histone RNPs from sea urchin embryos. J. Cell Biol., 91:363a.

Havaranis, A. S. and Heywood, S. M. (1981). Cytoplasmic utilization of liposome-encapsulated myosin heavy chain messenger ribonucleoprotein particles during muscle cell differentiation. Proc. Nat. Acad. Sci., U.S., 78:6898.

Heywood, S. M., Kennedy, D. S., Bester, A. J. (1975b). Studies concerning the mechanism by which translational control RNA regulates protein synthesis in embryonic muscle. Eur. J. Biochem. 58:587.

Heywood, S. M., Kennedy, D. S., Bester, A. J. (1975a). Stored myosin messenger in embryonic chick muscle. FEBS Lett., 53:69.

Heywood, S. M., Kennedy, D. S. (1976). Purification of myosin translational control RNA and its interaction with myosin messenger RNA. Biochemistry, 15:3314.

Lerner, M. R., Steitz, J. A. (1979). Antibodies to small nuclear RNAs complexed with proteins are produced by patients with systemic lupus erythemafocus. Proc. Nat. Acad. Sci., U.S., 76:5495

Mroczkowski, B. (1982). Translational control RNAs associated with messenger ribonucleoproteins in embryonic muscle. Ph.D. Thesis, Univ. of Conn., Storrs, CT 06268.

Mukherjee, A., Sarkar, S. (1981). The translational inhibitor 10S cytoplasmic ribonucleoprotein of embryonic muscle. J. Biol. Chem., 256:11301.

O'Loughlin, J., Lehr, L., Havaranis, A., Heywood, S. M. (1981). Encapsulation of core eIF3, regulatory components of eIF3 and mRNA into liposomes, and their subsequent uptake into myogenic cells in culture. J. Cell Biol. 90:160.

Siegel, E. (1981). Characterization of translational control RNA isolated from embryonic chick muscle. Ph.D. Thesis, Univ. of Conn., Storrs, CT 06268.

Limb Development and Regeneration
Part B, pages 417–428
© **1983 Alan R. Liss, Inc., 150 Fifth Avenue, New York, NY 10011**

MYOSIN LIGHT CHAIN EXPRESSION IN THE DEVELOPING AVIAN
HINDLIMB

Michael T. Crow, Pamela S. Olson, Sandra B. Conlon
and Frank E. Stockdale

Department of Medicine
Stanford School of Medicine
Stanford, California 94305

Introduction

Most vertebrate skeletal muscles are composed of a mosaic of
fibers exhibiting a broad range of metabolic and contractile
properties. On a molecular level, this heterogeneity in physiological
properties is thought to be reflected in the metabolic and contractile
protein isozyme composition of individual fibers (Barnard et al., 1971;
Pette and Schnez, 1977). The metabolic isozyme composition of a
given fiber is not invariant but is subject to modulation, such as that
induced by exercise (Holloszy and Booth, 1976). On the other hand,
the contractile protein isozymes form a relatively stable and
characteristic pattern for each fiber type (Weeds, 1976). The major
component of the contractile apparatus which is present in fiber-
specific isoforms is myosin. This molecule is a large molecular
weight protein consisting of 6 subunits: two 200,000 dalton heavy
chains and 4 lower molecular weight light chains (Gazith et al., 1970;
Weeds and Lowey, 1971). Isozymes of both the heavy and light chains
exist (Lowey and Risby, 1971; Rushbrook and Stracher, 1979; Whalen
et al., 1981). In the adult, two such isozyme systems are recognized
and designated as "fast" and "slow."

Functional adult skeletal muscle fibers arise from the maturation
of multinucleated muscle fibers (myotubes) which, in turn, are
derived from the fusion of many mononucleated precursor cell types
(myoblasts). One of the central questions in the biology of muscle
development is whether the heterogeneity of muscle fiber types seen
in the adult can be traced to heterogeneity in the myoblast types. In
other words, does the diversity of mature muscle fiber types derive
from two or more myogenic cell lineages (myoblast types) or from

extrinsic modulation of a single cell lineage or myoblast type. The two hypotheses are not mutually exclusive; multiple cell lineages may exist which differ in their response to modulation by extrinsic factors.

Monoclonal Antibodies as Probes for Isozyme Expression

One approach to identifying whether multiple myogenic precursor cell types do exist is to study the expression of fiber type-specific muscle isozymes during development. In this report we will review previous work as well as our own data on the expression of myosin isozymes in the embryo and adult. Such an approach requires that probes of high sensitivity and specificity exist for the accurate quantitation and localization of these isozyme types. In the past, these requirements were fulfilled by antisera to the myosin molecule. The use of antisera, however, suffers from two major drawbacks. The first is that, in order to obtain fast or slow myosin specific antibodies, antibodies to common and heterologous determinants must be removed by absorption. This requires a priori knowledge of the isozymes present in the tissues used for such absorption. The second drawback of this approach has only recently been realized. It is now clear that developing muscles contain myosin heavy chain isoforms that are distinct from any previously described adult isoform (Rushbrook and Stracher, 1979; Whalen et al., 1979;1981). Reaction of developing muscle with adult heavy chain antisera may be misleading since these antibodies may or may not cross react with these "embryonic" isoforms.

We have avoided these particular issues by employing monoclonal antibodies to the myosin light chains in our study of developmental expression. The approach is based on the assumption that expression of specific myosin light chain types functions as a useful marker for gene expression. There are three principal advantages to this approach over that employing antisera to the entire molecule. First, it avoids the complications associated with tentative "embryonic" isoforms of the heavy chain. The light chains synthesized by developing avian muscle are identical in all respects to those of the adult. This conclusion is based on their immunological cross-reactivity with adult light chain antibodies (Gauthier et al, .1979; Stockdale et al., 1982; Crow et al., 1982), their behavior in various electrophoretic systems (Keller and Emerson, 1981; Stockdale et al, 1981) and the fact that the light chain genes expressed by cultured embryonic avian muscle encode adult light chain proteins (Hastings and Emerson, 1982). Secondly, the light chains form a characteristic and distinct pattern on gel electrophoresis which

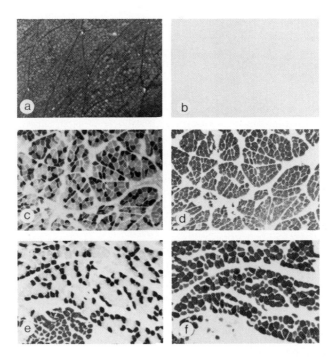

Figure I. Immunocytochemical staining of serial cross-sections of adult chicken skeletal muscle with anti-fast light chain (F_{310}) and anti-slow light chain (S_{21}) monoclonal antibodies. Frozen serial sections of adult pectoralis major (a,b), anterior latissimus dorsi (c,d), and medial adductor (e,f) muscles were cut at 10μ, incubated with antibody, and visualized as described by Billeter et al. (1980). In panels a, c, and e, sections were reacted with F_{310}, while panels b, d, and f demonstrate the staining pattern with S_{21}.

allows for precise characterization of the specificity of the antibodies using gel replication and blotting techniques ("Western" blotting; Burnette, 1981). Finally, the antibodies employed are, by definition, monospecific; antibody specificity can not be improved by heterologous immunoabsorption.

The specificity of the antibodies used in this study was determined by reaction on "Western" blots of adult muscle extracts and by immunocytochemical staining of adult muscle tissue sections (Stockdale et al., 1982; Crow et al, 1982;). The reaction of two of these antibodies with frozen serial sections of adult chicken skeletal muscle are shown in Figure 1. The muscles chosen were the pectoralis major (PM), the anterior latissimus dorsi (ALD), and the medial adductor (MA). Previous immunocytochemical and biochemical evidence suggested that these muscles were composed predominantly of fast, slow, and mixed fiber types, respectively. The panels to the left (Figure 1a, c, and e) show the staining pattern with an anti-fast light chain monoclonal antibody (F_{310}), while the panels on the right (Figure 1b, d, and f) show the staining pattern obtained with an anti-slow light chain monoclonal antibody (S_{21}). Three types of antibody staining are evident. Some fibers stain exclusively with F_{310}. The pectoralis major (Figure 1a and b) is a good example of a muscle composed exclusively of this fiber type . On the other hand, the fibers of the ALD as well as a number of fibers in the medial adductor react with both light chain antibodies. The relative intensities of staining vary among the fibers suggesting a spectrum of fiber types. The identification of a large number of adult fibers containing both fast and slow light chains was an unexpected observation. Previous typing with antisera had suggested that dual isozyme expression in the adult was rare (Gauthier et al, 1979).

The third group of fibers shown in Figure 1 are those that stain exclusively with anti-slow light chain antibody, S_{21}. In addition to their distinctive staining pattern, these fibers are larger in diameter than the fibers that stain with either F_{310} or both F_{310} and S_{21} (in the case of the medial adductor, twice as large). These fibers are not found in the more commonly studied adult chicken muscles of the pectoral region, such as the ALD. While this muscle is generally classified as a homogenous slow muscle, the fibers present in this muscle express both slow and variable amounts of fast light chains (see Figure 1c,d).

Myosin Light Chain Expression during Development

There is a wealth of conflicting information on isozyme expression in developing muscle. While some of the discrepancies may be related to the uncertainty of cross-reaction with adult myosin antisera, quantitative differences in the affinity of the antisera for each isozyme type may also limit definitive interpretation. To quantitate the absolute and relative amounts of fast and slow light chains, competitive inhibition immunoassays were developed based on the specificity and the physical requirements for antigen presentation for three monoclonal antibodies. Monoclonal antibodies T_{76} and F_{310} recognize the total and fast alkali light chains of myosin,

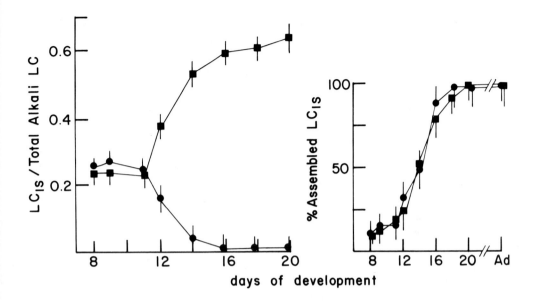

Figure 2. Slow Light Chain Content of Developing Medial Adductor (■) and Pectoralis Major (●). Extracts were solubilzed in either SDS lysis buffer (Laemmli, 1970) and diluted with 1% Triton-X100 and 1% deoxycholate to measure both free and associated light chains or in high salt buffer to measure associated light chains alone. On the left (2a), the total (both free and associated) slow light chain content relative to total alkali light chains is plotted as a function of development. On the right (2b), the percentage of slow light chains associated with myosin heavy chains is also plotted as a function of development.

respectively, either free in solution or dissociated from the heavy chain. Antibody S21, on the other hand, recognizes the slow alkali light chains of myosin only when they are associated with the myosin heavy chain. Slow light chain content could then be measured in one of two ways to distinguish between slow light chains that are or are not associated with myosin heavy chains: free and associated (total) slow light chains are determined by subtracting the results of assay with F310 (fast light chain) from the assay results of T76 (total light chain) following treatment of the extract with SDS and detergents. Slow light chains associated with heavy chains were determined by direct immunoassay with S21 in extracts solubilized with high salt.

The results of such determinations for the adductor muscles of the hindlimb are shown in Figure 2. In figure 2a, the time course of expression of both free and associated slow light chain relative to total light chain content is shown. For comparison and by way of contrast, the results of assay with the PM are also shown. In both muscle groups, total light chain content increased approximately 40 fold over the ages studied (8 days of incubation through the adult). At early stages of development (8-11 days), these two muscle groups were indistinguishable on the basis of light chain composition. Slow light chain content was similar in both muscles (solid lines) and constant over this period, comprising approximately 22% of the total light chain content. Beginning at day 12, slow light chain content rose in the adductors and fell in the pectoral extracts. By 16 days of incubation, these muscles had assumed a light chain composition similar but not identical to that of the adult.

Figure 2b shows the percentage of slow light chains that are associated with heavy chains as a function of the developmental stage. Between 8 and 12 days of incubation, the majority of the slow light chains were not associated with myosin heavy chains in either muscle. The percent of slow light chains that were associated with heavy chains increased progressively over this period from 2.5% at day 8 to approximately 20% at day 12 of incubation. From that time forward, the percentage of slow light chains associated with myosin heavy chains increased in both muscle groups even though the total slow light chain pool had decreased in the presumptive fast-twitch pectoral muscle and increased in the predominantly slow medial adductor. The data on total slow light chain content in pectoral muscle is consistent with previous observations using 2-dimensional gel analysis of developing muscle (Stockdale et al., 1981) and substantiates the contention that early in muscle development both fast and slow light chains are present.

Light Chain Expression in Early Muscle Fibers

While these results provide additional insights into isozyme expression during development, they do not answer the question of whether in early development both isozyme types are expressed within the same fiber or separate fibers. Ashmore et al. (1972) and Kelly and Zachs (1969) have described two early fiber populations in developing mammalian muscles. These populations differed both morphologically and biochemically. The first group of fibers, β fibers or primary myotubes, are characterized by a large diameter and a central core devoid of myofibrils. They appear "doughnut" shaped in transverse sections stained with antisera to adult myosin. At early stages of muscle development, the first fibers visible are the β type

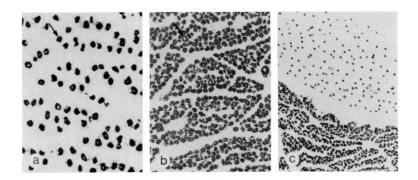

Figure 3. Primary and Secondary Myotubes in the Chick. (a) and (b): transverse serial sections of 12 day vastus medialis stained with antibody S_{21} (a) and antibody F_{310} (b). The arrows mark the location of a "primary" type fiber. (c): low power view of the 12 day medial (bottom) and lateral (upper) adductors stained with antibody S_{21}.

fibers. Rubinstein and Kelly (1981) have recently reported that these fibers initially only react with antisera to adult fast myosin. As development proceeds in the rat, these fibers react with antisera to both adult fast and slow myosin antisera. Gauthier et al. (1982) also find that myotubes with primary-type morphology react with antisera to both adult fast and slow myosin.

Succeeding generations of myotubes appear in close proximity to the primary myotubes. These have been termed α fibers (Ashmore et al., 1972) or secondary myotubes (Kelly and Zachs, 1969). These fibers are initially smaller in diameter than primary myotubes and lack a central vacuole. They vary in size as development proceeds and stain exclusively with fast heavy chain antisera throughout most of their development (Rubinstein and Kelly, 1981).

Myotube types exhibiting morphology similiar to the primary and secondary myotubes of mammals also exist in developing chicken muscle, although the period over which these distinctions are apparent is much shorter. Figure 3 shows the staining pattern of a cluster of such primary and secondary myotubes from the vastus medialis and adductor muscle group of the thigh of a 12 day chick embryo. As in mammalian development, the first fibers to appear are large in diameter and stain only along the perimeter of the fiber.

At the stage shown in Figure 3, these fibers react with monoclonal antibodies to both slow (S_21) or fast (F_310) light chains. In contrast, the smaller "secondary" fibers present in the field stain uniformly across their entire cross-section only with anti-fast light chain antibody. Figure 3c shows the distribution of these morphological fiber types in the adductor muscle group of the thigh at day 12 of development. In the presumptive slow medial adductor muscle, the majority of myotubes are of the primary type. In the presumptive fast lateral adductor there are few primary myotubes; instead, the majority of myotubes in this muscle are typical of secondary myotubes.

Figure 4 shows the staining pattern for the developing medial and lateral adductor muscles of the thigh. The lateral adductor in the 13 day embryo is composed principally of small diameter myotubes which react with fast light chain antibody--myotubes which on morphological grounds would be considered secondary myotubes. A small number of large diameter, centrally vacuolated myotubes which stain with fast as well as slow myosin light chain antibody are widely, but uniformly dispersed in this muscle. These correspond to primary myotubes. The medial adductor, on the other hand, is composed of variable sized myotubes which appear to lack central vacuoles but stain with both fast and slow light chain antibody. As development proceeds from day 13 to 18 in the lateral adductor, there is a loss of the central vacuole and a decrease in fast myosin light chain antibody staining. In the medial adductor, there is a progressive alteration in the intensity of reaction of the myotubes with antibodies to fast and slow light chains over this same period.

Conclusions

The results of the competitive immunoassays show that between 8 and 12 days of development, muscle groups express both fast and slow myosin light chains, regardless of their adult pattern of expression or anatomical location. Adult patterns of light chain expression are approached by 16 days of development. Immunocytochemistry with these same antibodies shows, however, that fast and slow light chains are not evenly distributed among all the fibers of these early muscle groups. By analogy with mammalian muscle development, there appear to be at least two myotube types which have been referred to as primary and secondary myotubes. In early chick muscle development, primary myotubes typically stain with both fast and slow light chain antibodies, while secondary myotubes stain with fast light chain antibody alone. These

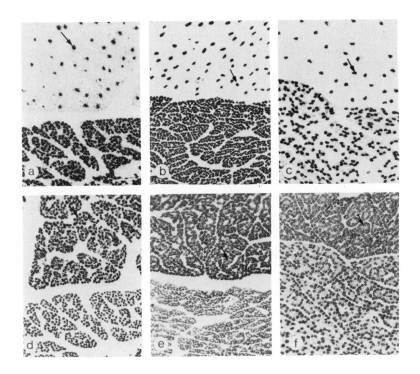

Figure 4. Immunocytochemical Detection of Myosin Light Chains in Developing Medial and Lateral Adductor Muscles. Panels a, b, and c show the staining pattern with slow light chain antibody, S21, for 13, 16, and 18 days of development, respectively. Panels d, e, and f show the staining pattern with fast light chain antibody, F310, at 13, 16, and 18 days of development, respectively. In each panel, the muscle in the lower half is the medial adductor, while the muscle in the upper half is the lateral adductor.

observations contradict the assertions made by some investigators that all the fibers of developing muscle synthesize both fast and slow isozyme types (Gauthier et al., 1978;1982).

Although the earliest stages of muscle development have not been well characterized, the pattern of myotube types found in the 12-13 day embryo suggest that the relative proportion of the two myotube types within the thigh muscles reflects the fiber type composition that will prevail in the adult: future slow muscles contain many more primary than secondary myotubes, while future fast muscles are composed predominantly of secondary myotubes. This is best illustrated by the development of two adjacent muscles in the thigh, the medial and lateral adductors. The medial adductor is a predominantly slow muscle in the adult and is composed in the 12 day embryo of myotubes all of which stain with both fast and slow light chain antibodies. The lateral adductor, on the other hand, is a predominantly fast muscle in the adult. There are few primary myotubes in this muscle at any of the stages shown. Instead, the majority of myotubes present early in its development are of the secondary type.

The complexity of muscle development is most apparent when analysis is performed at the cellular level. While each developing thigh muscle analysed for myosin composition appears qualitatively the same in that all contain both fast and slow light chains, each has a distinctive fiber composition. It is not clear whether the observed distinctions between primary and secondary myotubes actually reflect the development of two separate lineages of myogenic precursors or simply the asynchrony of muscle fiber development.

Acknowledgements

These investigations were supported by a grant from the National Institutes of Health (AG-02822) and a grant from the Muscular Dystrophy Association of America. M.T.C. is a Muscular Dystrophy postdoctoral fellow. We thank Gloria Garcia for clerical assistance in the preparation of this manusript.

References

Ashmore, C.R., D.W.Robinson, P.Rattray, and L. Doerr (1972). Biphasic development of muscles fibers in the fetal lamb. Exp. Neurol. 37: 241.

Barnard, R.J., V.R.Edgerton, T. Furokawa, and J.B.Peter (1971). Histochemical, biochemical, and contractile properties of red, white, and intermediate fibers. Am. J. Physiol. 220: 410.

Billeter, R., H.Weber, H. Lutz, H.Howard, H.M.Eppenberger, and E. Jenny (1980). Myosin types in human skeletal muscle fibers. Histochem. 65: 249.

Burnette, W.N. (1981). "Western Blotting." Electrophoretic transfer proteins from sodium dodecyl sulfate - polyacrylamide gels to unmodified nitrocellulose and radiographic detection with antibody and radioiodinated protein A. Anal. Biochem. 112:195.

Crow, M.T., P.S. Olson, and F.E. Stockdale (1982). The myosin light chains of developing avian muscle: Detection with monoclonal antibodies and quantitation by competitive immunoassay. J. Cell Biol. (submitted).

Gauthier, G.F. and S. Lowey (1979). Distribution of myosin isoenzymes among skeletal muscle fiber types. J. Cell Biol. 81:10.

Gauthier, G.F., S. Lowey, and A. Hobbs (1978). Fast and slow myosin in developing muscle fibers. Nature 274:25.

Gauthier, G.F., S. Lowey, P.A. Benfield, and A.W. Hobbs (1982). Distribution and properties of myosin isozymes in developing avian and mammalian skeletal muscle fibers. J. Cell Biol. 92:471.

Gazith, J., S. Himmelfarb, and W.F. Harrington (1970). Studies on the subunit structure of myosin. J. Biol. Chem. 245:15.

Hastings, K.E.M. and C.P. Emerson (1982). cDNA clone analysis of six co-regulated mRNAS encoding skeletal muscle contractile proteins. Proc. Natl. Acad. Sci. USA 79:1553.

Holloszy, J.O. and F.W. Booth (1976). Biochemical adaptations to endurane exercise in muscle. Ann. Rev. Physiol. 38:213.

Keller, L.R. and C.P. Emerson (1980). Synthesis of adult myosin light chains by embryonic muscle cultures. Proc. Natl. Acad. Sci. USA 77:1020.

Kelly, A. and S. Zachs (1969). The histogenesis of rat intercostal muscles. J. Cell Biol. 42: 154.

Laemmli, U.K. (1970). Cleavage of structural proteins during the assembly of the head of bacteriophage T4. Nature 227:680.

Lowey, S. and D. Risby (1971). Light chains from fast and slow muscle myosins. Nature 234:81.

Pette, D. and V. Schnez (1977). Myosin light chain patterns of individual fast and slow-twitch fibers of rabbit muscles. Histochem. 54:97.

Rubinstein, N.A. and A.M. Kelly (1981). Development of muscle fiber specialization in the rat hindlimb. J. Cell Biol. 90:128.

Rushbrook, J.I. and A. Stracher (1979). Comparison of adult, embryonic and dystrophic myosin heavy chains from chicken muscle by sodium dodecyl sulfate polyacrylamide gel electrophoresis and peptide mapping. Proc. Natl. Acad. Sci. USA 76:4331.

Stockdale, F.E., M.T. Crow, and P.S. Olson (1982). Myosin light chain isozymes in developing avian skeletal muscle cells in ovo and cell culture. Cold Spring Harbor Symp. on Muscle Development, in press.

Stockdale, F.E., N. Raman, and H. Baden (1981). Myosin light chains and the developmental origin of fast muscle. Proc. Natl. Acad. Sci. USA 75:931.

Weeds, A.G. (1976). Light chains from slow-twitch muscle myosin.
  Eur. J. Biochem. 66:157.
Weeds, A.G. and S. Lowey (1971). Substructure of the myosin
  molecule II. The light chains of myosin. J. Mol. Biol. 61:701.
Whalen, R.G., K. Schwartz, P. Bouveret, S.M. Sell, and F. Gros
  (1979). Contractile protein isozymes in muscle development:
  Identification of an embryonic form of myosin heavy chain.
  Proc. Natl. Acad. Sci. USA 76:5197.
Whalen, R.G., S.M. Sell, G.S. Butler-Browne, K. Schwartz, P.
  Bouveret, and I. Pinset-Harstrom (1981). Three myosin heavy
  chain isozymes appear sequentially in rat muscle development.
  Nature 292:805.

Index

PROGRESS IN CLINICAL AND BIOLOGICAL RESEARCH

Series Editors

Nathan Back
George J. Brewer
Vincent P. Eijsvoogel
Robert Grover

Kurt Hirschhorn
Seymour S. Kety
Sidney Udenfriend
Jonathan W. Uhr